W9-COY-902

Apraxia: The Neuropsychology of Action

edited by

Leslie J. Gonzalez Rothi and Kenneth M. Heilman

Department of Veterans Affairs Medical Center and Department of Neurology, University of Florida, Gainesville, FL 32610, USA

Psychology Press
An imprint of Erlbaum (UK) Taylor & Francis

Psychology Press, Publishers
27 Church Road
Hove
East Sussex, BN3 2FA
UK

British Library Cataloguing in Publication Data
A catalogue record for this book is available from the British Library

ISBN 0-86377-743-0

Typeset by Gilbert Composing Services, Leighton Buzzard, Bedfordshire
Printed and bound in the United Kingdom by TJ Press (Padstow) Ltd.

Contents

Chapter 1. Introduction to Limb Apraxia **1**
Leslie J. Gonzalez Rothi and Kenneth M. Heilman

Chapter 2. Limb Apraxia: A Look Back **7**
Kenneth M. Heilman and Leslie J.Gonzalez Rothi

Chapter 3. Handedness **19**
Kenneth M. Heilman

Chapter 4. A Cognitive Neuropsychological Model of Limb Praxis and Apraxia **29**
Leslie J. Gonzalez Rothi, Cynthia Ochipa, and Kenneth M. Heilman

Chapter 5. Conceptual Praxis **51**
Anastasia M. Raymer and Cynthia Ochipa

Chapter 6. Limb Praxis Assessment **61**
Leslie J. Gonzalez Rothi, Anastasia M. Raymer and Kenneth M. Heilman

Chapter 7. Management and Treatment of Limb Apraxia **75**
Lynn M. Maher and Cynthia Ochipa

Chapter 8. Kinematic Approaches to the Study of Apraxic Disorders **93**
Howard Poizner, Alma S. Merians, Mary Ann Clark, Leslie J. Gonzalez Rothi, and Kenneth M. Heilman

Chapter 9. Representations of Actions in Ideomotor Limb Apraxia: Clues from Motor Programming and Control **111**
Deborah L. Harrington and Kathleen Y. Haaland

Chapter 10. Disorders of Writing **149**
Steven Z. Rapscak

Chapter 11. Apraxia of Speech: Another Form of Praxis Disruption **173**
Paula A. Square, Eric A. Roy, and Ruth E. Martin

Chapter 12. Subcortical Limb Apraxia **207**
Bruce Crosson

Chapter 13. Developmental Dyspraxia **245**
Mary K. Morris

Chapter 14. Naturalistic Action **269**
Myrna F. Schwartz and Laurel J. Buxbaum

Glossary **291**

Author Index **299**

Subject Index **309**

Contributors

Laurel J. Buxbaum, Moss Rehabilitation Research Institute, Philadelphia, PA, USA.

MaryAnn Clark, Seton Hall University, Newark, NJ, USA.

Bruce Crosson, Department of Clinical and Health Psychology, University of Florida, Gainesville, FL, USA.

Kathleen Y. Haaland, Psychology Service, Department of Veterans Affairs Medical Center, and Departments of Psychiatry, Psychology, and Neurology, University of New Mexico, Albuquerque, NM, USA.

Deborah L. Harrington, Psychology Service, Department of Veterans Affairs Medical Center, and Departments of Psychology and Neurology, University of New Mexico, Albuquerque, NM, USA.

Kenneth M. Heilman, Department of Neurology, University of Florida, and Neurology Service, Department of Veterans Affairs Medical Center, Gainesville, FL, USA.

Lynn M. Maher, Department of Educational Psychology and Special Education, Georgia State University, Atlanta, GA, USA.

Ruth E. Martin, Department of Communicative Disorders, University of Western Ontario, London, Ontario, Canada.

Alma S. Merians, University of Medicine and Dentistry of New Jersey, Newark, NJ, USA.

Mary K. Morris, Department of Psychology, Georgia State University, Atlanta, GA, USA.

Cynthia Ochipa, Audiology and Speech Pathology Service, James A. Haley Department of Veterans Affairs Medical Center, Tampa, FL, USA.

Howard Poizner, Center for Molecular and Behavioral Neuroscience, Rutgers University, Newark, NJ, USA.

Anastasia M. Raymer, Department of Neurology, University of Florida, Gainesville, FL, USA.

Leslie J. Gonzalez Rothi, Audiology and Speech Pathology Service, Department of Veterans Affairs Medical Center and Department of Neurology, University of Florida, Gainesville, FL, USA.

Eric A. Roy, Department of Kinesiology, University of Waterloo, Waterloo, Ontario, Canada.

Myrna F. Schwartz, Moss Rehabilitation Research Institute, Philadelphia, PA, USA.

Paula A. Square, Graduate Department of Speech Pathology, University of Toronto, Toronto, Ontario, Canada.

Brain Damage, Behaviour and Cognition
Developments in Clinical Neuropsychology

Series Editors
Chris Code, University of Sydney, Australia
Glyn Humphreys, University of Birmingham, UK
Dave Müller, University College Suffolk, UK

Published titles

Cognitive Rehabilitation Using Microcomputers
Veronica A. Bradley, John L. Welch and Clive E. Skilbeck

The Characteristics of Aphasia
Chris Code (Ed.)

Classic Cases in Neuropsychology
Chris Code, Claus-W. Wallesch, Yves Joanette, and André Roch Lecours (Eds)

The Neuropsychology of Schizophrenia
Anthony S. David and John C. Cutting (Eds)

Neuropsychology and the Dementias
Siobhan Hart and James M. Semple

Clinical Neuropsychology of Alcoholism
Robert G. Knight and Barry E. Longmore

Acquired Neurological Speech/Language Disorders in Childhood
Bruce E. Murdoch (Ed.)

Neuropsychology of the Amnesic Syndrome
Alan J. Parkin and Nicholas R.C. Leng

Clinical and Neuropsychological Aspects of Closed Head Injury
John T.E. Richardson

Unilateral Neglect: Clinical and Experimental Studies
Ian H. Robertson and J.C. Marshall (Eds.)

Apraxia: The Neuropsychology of Action
Leslie J. Gonzalez Rothi and Kenneth Heilman (Eds.)

Acquired Apraxia of Speech in Aphasic Adults
Paula A. Square (Ed.)

Developmental Cognitive Neuropsychology
Christine Temple

Cognitive Rehabilitation in Perspective
Rodger Wood and Ian Fussey (Eds.)

Series Preface

From being an area primarily on the periphery of mainstream behavioural and cognitive science, neuropsychology has developed in recent years into an area of central concern for a range of disciplines. We are witnessing not only a revolution in the way in which brain–behaviour–cognition relationships are viewed, but a widening of interest concerning developments in neuropsychology on the part of a range of workers in a variety of fields. Major advances in brain-imaging techniques and the cognitive modelling of the impairments following brain damage promise a wider understanding of the nature of the representation of cognition and behaviour in the damaged and undamaged brain.

Neuropsychology is now centrally important for those working with brain-damaged people, but the very rate of expansion in the area makes it difficult to keep up with findings from current research. The aim of the *Brain Damage, Behaviour and Cognition* series is to publish a wide range of books which present comprehensive and up-to-date overviews of current developments in specific areas of interest.

These books will be of particular interest to those working with the brain-damaged. It is the editors' intention that undergraduates, postgraduates, clinicians and researchers in psychology, speech pathology and medicine will find this series a useful source of information on important current developments. The authors and editors of the books in this series are experts in their respective fields, working at the forefront of contemporary research. They have produced texts which are accessible and scholarly. We thank them for their contribution and their hard work in fulfilling the aims of the series.

CC and GH
Sydney, Australia and Birmingham, UK
Series Editors

1 Introduction to Limb Apraxia

Leslie J. Gonzalez Rothi and Kenneth M. Heilman

A DEFINITION OF APRAXIA

The term apraxia was coined by Steinthal in 1871. During the last century it has been used to describe a wide variety of neurologically induced, acquired, and developmental disorders including buccofacial apraxia, constructional apraxia, dressing apraxia, gait apraxia, gaze apraxia, limb apraxia, speech apraxia, truncal apraxia, and swallowing apraxia. Although the term apraxia is applied in each of these cases (and others), many of these disorders seem to be unrelated neurologically and psychologically. De Ajuriaguerra and Tissot (1969) therefore suggested that it may be necessary to abandon the general concept of apraxia as a class of disorders in favour of a description of various specific apraxias. But the term apraxia does have functional implications and we wondered about the attributes it implied to us. When Steinthal initially used this term apraxia it was not clear whether it reflected motor, sensory or gnosic deficits. Liepmann's (1977) earliest writings at the turn of the century suggested that the term apraxia should be used when the genesis of the disorder was motor in nature.

The term *apraxia* is sometimes used to indicate a disorder of volitional movement where nonvolitional movement is spared. However, volition is operationally difficult to define and discriminate. Others describe apraxia as a disorder of movement from memory. For example, when a person is asked to "pretend to blow out a match" they fail to do so, but when a lit match is placed before them they blow it out successfully and efficiently because the presence of the lit match "made it possible to retrieve the movement memories." These attributes of the disorder (i.e. failure with productions that are volitional or produced from memory), while probably the most consistently described attributes, are inconsistent with the movement behaviour found in some "apraxia" syndromes such as a few of those mentioned earlier. In other instances the "able/not able" distinction is less than absolute. For example, the long-held assumption that apraxic patients are only deficient in performing limb gestures to verbal command while they remain able to perform normally with actual tools and objects appears not to be correct. Whereas patients with limb apraxia may do better when using actual tools than they do with pantomime, they may also err when actually holding tools and objects (De Renzi, 1985) or when using tools and objects in their natural context (Foundas, Macauley, Raymer, Maher, Heilman, & Rothi, 1995).

The term *apraxia* also seems to have been used less discriminantly to refer to disorders of action where sensory-perceptual disorders have not been completely ruled out. One example is "constructional apraxia." Since the constructive behaviour of these cases may not be entirely explained by the movement aspect of the deficit, the more accurate term of "visuoconstructive disability" is now used (Benton & Tranel, 1993).

Hughlings Jackson (1884) proposed that the central nervous system was hierarchically organised. At the lowest level the behavioural repertoire is limited and the responses are stereotypic. As the level heightens the behavioural repertoire becomes more complex and less stereotypic. Whereas the behaviours mediated at the lower levels cannot be modified and appear to be automatic or involuntary, the behaviours mediated at the highest levels are learned, can be modified, and appear to be voluntary. For an example at the lower level, the spinal cord is programmed to send an efferent volley of action potentials to a muscle when that muscle's tendon is rapidly stretched. It is this program that gives us deep tendon reflexes. This simplest of motor programs cannot be modified by learning, is automatic, and is involuntary. At a higher level the brain stem, together with the basal ganglia, mediates righting reflexes such that, when there is rapid change in one's position with respect to gravity, postural compensations are implemented. This principle is perhaps best illustrated when sitting on a chair which is tilting backward. If the chair started to fall backward one would automatically extend the lower extremities at the knee, flex the lower extremities at the hip, and lean forward with arms outstretched and positioned in front of the trunk. Although one can superimpose voluntary behaviours on these brain stem-, basal ganglia- and spinal cord-mediated behaviours, these automatic behaviours are not learned, cannot be modified, and therefore are frequently termed "reflexes." The cerebral cortex is capable of programming both simple (e.g. finger flexion) and complex (e.g. using a pair of scissors to cut a piece of paper) behaviours. Almost all behaviours programmed by the cortex are voluntary (not automatic). In contrast to the primitive reflexes which may be motorically quite complex but not learned, the complex behaviours mediated by the cortex are almost always learned. *A loss (as the result of neurologic disorder) of the ability to perform these learned voluntary actions is what we term apraxia.*

Hughlings Jackson (1878) noted that, even when a specific body part may be unable to carry out a volitional act, that same body part may still be able to make the same movements when carrying out nonintentional or reflexive acts. The opposite may also occur such that an individual who can no longer perform an automatic involuntary behaviour may be able to perform voluntary behaviours with the same body parts. Although some term the loss of the ability to perform automatic behaviours apraxia (e.g. optic and swallowing apraxia), we use the term apraxia to denote a loss of learned behaviours. Therefore, we define apraxia simply as a *neurological disorder of learned purposive movement skill that is not explained by deficits of elemental motor or sensory systems.*

RELEVANCE

Almost 20 years ago we began our collaboration in research in limb apraxia. We have been asked many times by students, colleagues, and, most acutely, by grant reviewers: "Why is the study of apraxia of interest?". While it is a commonly held notion that apraxia is of theoretical interest without being of any practical import, we disagree. Although we did not have experimental data to support our position, it was not uncommon for patients' families to tell us that they had to remove access to dangerous tools and objects from the patients' environment because they were concerned that these patients might hurt themselves. In addition, because we see many patients who are aphasic, we commonly look to gesture in the form of Amer-ind (Skelly, 1979) as an alternative form of communication. Amer-ind is a gestural system that can be performed with a single hand. Amer-ind gestures are relatively transparent in terms of meaning and can be interpreted by most untrained viewers. However, because limb apraxia commonly co-occurs (Kertesz, Ferro, & Shewan, 1984) with aphasia, we found limited success with the ability of patients with both disorders to acquire this gestural system. Most recently, the work of Schwartz and colleagues (described in Chapter 14 of this book) as well as our own work looking at the ability of patients with limb apraxia to use tools and objects in natural contexts (Foundas et al., 1995) have confirmed our impression that apraxia is not a trivial behavioural disorder only of theoretical interest, but instead apraxia is a common and disabling disorder.

DOMAIN OF INTEREST

Unfortunately, interest in apraxia has been limited to a small group of neurologists, psychologists, and speech–language pathologists. However, because apraxia is of both theoretical and practical importance, we believe that many professions (from anthropologists to occupational therapists) should find this topic of interest as well. For example, several phylogenetic levels of animals are known to use tools and to learn to use body movements meaningfully and skilfully. Tool manufacture and complex tool use, while noted in rare examples, are less well developed in animals than man. The development of tool use in man is, of course, of interest anthropologically. Therefore what is learned from comparative psychological and anthropological studies of the development of tool use and "gesture" across species and across human history with specific reference to changes in relevant brain structures should be consistent or compatible with what is known about the cognitive neuropsychology of human praxis using an ablation paradigm of study.

Additionally, as pointed out by Morris in Chapter 13, how children learn to use tools and gesture with each hand is poorly understood. Knowledge of how the praxis system is acquired should inform us not only of the neural substrates of skill acquisition but also about which portions of a componential system of limb praxis are modular and which are not by virtue of their dependence on development of others. Studying developmental models of normal praxis may also help us understand developmental disorders of praxis.

Neuropsychologically, the study of how one loses a previously attained level of movement skill as the result of brain impairments allows us to study how the adult human brain instantiates planned movements. The principles of movement processing, such as those learned by use of the ablation paradigm of neuropsychology, must be reconciled with those learned by using the paradigms of cognitive psychology, such as the study of normal reaching patterns, etc. (for example, see Chapter 9 by Harrington and Haaland). In turn, the interest of the physician is to use the knowledge gained from neuropsychological evaluation to assist in diagnosis, lesion localisation and patient care.

Finally, information regarding praxis processing and deficiencies should be of interest to rehabilitation clinicians such as speech–language pathologists, clinical neuropsychologists, occupational therapists and physical therapists for a number of reasons. For one, these clinicians need to advise their clients of potential environmental dangers. In addition, there is a need to develop efficacious and theoretically informed treatment and rehabilitative strategies (see Chapter 7, by Maher and Ochipa, for examples).

THE BOOK

This book is a compilation of the works of many clinical investigators working on different praxis problems. We are indebted to these authors for their contributions, their guidance, and their support. While certainly not an exhaustive review of the various perspectives currently available in the study of limb apraxia, we have included these 14 chapters because we feel they provide a reasonably cohesive and compatible statement collectively. In Chapter 2, Heilman and Rothi provide a selected history of the re-emergence of interest in the psychology of limb apraxia largely as a result of Geschwind's articles of 1965 entitled "Disconnexion syndromes in animals and man." Serving as one of Heilman's professors during his clinical training, Geschwind was influential in encouraging Heilman to read Liepmann's writings of the turn of the century regarding limb apraxia. Because hand preference and hand skill are so closely linked functionally, theories accounting for hand preference are reviewed, prior to a discussion of the mechanism of hand skill (limb praxis) in subsequent chapters. Then, utilising the discoveries of Liepmann (and the reformulations of Geschwind) and recruiting parallels found in the cognitive neuropsychology of language, Rothi, Ochipa, and Heilman (1991) proposed a cognitive neuropsychological model of limb praxis and cited cases in the literature supporting its structure. An updated, but fundamentally unaltered reiteration of that 1991 paper is presented in Chapter 4 and forms the basis of many of the subsequent chapters of this book. For example, Chapter 5 by Raymer and Ochipa expands upon the issue of action semantics raised in Chapter 4 by Rothi, Ochipa, and Heilman and enlists many of the same methodologies that have become the hallmark of the cognitive neuropsychological approach. Similarly, in Chapter 6 Rothi, Raymer, and Heilman provide the interested clinician and researcher with attributes of test construction that would allow one to assess a patient in a manner compatible with the fractionation of praxis functions implied by the model described in Chapters 4 and embellished in Chapter 5. While relatively little is known about the neurology of the praxis system based on this functional approach, Crosson in Chapter 12 embraces the challenge of parallels between language and praxis modelled in Chapter 4 and applies these notions to what the role of the subcortical structures might be in praxis programming. As a result, seven hypotheses are generated to challenge future research. Poizner and colleagues in Chapter 8 describe a technology used to analyse the spatiotemporal characteristics of skilled limb movement as applied to patients with ideomotor limb apraxia and, with this technology, confirms Liepmann's hypothesis of apraxia resulting from a failure of stored movement representations. Explicating the nature of information in and structure of these "movement representations" as proposed by Liepmann is the challenge approached in Chapter 9 by Harrington and Haaland, whereas a higher order perspective on praxis processing is the perspective provided in Chapter 14 by Schwartz and Buxbaum in their work on action pragmatics or how actions are used in natural contexts. Finally, Chapters 10 and 11 represent interesting reviews of praxis systems closely linked to limb praxis specifically: notably, writing and speaking. For example, it is a commonly held (but inaccurate) notion that failure to write legibly as the result of brain damage is the result of a coexisting limb apraxia (Zangwill, 1954). In Chapter 10 Rapcsak reviews the cognitive neuropsychological model specific to peripheral agraphias that is compatible but distinctive from that which is proposed for limb apraxia. Regarding apraxia of speech, Square, Roy and Martin review in Chapter 11 the controversy that continues to rage over the existence of a defect of motor planning of speech and its potential relationship to limb and buccofacial apraxias.

Thus, the chapters that follow provide the reader with a perspective of limb apraxia that expands upon Hugo Liepmann's vision that is almost 100 years old. The reader may be a clinician looking for theoretical underpinnings that might provide greater precision to clinical interventions for this disorder while the reader who is a researcher of normal motor skill acquisition may find concurrence in principles provided by a look at pathology and information emanating from the ablation paradigm.

REFERENCES

Ajuriaguerra, J. de, & Tissot, R. (1969). The apraxias. In P. J. Vincken & G. W. Bruyn (Eds.), *Handbook of Clinical Neurology.* Amsterdam: North Holland.

Benton, A. & Tranel, D. (1993). Visuoperceptual, visuospatial, and visuoconstructive disorders. In K. M. Heilman & E. Valenstein (Eds.), *Clinical Neuropsychology,* pp. 165–213. New York: Oxford University Press, pp. 165–213.

De Renzi, E. (1985). Methods of limb apraxia examination and their bearing on the interpretation of the disorder. In E. A. Roy (Ed.), *Neuropsychological Studies of Apraxia and Related Disorders,* (pp. 45–64). Amsterdam: Elsevier Science.

Foundas, A., Macauley, B. L., Raymer, A. M., Maher, L. M., Heilman, K. M., & Rothi, L. J. G. (1995). Ecological implications of limb apraxia: Evidence from mealtime behavior. *Journal of the International Neuropsychological Society, 1,* 62–66.

Geschwind, N. (1965). Disconnexion syndromes in animals and man. *Brain, 88,* 237–294, 585–644.

Hughlings Jackson, J. (1878). Remarks on non-protrusion of the tongue in some cases of aphasia. *Lancet, 1,* 716–717.

Hughlings Jackson J. (1884). The Croonian lectures: Evolution and dissolution of the nervous system. *British Medical Journal, 1,* 591–593, 660–663, 703–707.

Kertesz, A., Ferro, J. M., & Shewan, C. M. (1984). Apraxia and aphasia: The functional-anatomical basis for their dissociation. *Neurology, 34,* 40–47.

Liepmann, H. (1977). The syndrome of apraxia (motor asymboly) based on a case of unilateral apraxia. (Translation from *Monatschrift für Psychiatrie und Neurologie,* 1900, *8,* 15–44.) In D. A. Rottenberg & F. H. Hockberg (Eds.), *Neurological Classics in Modern Translation.* New York: Macmillan.

Rothi, L.J.G., Ochipa, C., & Heilman, K.M. (1991). A cognitive neuropsychological model of limb praxis. *Cognitive Neuropsychology, 8,* 443–458.

Skelly, M. (1979). *Amer-Ind gestural code based on universal American Indian hand talk.* New York: Elsevier.

Steinthal, P. (1871). *Abriss der Sprachwissenschaft.* Berlin.

Zangwill, O. L. (1954). Agraphia due to a left parietal glioma in a left-handed man. *Brain, 77,* 510–520.

2 Limb Apraxia: A Look Back

Kenneth M. Heilman and Leslie J.Gonzalez Rothi

INTRODUCTION

This chapter will briefly present some of the history of limb apraxia. Many of the specific historical details can also be found in the subsequent chapters (most especially Chapter 4 by Rothi and Heilman). Apraxia (loss of skilled movements) is a term that has been used to describe a wide variety of neurobehavioural disorders. Many of these disorders, however, are not disorders of learned skilled movements (e.g. apraxia of gaze, gait apraxia). In other instances, the deficit does involve deficient productions of learned or skilled movements but involves body parts other than the upper limb (i.e. buccofacial apraxia). Although what has been termed dressing apraxia and constructional apraxia may involve the limb, they will not be discussed because many of the behaviour deficits associated with these disorders are induced by either visuospatial disorders or neglect. In summary, although there are many disorders that are labelled apraxia, each has a unique underlying mechanism and this historical review will be limited to a discussion of limb apraxia.

The limb apraxia that Liepmann termed "ideomotor apraxia" has received the most interest, and the major portion of our chapter will discuss that disorder. However, we will also briefly discuss what Liepmann termed ideational apraxia. Liepmann described a third type of apraxia that he termed limb-kinetic or melokinetic apraxia. This disorder is similar to what Kleist (1907) termed innervatory apraxia. In this disorder there is an inability to make fine, precise, independent finger movements of the hand contralateral to a cerebral lesion. Limb-kinetic apraxia is probably related to injury to the cortico-spinal neurons or their projections to the spinal cord. Although one needs to be able to perform precise independent finger movements to perform skilled acts, limb-kinetic apraxia is not primarily a disorder of learned skilled movement but more of an elemental motor disorder. Therefore limb-kinetic apraxia will not be discussed further in this chapter.

Hughlings Jackson (1932) described a patient who was unable to perform buccofacial movements to command. For example, when asked to protrude his tongue, the patient was unable to do so. However, the patient was able to understand the command and there was no evidence of weakness that could account for the inability to protrude the tongue. Whereas this report may have been the first description of apraxia, it was Steinthal in 1871 who first used the term apraxia. Steinthal used this term for defects in "... the relationship between movements and the objects with which the movements were concerned" (Hécaen & Rondot, 1985). Although Finkelnburg (1873) noted that aphasics had an impaired ability to gesture, he attributed these defects to asymbolia. That is, he thought that the core deficit in aphasia is the inability to comprehend or express symbols. According to Finkelnburg, whereas aphasia is the verbal inability to express symbols apraxia is the inability to produce gestural symbols.

Despite these 19th-century descriptions of disordered action, not many important advances were made until the beginning of the 20th century and the work of Hugo Liepmann. Liepmann was a student of Karl Wernicke. Karl Wernicke not only was responsible for describing new types of aphasia (e.g. Wernicke's aphasia) but he also introduced explanations of these aphasic syndromes in what now is termed information-processing systems models and modular networks. In these networks each module contains different types of information or what is now termed "representations." Consistent with the influence of his mentor, Liepmann was the first to apply this form of cognitive neuropsychological analysis to the behavioural disorder of apraxia.

IDEOMOTOR APRAXIA

Liepmann's (1900) first major manuscript on apraxia attempted to demonstrate that apraxia is not a disorder of symbolic behaviour, an elemental motor disorder, or a disorder of recognition (agnosia). His patient, M.T., was apraxic with only his right arm. When forced to use the left arm the patient was not apraxic. Liepmann argued that if the defect was that of asymbolia the patient should have made the same errors with both arms. Subsequently, Liepmann and others demonstrated that apraxia may occur in the absence of aphasia and aphasia may occur in the absence of apraxia. In addition, Goodglass and Kaplan (1963) demonstrated that when apraxia and aphasia occur together the severity of the aphasia does not predict the severity of the apraxia or vice versa. These results are not compatible with a causal relationship between aphasia, apraxia, and asymbolia.

Kimura and Archibald (1974) proposed a unitary mechanism that may account for both aphasia and apraxia with left-hemisphere lesions. Kimura and Archibald noted that the defects in gesture associated with left hemisphere damage were primarily for complex gestures that are comprised of multiple movements. These authors studied right- and left-hemisphere damaged subjects performing simple or complex hand gestures. When compared to the right-hemisphere damaged subjects, those with left-hemisphere damage were impaired on those complex gestures that required sequences. Kimura and Archibald (1974) noted that the sequencing defect found in apraxia may also be found in aphasia and posited that left-hemisphere dominance for sequencing may account for the left hemisphere's ability to mediate both praxis and speech–language. Although sequencing is important for both of these behaviours, apraxia is associated with spatial errors as well as other types of temporal errors (see Chapter 6). In addition, as discussed, apraxia can be seen in the absence of aphasia and aphasia in the absence of apraxia. For example, we reported a left-handed patient who had a large right-hemisphere stroke. Although this patient was severely apraxic, he was not aphasic. Using selective hemisphere anaesthesia (the Wada test), it has been repeatedly demonstrated that in most left handers it is the left-hemisphere that is dominant for language. Since our patient had a large right-hemisphere lesion and was not aphasic, it was probably his left hemisphere that mediated language. However, because he was apraxic it was probably his right-hemisphere that mediated praxis. This case provides further evidence that there is not a unitary hemispheric mechanism that is impaired in both aphasia and apraxia (Heilman, Coyle, Gonyea, & Geschwind, 1973).

The patient reported by Liepmann in 1900 was a complex case. The patient's hand preference was not clearly defined and the patient had syphilis with multifocal brain pathology. Liepmann and Maas (1907) subsequently described another patient (Ochs), who was hemiparetic with his right arm and hand and who could not correctly carry out commands with his left hand. On postmortem examination, Ochs was found to have a lesion of his left basis pontis that accounted for his right hemiparesis and a lesion of his corpus callosum. Liepmann was aware of Broca's report that demonstrated, in right handers, that language is mediated by the left hemisphere. Liepmann and Maas could have explained their patient's left-hand apraxia by suggesting that the left hemisphere's language systems were disconnected for the right hemisphere's motor systems that control the right hand (i.e. a right hemisphere verbal comprehension deficit). Unlike verbal commands, gesture imitation and using actual tools and objects do not require language. Therefore, Liepmann and Maas felt their patient's left-arm apraxia could not be explained by a language motor disconnection. The right hemisphere's visual, somesthetic and motor systems were all intact and the right hemisphere should have been able to mediate both gesture imitation and actual tool and object use. That the callosal disconnection impaired these functions suggests that the left hemisphere of this patient contained some other form of nonlanguage knowledge that is critical for being able to perform these movement tasks. Liepmann and Maas posited that the left hemisphere of right handers mediates not only language but also contains movement formulas that store the knowledge of how to control purposeful skilled movements.

Liepmann also recognised that apraxia may be associated with hemispheric lesions that do not involve the corpus callosum. In 1905, he reported the data on 83 right-handed patients with either a right or left hemiplegia. He tested these cases by asking them to pantomime to command and to imitate both transitive and intransitive gestures. He also had these patients perform with actual tools and objects. Liepmann found that none of his patients with left hemiplegia (right-hemisphere lesions) were apraxic. However, about one-half of the subjects with right-hemiplegia (left-hemisphere lesions) were apraxic even when using their nonparetic left arm and hand. Based on this study, Liepmann concluded that the left sensory motor cortex dominates the right in the control of movement for both hands and the left hemisphere controls the ipsilateral left hand via the corpus callosum.

In addition to suggesting that the left hemisphere of right handers plays a dominant role in the mediation of skilled movements, independent of the hand used to carry out these movements, Liepmann also suggested that the left hemisphere of right handers contains an action system with several major, distinct components. These components, or what now may be termed modules, included the movement formula (time–space sequence memories of learned, skilled, purposeful movements) and a mechanism that allowed the person to realise a movement's innervation. Liepmann suggested that destruction of these movement representations was associated with lesions that were in the posterior portions of the left hemisphere and usually involved the temporal-parietal interface. In contrast, the innervatory pattern system transformed the movement representations, which were primarily represented in sensory modalities, into motor innervatory patterns. However, Liepmann recognised that correctly using a limb required interactions as well and "co-operation of innervatory and extra-innervatory areas" of the brain. Liepmann did not state the location of this innervatory system in the brain but implied that it may be in the frontal lobes.

Geschwind noted in his introduction to his classic 1965 *Brain* paper, "Disconnexion syndromes in animals and man," that after Liepmann's series of manuscripts there was little interest in apraxia. Geschwind attributed this loss of interest to the growth of "holistically oriented" "neurologists such as Head, Marie, and von Monakow and to "holistically oriented" psychologists such as Lashley and the Gestalt school. In addition, many of Liepmann's theories were based on the patient with a callosal lesion (Liepmann & Maas, 1907) and could not be replicated by Akelaitis (1944), who performed callosal disconnections in patients with intractable epilepsy.

Geschwind's 1965 *Brain* paper renewed interest in apraxia. Geschwind and Kaplan (1962) reported on a patient with a naturally occurring callosal lesion. Like the patient reported by Liepmann, their patient was unable to gesture to command with his left hand. When attempting to write with his left hand, the patient would be able to write legible letters but frequently either wrote an incorrect word or misspelled the word. Because the patient could copy writing and imitate gestures with his left hand, Geschwind and Kaplan concluded that the patient's defects could not be attributed to an elemental motor deficit. When the patient used actual objects or tools with his left hand, his performance was also excellent. His right hand performance was normal in all conditions. Geschwind and Kaplan interpreted these findings as a disconnection of the right motor cortex from the speech areas. Furthermore, the preserved ability of the left hand to copy writing, imitate gestures, and use actual objects and tools suggested to Geschwind and Kaplan that the right hemisphere was capable of mediating these activities by itself and without information from the left hemisphere. Whereas Geschwind and Kaplan's patient could imitate and use actual tools and objects, they were aware that Liepmann and Maas' patient, Ochs, could not, suggesting that these two patients with callosal lesions may have had differences in their brain organisation.

In regard to the lesions in the region of the left supramarginal gyrus that are associated with apraxia of both the right and left limbs, Geschwind interpreted Liepmann as suggesting that such lesions disconnect Wernicke's area, important for speech comprehension, from the more anterior left motor association cortex. The motor association cortex on the left side projects to the motor cortex on the left side and to the motor association cortex on the right side, which in turn projects to the motor cortex in the right hemisphere. Patients with apraxia from supramarginal lesions are usually unable to imitate correctly. Geschwind thought that imitation was impaired because the lesions were deep and interrupted visuomotor connections. Although Geschwind considered the postulate that the supramarginal gyrus may contain the "memories for movement," he rejected this hypothesis because these representations should be able to cross the callosum and gain access to the right hemisphere's motor systems. If these movement memories represented in the supramarginal gyrus of the left hemisphere are able to reach the right hemisphere, it could not account for why patients with more anterior left hemisphere lesions are apraxic when using their left hand (sympathetic apraxia).

In 1983, Watson and Heilman reported a woman with a naturally occurring callosal lesion who, unlike Liepmann's patient, was not hemiparetic. Although she performed gesture to command flawlessly with her right hand, she was unable to gesture correctly to command with her left. However, unlike the patients reported by Geschwind and Kaplan, those reported by Akelaitis and coworkers, and those reported by Sperry, Gazzaniga, and Bogen (1969), Watson and Heilman's patient was unable either to imitate correctly or use actual objects with her left arm and hand. The right hemisphere of this patient had intact visuomotor and tactile-motor connections and neither imitation nor object use requires input from the speech language areas. The observation that a callosal disconnection in this patient induced behavioural deficits that could not be accounted for by a language motor disconnection suggested that Liepmann's original account could be correct and that, in some right handers, movement representations are lateralised to the left hemisphere and that a callosal disconnection in these patients not only deprives the right hemisphere of verbal information but also deprives the right hemisphere of the temporal and spatial knowledge needed to perform learned, skilled movements.

Although Watson and Heilman's observations on their patient with callosal disconnection has been replicated (e.g. see Graff-Radford, Welsh, & Godersky, 1987) it is not entirely clear why some patients with callosal disconnection can imitate and use actual objects with their left hand and others cannot. Although intractable seizures may be associated with anomalies of brain organisation, Geschwind and Kaplan's patient could imitate and use actual objects and had a naturally occurring callosal disconnection. In addition, one of us (K.M.H) had the opportunity to examine a patient with a surgical disconnection for intractable seizures who was unable to imitate and use actual objects correctly with her left hand. These observations and the knowledge that there are many patients who, in spite of sustaining large cerebral infarctions in the distribution of the middle cerebral artery, are not apraxic would suggest that in some right handers these movement memories may be bilaterally represented.

Although Geschwind thought that lesions in the region of the supramarginal gyrus induce apraxia because they involve the white matter below the cortex and therefore disconnect both Wernicke's area and visual association cortex from premotor cortex, Geschwind was aware, as was Liepmann, that there were patients who also had difficulty when using actual objects or tools. To resolve this dilemma, Geschwind reported that Liepmann asserted that even the manual handling of tools and objects was frequently learned visually. However, Geschwind could not account for why the visuomotor pathways of the left hemisphere would be dominant.

The hypothesis that may explain why right-handed patients with left parietal lesions cannot correctly gesture to command, imitate, or use actual objects and tools is that the movement representations posited by Liepmann are stored in the left inferior parietal lobule. When one is asked to pantomime a gesture, imitate a gesture, or actually work with an object or tool, one must activate these representations. Heilman, Rothi, and Valenstein (1982) and Rothi, Heilman, and Watson (1985) provided evidence that these representations are stored in the parietal lobe contralateral to the preferred hand. These activated representations help program the motor association or premotor cortex that in turn selectively activates the motor cortex.

One of the problems that Geschwind had with the postulate that movement representations were stored in the left parietal lobe was that, although lesions of the right parietal lobe did not cause apraxia of the left hand, left frontal lesions did cause apraxia of both hands. Geschwind thought that if these motor representations were stored in the left parietal lobe, then frontal lesions in the left hemisphere should not be associated with sympathetic apraxia (apraxia of the left hand from left-hemisphere lesions) because the left and right parietal lobe have connections through the corpus callosum and the stored motor memories should be able to gain access to the right hemisphere via these posterior callosal connections. Left-hemisphere lesions anterior to the parietal lobe can also induce apraxia. If movement representations are stored in the inferior parietal lobe, why would anterior left-hemisphere lesions be associated with apraxia of the left as well as the right hand? Perhaps the left premotor cortex also contains a special neural apparatus important in programming skilled movements. Liepmann suggested that movement representations had to be transformed into innervatory patterns. Unfortunately, Liepmann never specified exactly where he thought the transformation from a movement representation to a motor pattern took place. Geschwind thought that the convexity premotor cortex was important for praxis. Although lesions in this region may be associated with apraxia, these lesions often involve other areas as well as subcortical white matter. Whereas the convexity premotor cortex may be important in adapting the motor program to environmental perturbations, there is little evidence that this area can develop innervatory patterns.

The premotor cortex in the medial portion of the frontal lobe is called the supplementary motor area (SMA). Whereas stimulation of this area produces complex movements that include the fingers, hand, and arm, stimulation of the primary motor cortex (Brodmann's 4) produces simple single movements (Penfield & Welch, 1951). The SMA receives projections from the parietal lobe and projects to the primary motor cortex. The SMA neurons discharge before neurons in the primary motor cortex discharge (Brinkman & Porter, 1979). Studies of cerebral blood flow demonstrate that when one makes simple finger movements, the contralateral primary motor cortex becomes activated. However, when one makes complex movements, not only is the primary cortex activated, but the SMA also becomes activated. Last, when individuals think about making a complex movement but do not make the movement, the SMA becomes activated but the motor cortex does not become activated. These anatomic and physiological findings suggest that the SMA would be an ideal area for movement representations to be transformed to innervatory patterns. In 1986, Watson, Fleet, Rothi and Heilman reported several right-handed patients who had lesions of the left medial frontal lobe that included the SMA. These patients demonstrated an ideomotor apraxia of both hands. Based on the experimental evidence that the parietal lobe may store movement representations and that the SMA may transcode these time–space or visual kinaesthetic representations into innervatory patterns, we have proposed a modular system that mediates learned skilled movements (Rothi, Ochipa, & Heilman, 1991). This praxis system is discussed in Chapter 4 .

DISASSOCIATION APRAXIA

Heilman (1973) reported three patients who were unable to pantomime to command. However, unlike patients with ideomotor apraxia who primarily make spatial and temporal errors (Rothi, Mack, Verfaellie, Brown, & Heilman, 1988 and Poizner, Mack, Verfaellie, Rothi, & Heilman, 1990) when asked to pantomime, these patients would hesitate to make any meaningful movement. In addition, unlike patients with ideomotor apraxia, when they were shown the tool or object, they were able to pantomime flawlessly and were also able to imitate flawlessly.

In Heilman's 1973 *Brain* article he termed this disorder "ideational apraxia," an unfortunate choice of terms for several reasons. Not only had this term been used for other types of behavioural disorders that we will discuss but also because he proposed in this article that the probable mechanism accounting for this disorder was a disconnection between the movement representations previously discussed and the verbal system that interprets the command.

The disassociation apraxia described by Heilman was similar to that of the callosal patients described by Geschwind and Kaplan (1962) and Gazzaniga, Bogen, and Sperry (1967). However, in the callosal patients the apraxia to verbal command was restricted to the left hand. In contrast, the patients with disassociation apraxia from intrahemispheric lesions are unable to gesture to command with either hand. Although the exact foci of lesions that induce this defect remains unknown, based on the clinical presentation Heilman suggested that the lesion was deep to the cortex in the region of the inferior parietal lobe. De Renzi, Faglioni, and Sorgato (1982) not only replicated Heilman's observations but also described individuals who could not correctly gesture to visual stimuli but would correctly gesture to verbal commands. He even described two patients who performed better to verbal commands and visual stimuli than they did when they were blindfolded and actually holding the tool or object.

IDEATIONAL APRAXIA

Pick (1905) described a patient with an aphasia who had some preserved comprehension. Pick's patient was unable to use correctly actual tools and objects and also had difficulty sequencing a series of acts with tools or objects that led to an action goal. Although a failure to recognise a tool or object (an agnosia) may interfere with a person's ability to use correctly tools or objects, Pick thought that his patient's errors were not related to an agnosia because the patient could name tools and objects that were incorrectly used. Therefore, from the time of

Pick's original description there was some conflict as to what actually constituted ideational apraxia, i.e. a defect in using actual tools vs. an inability to perform a sequence of acts leading to an action goal. Liepmann (1920) defined ideational apraxia as an impairment on tasks that require a sequence of serial acts with objects or tools, and modern investigators such as Poeck (1983) have continued to adhere to this definition.

Many clinicians and investigators thought that ideomotor apraxia was limited to tests of pantomime and imitation but that, in the natural environment with the use of real tools, patients with ideomotor apraxia did not make errors. De Renzi and Lucchelli (1968), therefore, functionally defined ideational apraxia as a failure to use correctly actual tools and objects, and these actual tool errors are distinct from errors made by patients with ideomotor apraxia who use actual tools and objects correctly. However, Zangwill (1960) noted that the failure to use properly actual tools may be related to a severe production disorder (ideomotor apraxia) rather than an ideational disorder. Subsequently, using quantitative methods, Poizner et al. (1990) demonstrated that patients with ideomotor apraxia do make timing and spatial errors even when using actual objects or tools. Although errors appeared to be less severe when using actual tools or objects than when pantomiming, the nature of the errors were similar in both conditions.

CONCEPTUAL APRAXIA

Roy and Square (1985) proposed that acting on the world may involve the operation of two systems: conceptual and production. Whereas patients with ideomotor apraxia make production errors (spatial and timing errors), patients with conceptual apraxia should make content errors. Ochipa, Rothi, and Heilman (1989) reported a patient who used tools inappropriately. For example, in the bathroom he used a tube of toothpaste to brush his teeth. On another occasion he used a comb. Although one may expect such defects from a patient with visual and tactile agnosia, this patient was able to name correctly the items he misused.

In a subsequent study, Ochipa, Rothi, and Heilman (1992) posited that there are four types of conceptual knowledge that are needed to interact correctly with objects in the environment. These include: (1) tool–action associations or the type of actions that are associated with using tools (e.g. twisting motions are associated with using a screw driver); (2) tool–object associations or the type of tool that works on a specific object (e.g. hammers are used to hit nails); (3) the mechanical advantage afforded by tools or mechanical knowledge (e.g. if a hammer is not available to pound in a nail to use a wrench rather than an ice pick); (4) tool fabrication or the ability to make a tool (e.g. ability to bend and use coat hanger to open a locked car door). Ochipa et al. (1992) studied patients with probable Alzheimer's disease and compared them to controls. They demonstrated that many of these patients did have conceptual apraxia and that these patients made all four types of conceptual errors. In some patients, conceptual apraxia was not associated with ideomotor apraxia or a semantic language impairment, suggesting that the systems that mediate conceptual praxis knowledge may be independent of the gesture production system and verbal semantics.

SUMMARY

Limb apraxia is a loss of the ability to perform learned skilled limb movements. Hugo Liepmann, at the beginning of the 20th century, was the first to perform systematic studies of limb apraxia. By studying a patient with a callosal disconnection, Liepmann and Maas demonstrated that apraxia was not a form of asymbolia and could be dissociated from disorders of language. In addition, he demonstrated that apraxia in right handers is most commonly associated with left-hemisphere lesions. Liepmann described three forms of apraxia: limb kinetic, ideomotor, and ideational. Patients with limb kinetic apraxia are unable to make precise, independent finger movements. Subsequent studies have demonstrated that this disorder is probably related to dysfunction of the corticospinal system. Patients with ideomotor apraxia make spatial and temporal errors even when performing with the hand (e.g. left) that is ipsilateral to their hemispheric lesion (e.g. left). The term ideational apraxia was used to describe both sequencing and conceptual errors. However, currently the term ideational apraxia is used to denote the inability of patients to perform a series of acts leading to a goal (e.g. making a sandwich) and the term conceptual apraxia is used to denote the disorder characterised by content errors (a loss of tool–action associative knowledge), a loss of tool–object associative knowledge, and other forms of mechanical knowledge.

After the First World War and the ascent of the "holistic" approach, there was a loss of interest in apraxia and the information-processing approach of Liepmann and his mentor, Karl Wernicke. However, in his classic article, Geschwind stimulated renewed interest. Geschwind primarily attributed ideomotor apraxia to disconnections (e.g. a language motor disconnection.) However, the disconnection hypotheses could not account for the inability of patients with left-hemisphere disease both to imitate and use actual objects with their left hand. Therefore, Heilman, Rothi and their coworkers proposed and provided evidence for a mixed representational–disconnection hypothesis. Using this hypothesis together with an information-processing approach, Rothi, Ochipa, Heilman, and their coworkers described several forms of apraxia (e.g. disassociation, conduction, and conceptual). A more detailed discussion of both ideomotor and these "new" forms of apraxia can be found in Chapters 4 and 5.

REFERENCES

Akelaitis, A.J. (1944). A study of gnosis, praxia, and language following section of the corpus callosum and arterior commisure. *Journal of Neurosurgery, 1*, 94–102.

Brinkman, C. & Porter, R. (1979). Supplementary motor area in the monkey: Activity of neurons during performance of a learned motor task. *Journal of Neurophysiology, 42*, 681–709.

De Renzi, E. & Lucchelli, F. (1968). Ideational apraxia. *Brain, 113*, 1173–1188.

De Renzi, E., Faglioni, P., & Sorgato, P. (1982). Modality-specific and supramodal mechanisms of apraxia. *Brain, 105*, 301–312.

Finkelnburg, F. (1873). Über Aphasie und Aysmobolie Nebst Versuch Elmer Theorie der Sprachbildung. *Archiv für Psychiatrie, 6*.

Gazzaniga, M., Bogen, J., & Sperry, R. (1967). Dyspraxia following diversion of the cerebral commissures. *Archives of Neurology, 16*, 606–612.

Geschwind, N. (1965). Disconnexion syndromes in animals and man. *Brain, 88*, 237–294.

Geschwind, N. & Kaplan, E. (1962). A human cerebral disconnection syndrome. *Neurology, 12*, 675–685.

Goodglass, H. & Kaplan, E. (1963). Disturbance of gesture and pantomime in aphasia. *Brain, 86*, 703–720.

Graff-Radford, N.R., Welsh, K., & Godersky, J. (1987). Callosal apraxia. *Neurology, 37*, 100–105.

Hécaen, H. & Rondot, P. (1985). Apraxia as a disorder of a system of signs. In E.A. Roy (Ed.), *Studies of Apraxia and Related Disorders*. North Holland: Elsevier.

Heilman, K.M. (1973). Ideational apraxia—A re-definition. *Brain, 96*, 861–864.

Heilman, K.M., Coyle, J.M., Gonyea, E.F., & Geschwind, N. (1973). Apraxia and agraphia in a left hander. *Brain, 96*, 21–28.

Heilman, K.M., Rothi, L.J., & Valenstein, E. (1982). Two forms of ideomotor apraxia. *Neurology, 32*, 342–346.

Hughlings Jackson, J. (1932). Remarks on non-protrusion of the tongue in some cases of aphasia. In J. Taylor (Ed.), *Selected Writings of John Hughlings Jackson*. London: Hodder & Stoughton.

Kimura, D. & Archibald, Y. (1974). Motor function of the left hemisphere. *Brain, 97*, 337–350.

Kleist, K. (1907). Apraxie. *Jarbuch für Psychiatrie und Neurologie, 28*, 46–112.

Liepmann, H. (1900). Das Krankheitshild der Apraxie (motorischen/Asymbolie). *Monatschrift für Psychiatrie und Neurologie, 8*, 15–44, 102–132, 182–197.

Liepmann, H. (1905). Die linke Hemisphare und das Handeln. *Münchener Medizinische Wochenschrift, 49*, 2322–2326, 2375–2378.

Liepmann, H. (1920). Apraxia. *Ergebnisse der Gesamten Medizin, 1*, 516–543.

Liepmann, H. & Maas, O. (1907). Fall von linksseitiger Agraphie und Apraxie bei rechsseitiger Lahmung. *Zeitschrift für Psychologie und Neurologie, 10*, 214–227.

Ochipa, C., Rothi, L.J.G., & Heilman, K.M. (1989). Ideational apraxia: A deficit in tool selection and use. *Annals of Neurology, 25*, 190–193.

Ochipa, C., Rothi, L.J.G., & Heilman, K.M. (1992). Conceptual apraxia in Alzheimer's disease. *Brain, 115*, 1061–1071.

Penfield, W. & Welch, K. (1951). The supplementary motor area of the cerebral cortex. *Archives of Neurology and Psychiatry, 66*, 289–317.

Pick, A. (1905). *Sudien uber Motorische Apraxia und ihre Mahestenhende Erscheinungen*. Leipzig: Deuticke.

Poeck, K. (1983). Ideational apraxia. *Journal of Neurology, 230*, 1–5.

Poizner, H., Mack, L., Verfaellie, M., Rothi, L.J.G., & Heilman, K.M. (1990). Three dimensional computer graphic analysis of apraxia. *Brain, 113*, 85–101.

Rothi, L.J.G., Heilman, K.M., & Watson, R.T. (1985). Pantomime comprehension and ideomotor apraxia. *Journal of Neurology, Neurosurgery, and Psychiatry, 48*, 207–210.

Rothi, L.J.G., Mack, L., Verfaellie, M., Brown, P., & Heilman, K.M. (1988). Ideomotor apraxia: error pattern analysis. *Aphasiology, 2*, 381–387.

Rothi, L.J.G., Ochipa, C., & Heilman, K.M. (1991). A cognitive neuropsychological model of limb praxia. *Cognitive Neuropsychology, 8*, 443–458.

Roy, E.A. & Square, P.A. (1985). Common considerations in the study of limb, verbal and oral apraxia. In E.A. Roy (Ed.), *Neuropsychological Studies of Apraxia*. New York: Elsevier.

Sperry, R.W., Gazzaniga, M.S., & Bogen, J.E. (1969). Interhemispheric relationships: The neocotical commisures: Syndromes of hemisphere disconnecion. In P.J. Vinken & G.W. Bruyn (Eds.), *Handbook of Clinical Neurology,* vol. 4, (273–290) North Holland: Amsterdam.

Steinthal, P. (1871). *Abriss der Sprach wissenschaft.* Berlin.

Watson, R.T. & Heilman, K.M. (1983). Callosal apraxia. *Brain, 106,* 391-403.

Watson, R.T., Fleet, W.S., Rothi, L.J.G., & Heilman, K.M. (1986). Apraxia and the supplementary motor area. *Archives of Neurology, 43*, 787–792.

Zangwill, O.L. (1960). L'apraxie ideatorie. *Nerve Neurology, 106*, 595–603.

3 Handedness

Kenneth M. Heilman

INTRODUCTION

The most blatant behavioural asymmetry observed in people is hand preference or handedness. Approximately 90 % of people prefer to use their right hand, and the remainder either prefer their left hand or do not have a hand preference. There have been many books, articles, chapters, and abstracts written about handedness and an in-depth review would require a multivolume book rather than a brief chapter. In this chapter we plan to focus on the brain mechanisms that may account for hand preference.

Hand preference implies that there is some relative advantage of using the hand on one side of the body as opposed to using the opposite hand. Since it is primarily the contralateral side of the brain that controls (via the spinal cord and nerves) the muscles of the hand and arm on each side of the body, hand preference implies functional brain asymmetries. Liepmann, whom we discuss in Chapters 2 and 4, thought that the instructions for learned skilled movements were stored in movement representations or movement formulae and that these representations were stored in the hemisphere that is opposite to the preferred hand. However, there are other theories of functional brain asymmetries that may account for hand preference. Although this chapter will focus on praxis asymmetries, we will first briefly discuss other theories of hand preference, including; (1) learning theory; (2) language lateralisation; (3) attentional and intentional asymmetries; (4) elemental motor asymmetries such as strength, speed and precision, and (5) finally praxis asymmetries.

LEARNING THEORIES

Once a person develops a hand preference in the absence of a neurological disease, it is rare for this preference ever to change. Watson (1919) proposed that learning theories could account for hand preference, and there is strong evidence that social pressures and culture may influence hand preference. In general, there are more items in the environment that are more convenient for right handers to use than there are for left handers to use. In many cultures there are also social stigmas attached to being left handed. Although environmental, social, and cultural factors may influence the overall prevalence of handedness, there is also strong evidence that these factors cannot fully account for hand preference. Therefore, this topic will not be discussed further in this chapter, and the reader is referred to Porac and Coren (1977) or Coren (1992) for reviews.

LANGUAGE LATERALISATION

Paul Broca (1861) described a patient with nonfluent aphasia that was associated with a left hemisphere anterior perisylvian lesion and subsequently described a series of eight patients who were aphasic. All had left-hemisphere lesions and all were right handed (Broca, 1863), suggesting to Broca that the hemisphere that mediates speech is opposite to the preferred hand. Multiple subsequent studies have replicated Broca's original observation that, right handedness appears to be closely associated with left-hemisphere dominance for language. Broca posited that, in contrast to right handers, left handers were right-hemisphere dominant for speech and language.

There are several reasons why language lateralisation may influence hand preference. The cortico-spinal tract, which is important for the mediation of fine distal movements, is primarily a crossed system such that the left hemisphere controls movements of the right hand and the right hemisphere controls movements of the left hand.

When writing, the cortico-spinal system must receive information from those parts of the brain that mediate language. Whereas the left hemispheric motor systems would have direct access to this information, the right hemisphere would only have indirect access via the corpus callosum. Supportive evidence for this postulate comes from the observation that patients with callosal lesions are unable to write correctly with their left hand while retaining the ability to write with the right hand.

Although writing may be an important factor in hand preference, it alone cannot account for hand preference because in most cases hand preference develops before the child learns to write and even illiterate people have hand preferences. However, many actions that we perform with our hands and arms are in response to some form of verbal input and the left hemisphere of right handers is not only dominant for speaking and writing but also mediates comprehension of spoken and written language.

In addition to being unable to write with their left hand, patients with callosal disconnection may also demonstrate what has been termed an "alien" hand. In the past decade many types of alien hands have been reported, and the definition of alien hand has varied. Some use this term to refer to any hand or arm movement that is not voluntary: for example, the abnormal movement associated with cortical basal ganglia degeneration. Others have reserved the term for movements of a limb that are purposeful and goal directed but are not consistent with a person's own conscious goals or intentions. For example, I examined a patient with a surgical callosal disconnection for intractable epilepsy. The patient told me that one day when she was getting dressed she had on a blue outfit and wanted to put on blue shoes. Whereas her right hand picked up a blue pair of shoes, her left hand picked up a pair of red shoes. It appears that this patient was aware of the intention of her left hemisphere (i.e. blue shoes) and not those of her disconnected right hemisphere (i.e. red shoes). Conscious awareness of one's intentions may be related to language and "inner speech." The corpus callosum myelinates fairly late in development and does not complete myelination until after a child develops language and speech. Therefore, right hand preference, even in people without callosal disconnection, may be related to the "privileged" access that the left hemisphere has to the right hand during development such that the right hand more easily carries out conscious intentions.

Studies of patients with unilateral hemispheric strokes and studies of patients undergoing selective hemispheric anaesthesia (Wada testing) for preoperative evaluation of intractable epilepsy have demonstrated that there are individuals who, in spite of having strong right-hand preference, have their language mediated by their right hemisphere. Perhaps even more dramatic is the observation that the majority of left handers have language mediated by their left hemisphere (see Porac & Coren, 1977, for a review). Therefore, whereas hemispheric language dominance may have an important influence on hand preference, because there are dissociations between language dominance and hand peference, the side of language dominance cannot solely account for hand preference.

INFLUENCE OF ATTENTION AND INTENTION

Hands interact with tools and objects in the environment. The environment contains many stimuli, some relevant and some irrelevant. Therefore, many tasks require the use of attentional systems to regulate the processing of incoming stimulus information. In addition, motor systems need to be prepared for responding to these environmental stimuli. This selective preparation for motor action we term intention.

Studies of brain-impaired individuals (Heilman, Watson, & Valenstein, 1993), normal individuals (Verfaellie, Bowers, & Heilman, 1988a; 1988b), and electrophysiological studies in monkeys (Bushnell, Goldberg, & Robinson,

1981; Goldberg & Bushnell, 1981) all suggest that spatially directed attention and intention are mediated by different neuronal systems. Although attention and intention are mediated by anatomically distributed systems, the parietal lobe appears to play a critical role in attention and the frontal lobes in intention (see Heilman et al., 1993 for a detailed review). There are, however, strong reciprocal connections between the frontal and parietal lobes, and behaviourally it appears that attention can influence intention and intention can influence attention (Verfaellie & Heilman, 1990).

In addition to the separability but interconnectiveness of these systems, there is also evidence that the attentional and intentional systems are asymmetrically organised in humans. For example, whereas the left hemisphere primarily attends to the right side of space, the right hemisphere attends to both sides of space (Heilman & Van Den Abell, 1980). Similarly, whereas the left hemisphere primarily prepares the right-sided limbs for movement, the right hemisphere prepares both sides for movement (Heilman & Van Den Abell, 1979).

To learn how attentional and intentional asymmetries and their interactions may influence hand preference, Verfaellie and Heilman (1990) studied right- and left-hand reaction times in normal individuals to stimuli presented in either right or left hemispace. Prior to presenting the imperative (reaction time) stimulus, they provided their participants either motor intentional or sensory attentional cues and measured how these cues influenced reaction times. When a cue directs attention to one side of space, the attentional systems of the hemisphere that is contralateral to the cue may be activated, and the activated attentional systems may activate the motor preparatory systems of either one or both hemispheres. Differences in reaction times between the two hands when attention is directed to the right vs. left side by an attentional cue may reveal asymmetrical activation of the right and left hemisphere's intentional systems. Similarly, when an intentional cue (which gives advanced warning as to which hand will likely be called upon to move) prepares either the right or left hand to respond, the intentional systems of the corresponding hemisphere may in turn activate the attentional systems of the same or both hemispheres.

Verfaellie and Heilman (1990) found that directing attention to the right side of space resulted in faster reaction times for the right than left hand. This finding suggests that the attentional systems of the left hemisphere primarily activate the intentional systems of the left hemisphere. In contrast, no differences between hands were found when attention was directed toward the left hemispace, suggesting that attentional systems of the right hemisphere activate the intentional systems of both hemispheres.

Because the cortical spinal system primarily projects to its contralateral hand, the left hand is optimally prepared to respond primarily when attention is directed to the left hemispace. In contrast, the right hand is prepared to respond to stimuli in either right or left hemispace. When the right hand was prepared to respond by an intentional cue, there was no difference in reaction times between responses to stimuli in right vs. left hemispace. However, when the left hand was prepared to respond, the responses were more rapid to stimuli presented in left than right hemispace. These findings suggest that the intentional systems of the right hemisphere primarily activate attentional systems of the right hemisphere, but the intentional systems of the left hemisphere activate the attentional systems of both hemispheres.

Overall these attentional and intentional asymmetries provide a right-hand advantage because attending to stimuli in either side of space prepares the right hand for action and when the right hand is prepared for action one can equally attend to either side of space.

ELEMENTAL MOTOR ASYMMETRIES

When selecting a hand to perform a unimanual task, we may select the hand based on the demands of the task. We may need the hand–arm that has the greater strength, or that can move more rapidly, or that can make more precise and accurate movements.

Many studies have demonstrated that the preferred hand is often the hand that is stronger, can move more rapidly, or is more accurate than the non-preferred hand (see Porac & Coren, 1977). Unfortunately the neuronal asymmetries that account for these performance asymmetries remain unknown. Whereas muscle mass may play a crucial role in determining strength, there is not an invariate relationship between muscle mass and strength. In addition, neurotrophic factors (e.g. the manner in which muscle is innervated) strongly influence muscle mass. Although muscle mass may play a role in strength, the patterns of neuronal activity that activate these muscles may also play a crucial role. Unfortunately, little is known about the physiologic asymmetries that may account for asymmetries of strength. Similarly, asymmetries of speed and precision may also be related to asymmetrical neuronal firing patterns. Asymmetrical patterns of neuronal innervation may also account for these elemental motor asymmetries.

Triggs, Calvanio, Macdonnell, Cros, and Chiappa (1994), using transcranial magnetic stimulation, found in right handers that the threshold for activation of muscles in the right arm was lower than the threshold for activation of the corresponding muscles in the left arm. Left handers showed the opposite pattern. Wassermann, McShane, Hallet, and Cohen (1992) found that a reduced threshold to magnetic stimulation was associated with larger cortical representations. Therefore, the motor neurons in the spinal cord (anterior horn cells) on the preferred side may receive input from more cortical spinal neurons than do the anterior horn cells on the opposite side of the cord. Foundas, Hong, Leonard, and Heilman (1995) using magnetic resonance images found that in right handers the hand–arm region of Brodmann area 4 was larger in the left than right hemisphere. Non-right handers did not show this asymmetry. In addition, Kertesz and Geschwind (1971) measured the cortical spinal tract at a level after this tract decussates. They found that, in right handers, the tract on the right side appeared to have more fibres than the tract on the left. However, in the study of Kertesz and Geschwind, participants who were left handed had a similar asymmetry. The innervatory asymmetry that accounts for elemental motor asymmetries may be more distal in the nervous system such that muscles on the preferred side may receive more direct input from the anterior horn cells than do the muscles on the non-preferred side. This would produce the situation where there are a smaller number of muscle fibres innervated by each anterior horn cell on the preferred side than there are on the nonpreferred side (i.e. the ratio of anterior horn cell to muscle fibres is higher on the preferred than nonpreferred side). To our knowledge this hypothesis has not been tested in right or left handers.

Although asymmetries of strength, speed, or precision may influence the hand that one selects for a specific task and thereby influence overall hand preference, there are people with right-hand preference who are stronger, faster, and more accurate with their left hand than their right hand and similarly, there are left handers who are stronger, faster, and more precise with their right than left hand (see Porac & Coren, 1977). Therefore, whereas elemental motor asymmetries may play an important role in hand preference, these cannot be the sole determinant of hand preference.

PRAXIS

Many skilled motor tasks that we perform require complex postures and movements where one or more joints have to be moved at different amplitudes while other joints have to be fixed. To perform skilled acts, these joint movements or fixations must take place in concert. Many of these joint movements or fixations must also take place in a sequential manner, and different portions of the movement must take place at different speeds.

Liepmann and Maas (1907) reported a patient with a right hemiplegia who, when attempting to perform skilled learned acts to command with his left hand, was impaired. On postmortem examination he was found to have a lesion of his pons that accounted for his right hemiplegia and an infarction of his corpus callosum. More recently, Watson and Heilman (1983) and Graff-Radford, Welsh, and Godersky (1987) also reported patients with callosal lesions who were unable to carry out skilled acts with their left hand but, because they were not hemiparetic, they could normally carry out skilled acts with their right hand. Since the time of Paul Broca and Karl Wernicke it has been repeatedly demonstrated that for most right handers it is the left hemisphere that mediates language. A callosal lesion, therefore, could have disconnected the left hemisphere that mediates language from the right hemisphere that contains the cortical spinal system that controls the left hand. However, these patients could also not correctly imitate skilled acts or correctly perform skilled acts when provided with the actual tool or object. A language motor disconnection could not account for the failure to imitate or use actual tools or objects because imitation and actual tool or object use do not require language. Liepmann and Maas (1907) posited that the left hemisphere not only mediates language but also contains what they termed "movement formulas." These movement formulas contained the time–space–form pictures of the skilled movement. Lesions of the corpus callosum, therefore, not only disconnect left-hemisphere mediated language systems from the right hemisphere, but also disconnect these movement representations.

The loss of the ability to perform learned skilled movements is termed apraxia. In right handers, apraxia is often caused by left-hemisphere lesions and apraxia is often associated with aphasia. The association of apraxia with aphasia has led to the proposal that both are primarily a defect of symbolisation, aphasia being a defect in verbal symbolisation and apraxia being a defect in nonverbal symbolisation. However, Goodglass and Kaplan (1963) demonstrated that there was not a direct relationship between the severity of apraxia and the severity of aphasia. Liepmann noted that only 14 of his 20 apraxic patients were aphasic, and Heilman (1975) reported that not all aphasic patients are apraxic. Because there is a disassociation between apraxia and aphasia, there is little evidence to support the hypothesis that apraxia is a disorder of symbolic behaviour.

Liepmann thought that hand preference was related to the lateralisation of these "movement formulae" or movement representations. If these movement representations are lateralised to one hemisphere and one were to perform a skilled learned unimanual task, the neural apparatus that controls the arm that is contralateral to these representations would have direct access to these representations and the arm ipsilateral to these representations would only have indirect access (via the corpus callosum). Similarly, when learning a new unimanual skill, the hand and arm that is contralateral to these movement representations would have direct access to the area of the brain that stores these representations and the hand and arm that is ipsilateral to these representations would only have indirect access. Taylor and Heilman (1980) studied the relationship between handedness and motor learning. If one has lateralised representations and is learning a new skill, the hand opposite these representations should not only acquire the skill more rapidly but also, when a skill is learned by the nonpreferred hand, the acquisition of the skill should benefit the preferred hand more than vice versa. Taylor and Heilman's results supported these predictions.

Further support of Liepmann's handedness hypothesis comes from observations of patients with focal lesions. Heilman, Coyle, Gonyea, and

Geschwind (1973) reported a patient who as a child preferred to use his left hand for skilled unimanual activities. However, in school the teachers required him to write with his right hand, and as an adult he continued to write with his right hand. When the patient developed a left hemiparesis from a large right-hemisphere cerebral infarction, he was surprised to learn that he could no longer write with his right hand. Agraphia may be related to either linguistic or praxic deficits. Because this patient could correctly spell words, his deficit appeared to be related to motor production deficit, possibly apraxic agraphia. When we tested his ability to pantomime other skilled acts he performed poorly, demonstrating an ideomotor apraxia. He was, however, not aphasic. Because he was not aphasic and had a large right hemispheric perisylvian lesion, it was probably his left hemisphere that was mediating language. His preferred hand, therefore, was opposite the hemisphere that contained his movement representations and not necessarily the hemisphere that mediated language.

Selective hemispheric anaesthesia, the Wada test, is used to determine language dominance in patients who are being considered for epilepsy surgery. Wada studies of left handers has revealed that the majority of left handers are left-hemisphere dominant for language and speech (Milner, 1974). The Wada studies in these left-handed patients reveal that language dominance cannot account for hand preference.

Similarly, although crossed aphasia (aphasia that is associated with a hemispheric lesion that is ipsilateral to the preferred hand) has been repeatedly reported, cases of crossed apraxia (apraxia that is associated with a hemispheric lesion that is ipsilateral to the preferred hand) are exceedingly rare. Whereas these observations would appear to support Liepmann's postulate that handedness is determined by the laterality of movement representations, Rapscak, Gonzalez-Rothi, and Heilman (1987) reported a patient who preferred to use his right hand, but developed apraxia and aphasia with a right-hemisphere lesion. Although this case would appear to refute Liepmann's hypothesis, Rothi, Ochipa, and Heilman (1991) put forth and provided support for a modular system that mediates praxis (see Chapter 4). Although the patient reported by Rapscak et al. (1987) was clearly apraxic, he may have been apraxic not because his lesion destroyed the movement formulae or praxic representations, but rather because he injured another portion of the modular praxis system (discussed in Chapter 4).

It is possible that the development of lateralised praxic representations is related to our dependence upon other nonpraxic functional asymmetries. For example, as we discussed in right handers, the left-hemisphere cortico-spinal system is larger than the right, and most right-handed subjects are more deft with their right than left hand. Therefore, because of asymmetrical deftness when learning a new skill, right handers may prefer to use their right hand and the use of this right hand that is controlled by and gives feedback to the left hemisphere is important in the development of left-hemisphere praxic representations. However, this postulate cannot explain why some people who prefer to use their right hand for learned skilled acts are more deft with their nonpreferred left hand.

Although the laterality of praxic representation may play an important role in determining hand preference, we believe that hand preference is influenced by many factors, including but not limited to those factors discussed in this brief chapter. This is why, perhaps, no one has been able to develop a model that can entirely account for the genetics of hand preference.

REFERENCES

Broca, P. (1861). Remarques sur le siège de la faculté du langage articulé, suivies d'une observation d'aphemie. *Bulletins de la Société Anatomique de Paris, 2*, 330–357.

Broca, P. (1863). Localisation des fonctions cérébrales siège du langage articule. *Bulletins de la Société Anthropologique de Paris, 4*, 200–204.

Bushnell, M.C., Goldberg, M.E., & Robinson, D.L. (1981). Behavioral enhancement of visual responses in monkeys' cerebral cortex I. *Journal of Neurophysiology, 46*, 755–772.

Coren, S. (1992). *The left-hander syndrome. The causes and consequences of left handedness.* New York: Free Press.

Foundas, A.L., Hong, K., Leonard, C.M., & Heilman, K.M. (November 1995). *Hand preference and MRI asymmetries of the human central sulcus.* Paper presented at Society for Neuroscience meeting, San Diego, CA.

Goldberg, M.E. & Bushnell, M.C. (1981). Behavioral enhancement of visual responses in monkey cerebral cortex II. *Journal of Neurophysiology, 46*, 773–787.

Goodglass, H. & Kaplan, E. (1963). Disturbance of gesture and pantomime in aphasia. *Brain, 86*, 703–720.

Graff-Radford, N.R., Welsh, K., & Godersky, J. (1987). Callosal apraxia. *Neurology, 37*, 100–105.

Heilman, K.M. (1975). A tapping test in apraxia. *Cortex, 11*, 259–263.

Heilman, K.M., Coyle, J.M., Gonyea, E.F., & Geschwind, N. (1973). Apraxia and agraphia in a left-hander. *Brain, 96*, 21–28.

Heilman, K.M. & Van Den Abell, T. (1979). Right hemispheric dominance for mediating cerebral activation. *Neuropsychologia, 17*, 315–321.

Heilman, K.M. & Van Den Abell, T. (1980). Right hemisphere dominance for attention: The mechanisms underlying hemispheric asymmetries of inattention (neglect). *Neurology, 30*, 327–330.

Heilman, K.M., Watson, R.T., & Valenstein, E. (1993). Neglect and related disorders. In K.M. Heilman & E. Valenstein (Eds.), *Clinical neuropsychology.* New York: Oxford University Press.

Kertesz, A. & Geschwind, N. (1971). Patterns of pyramidal decussation and their relationship to handedness. *Archives of Neurology, 24*, 326–332.

Liepmann, H. & Maas, O. (1907). Fall von linksseitiger Agraphie und Apraxie bei rechsseltiger Lahmung. *Zeitschrift für Psychologie und Neurologie, 10*, 214–227.

Milner, B. (1974). Hemispheric specialization: Scope and limits. In T.O. Schmitt & F.G. Worden (Eds.), *The neurosciences: Third study program.* Cambridge, MA: MIT Press.

Porac, C., & Coren, S. (1977). *Lateral preferences and human behavior.* New York: Springer-Verlag.

Rapscak, S.Z., Gonzalez-Rothi, L.J., & Heilman, K.M. (1987). Apraxia in a patient with atypical cerebral dominance. *Brain and Cognition, 6*, 450–463.

Rothi, L.G., Ochipa, C., & Heilman, K.M. (1991). A cognitive neuropsychological model of limb praxis. *Cognitive Neuropsychology, 8*, 443–458.

Taylor, G. & Heilman, K.M. (1980). Left-hemisphere motor dominance in right handers. *Cortex, 16*, 587–603.

Triggs, W.J., Calvanio, R., Macdonell, R.A.L., Cros, D., & Chiappa, K.H. (1994). Physiological motor asymmetry in human handedness: Evidence from transcranial magnetic stimulation. *Brain Research, 636*, 270–276.

Verfaellie, M., Bowers, D., & Heilman, K.M. (1988a). Hemispheric asymmetries in mediating intention, but not selective attention. *Neuropsychologia, 26*, 521–532.

Verfaellie, M., Bowers, D., & Heilman, K.M. (1988b). Attentional factors in the occurrence of stimulus-response compatibility effects. *Neuropsychologia, 26*, 435–444.

Verfaellie, M. & Heilman, K.M. (1990). Hemispheric asymmetries in attentional control: Implications for hand preference in sensorimotor tasks. *Brain and Cognition, 14*, 70-80.

Wassermann, E.M., McShane, L.M., Hallett, M., & Cohen, L.G. (1992). Noninvasive mapping of muscle representations in human motor cortex. *Electroencephalography and Clinical Neurophysiology, 85*, 1–8.

Watson, J.B. (1919). *Psychology from the standpoint of a behaviorist*. Philadelphia: Lippincott.

Watson, R.T. & Heilman, K.M. (1983). Callosal apraxia. *Brain, 106*, 391–403.

4 A Cognitive Neuropsychological Model of Limb Praxis and Apraxia

Leslie J. Gonzalez Rothi, Cynthia Ochipa, and
Kenneth M. Heilman

INTRODUCTION

The cognitive neuropsychological approach is one that draws liberally on empirical analyses of the behavioural components developed in cognitive psychology; at the same time it is an endeavor to understand the abberent behaviour resulting from brain injury. This chapter is an attempt to further develop a cognitive neuropsychological model of the processing system that mediates normal limb praxis and helps us account for apraxia.

Coltheart (1984) has noted that emphasis on the simultaneous explanation of normal as well as neuropsychologically abnormal information processing found in cognitive neuropsychology is not of recent origin. In fact, this approach characterised the work of neurologists of the late 19th and early 20th centuries such as Wernicke, Lichtheim, and Liepmann, whom Head (1926) subsequently dismissed as the "diagram makers". Because Liepmann contributed most significantly to our understanding of apraxia and because we find the beginnings of our present day model of limb praxis processing in his work, we will begin our discussion with a brief review of his contributions.

Asymbolia versus Apraxia

Steinthal (1871) was the first to use the term apraxia to refer exclusively to a collection of disorders where the common feature was the inability to perform motor activity correctly on command; however, the presumed explanation for the mechanism of this disorder was quite different than our understanding today.

One thought was that disorders of skilled movement were the result of an inability to recognise the implement (tool or object) associated with a particular movement: in other words, a form of agnosia. But how an agnosia could induce an inability to perform skilled movements to verbal command remained unanswered. This inability to perform purposeful movements in response to commands was also described in the late 1800s (and for many years to come) only superficially as an aside to the presence of aphasia. In 1870, Finkelnburg (Duffy & Liles, 1979) referred to aphasia as "asymbolia" reflecting a deficit in the use of all symbols including gesture as well as verbal language. Again, one unanswered question was how an "asymbolia" explanation of apraxia could explain failure to perform skilled movements in an imitation task.

In 1900 Liepmann (1977) described the case of a right-handed civil servant who suddenly developed limb apraxia as the result of "apoplexy," stating that the case presentation needed no justification because the patient's deficits were so extraordinary that the importance of the case was readily evident. It was Liepmann's intent, by carefully describing the behavioural exam of this case, to establish apraxia as a disorder of motor planning rather than a manifestation of a recognition disorder such as agnosia or of asymbolia.

That aphasia and apraxia did not represent an underlying "asymbolia" was further addressed by Liepmann (1980) in 1905 who reviewed the gesture performance (on a number of tasks) of 83 cases classified as either left- or right-hemisphere lesioned on the basis of side of hemiplegia. He noted that of 42 left-paralytic patients, it was "rare" to find any who had difficulty with performance on the gesture tasks. In contrast, with the 41 right-paralytic cases studied, he noted "definite disturbances" of praxis performance in no less than 20 of the patients. Liepmann noted that with these 20 patients who were aphasic as well as apraxic, "...it makes no difference if one demonstrates the required movement for them. They are not able to succeed even when speech has had no influence on the information." Thus, Liepmann implied that the praxis and language disorders did not emanate from a common cause nor did the language disorder in these 20 patients induce the praxis failure. The controversy of the relationship between apraxia and aphasia continues today however (see Duffy & Duffy, 1990; Duffy, Duffy, & Pearson, 1975; Duffy, Watt & Duffy, 1981; Kempler, 1988; Square-Storer, Roy, & Hogg, 1990).

The results of his study of the 83 paralytic patients led Liepmann to postulate what is today credited as the original description of the mechanism underlying limb apraxia. He proposed that in right handers the guidance of skilled movements of both the left- and right-sided limbs was predominately the responsibility of the left hemisphere. More specifically, he suggested that the left sensorimotor cortex guides not only the "arm center" of the right arm but also the left arm (right hemisphere) via the corpus callosum. Therefore, an apraxia could occur from a lesion of the left sensorimotor cortex that would affect both right and left hands (possibly masked in the right hand by hemiparesis). In comparison, with a corpus callosum lesion, apraxia would also occur but only of the left hand. Just such a case was reported by Liepmann and Maas in 1907 and in recent times by others (Boldrini, Zanella, Cantagallo, & Basaglia, 1992; Geschwind & Kaplan, 1962; Tanaka, Iwasa, & Obayashi, 1990; Watson & Heilman, 1983).

In addition to proposing the novel postulate of left-hemisphere predominance in skilled movement planning of right handers and of specialisation of function that is inherent in his proposals, Liepmann proposed a "vertical schema" that included distinct functional components dedicated to praxis processing. He believed that the acquisition of skilled limb movements required the acquisition of "movement formulae," of "innervatory patterns" and of "kinetic memories" for overlearned movement segments.

In 1905 Liepmann (1980) proposed that these "movement formulae" contained the "time–space–form picture of the movement" or "space-time sequences." He defined them as "...general knowledge of the course of the procedure to be realized." Though sensory in nature and most commonly visual, this form of knowledge might also be represented in other sensory modalities when inherently necessary to the target action. The "innervatory patterns," which were proposed by Liepmann to be acquired through practice, were described as providing efficiency in the "transform(ation of)...movement formulae promptly and precisely into innervation" yielding a "...position(ing) of the limbs according to directional ideas." Finally, the kinetic memories referred to by Liepmann (1980) involved ". . . a functional linkage tak(ing) place between innervations...which runs its course without innervation from orientation and visual images, through short-circuiting." We (Rothi & Heilman, 1996) have noted that "an obvious obligation of this kinetic memory portion of the praxis system is that these memories require highly practised and familiar movement associations. Therefore, the kinetic memory portion may or may not contribute to the implementation of any particular action plan depending on that plan's component parts."

Disconnection Model of Apraxia

Geschwind (1965), like Liepmann, believed that the left hemisphere of the right-handed individual was responsible for the performance of skilled movements of both hands. Geschwind proposed a neuronal circuit for the mediation of skilled movements that is similar to that proposed by Wernicke (1874) to account for speech. He suggested that pantomime to command requires information to flow sequentially from the auditory pathway to Heschyl's gyrus, and in turn to the posterior superior temporal lobe (Wernicke's area). Information subsequently flows to the motor association cortex for control of the right hand. For left-handed movements, information flows to the right hemisphere premotor area via the corpus callosum. According to Geschwind's (1965) schema, lesions of the supramarginal gyrus or arcuate fasciculus would result in an apraxia by disconnecting the posterior language areas from the anterior motor association area. Because Wernicke's area is spared, patients with such lesions should be able to comprehend commands, but should not be able to perform skilled movements to verbal command. However, if the disconnection is limited to the association between Wernicke's area and the motor association cortex, patients should be able to imitate gestures; they cannot, however. To explain this discrepancy, Geschwind noted that fibres passing from the visual association cortex to the premotor cortex also course anteriorly through the arcuate fasciculus. He proposed that the left hemisphere arcuate fasciculus was dominant for these visuomotor connections. Thus, a lesion of the arcuate fasciculus would disrupt both the auditory and visual pathways and would result in impaired performance to command and to imitation. In keeping with the disconnection hypothesis is the observation that actual implement use often appears normal. However, performance with actual implements may be better than performance to command or imitation because use of actual implements reduces the number of possible errors and also provides strong cues (i.e. affordances). In addition, even though the use of actual implements is commonly reported to be performed better than pantomime, several studies (De Renzi, Faglioni, Scarpa, & Crisi, 1986; De Renzi, Faglioni, & Sorgato, 1982; De Renzi & Lucchelli, 1988; Pilgrim & Humphreys, 1991; Poizner, Soechting, Bracewell, Rothi, & Heilman, 1989; Watson, Fleet, Rothi, & Heilman, 1986) demonstrated that performance with actual implements is impaired. Because there is little evidence to suggest that the arcuate fasciculus also carries somaesthetic information to the premotor area, the disconnection hypothesis could not by itself account for impaired implement use. Geschwind (1965) was aware of this conundrum and suggested that perhaps somaesthetic information reaches the premotor areas by using visual pathways.

A REPRESENTATIONAL MODEL OF APRAXIA

Although the studies of Liepmann, as well as Geschwind's extension and interpretation of Liepmann's work, were the critical initial steps in developing a model of praxis processing, many questions about this model remain. The development of new constructs and the use of newer research strategies developed by cognitive psychologists to study normal language processing have been applied to studies of aphasic patients. This cognitive neuropsychological approach led to studies that not only informed us as to the pathophysiology of aphasic behaviours but also led to the development of new models of normal language function. Unfortunately, similar collaboration between those who perform research on apraxia and those who perform research on normal motor control has been somewhat limited (but see Harrington and Haaland chapter, or Paillard, 1982, for notable exceptions). Therefore, the remainder of this chapter is devoted to developing a composite model of praxis processing and providing support for this model by describing a series of dissociations noted in the performance of apraxic patients. Hopefully, this model will not only account for much of the present apraxia data, but more importantly will generate an empirical link to experimental studies of motor control in normal individuals and provide an impetus for further cognitive neuropsychological research with apraxic patients.

A number of praxis dissociations have been noted in the performance of neurologically impaired patients, including the separability of praxis production from praxis reception, the selectivity of input modalities of the praxis system, an "assembled" (not addressed) route of praxis imitation, and the possible fragmentation of semantics into an action and a nonaction system. Each of these dissociations will be discussed separately and noted in a series of figures that will culminate in a cumulative model of praxis processing. We begin with Liepmann's notion of "movement formulae" that are stored and addressed.

Praxis Production Versus Praxis Reception: An Elaboration of Liepmann's Work

To acquire skilled motor behaviour implies that the central nervous system stores information that the individual has previously experienced and that this stored information expedites future behaviour. Therefore, rather than portions of the process being reconstructed *de novo* with each experience, they may be called up from memory and reutilised. We use the term "processing-advantage" to describe the assistance provided by a system that can be called upon to reconstitute previously constructed programs.

Based on the assumption that the movement formulae or memories described by Liepmann are represented in the human brain and, by definition, that they provide a processing-advantage when performing motor acts with which the individual has prior experience, Heilman, Rothi, and Valenstein (1982) and Rothi, Heilman, and Watson (1985) proposed that there are at least two types of mechanisms that could account for the performance deficit associated with apraxia. First, a degraded memory trace would yield difficulties with both the reception and production of gesture and, second, a memory egress disorder would yield only a gesture production deficit.

Liepmann (1980) referred to disorders of his proposed movement formulae as resulting from left posterior lesions that involved the temporal-parietal interface and made a vague reference to a past undocumented presentation where he had suggested that the left supramarginal gyrus supported these formulae. Heilman et al. (1982) quite specifically proposed that these movement memories are stored in the left inferior parietal lobe. Therefore, a degraded memory trace as described above would be found in patients with left parietal injury and a difficulty in memory egress alone would not. Heilman et al. (1982) tested 20 patients with unilateral left-hemisphere lesions using a gesture-to-command task and also a gesture discrimination paradigm. On the discrimination task, each subject was shown a film of a man performing pantomimes. On each of 32 trials, the patient was asked to select from three choices which pantomime best represented a target pantomime named by the examiner. Of the patients found to be apraxic, we found that there was a group that could discriminate gestures and another group that were deficient in gesture discrimination. Although these apraxic groups did not differ in the degree of gesture production deficit, they did differ in the localisation of the lesion producing their deficits. Specifically, those apraxic patients with lesions that involved the left parietal lobe had gesture discrimination difficulties whereas those with apraxia from lesions that were more anterior and did not involve the left parietal lobe did not have a gesture discrimination problem. These findings are consistent with Liepmann's proposal that there are "movement formulae" or memories that guide the skilled purposive movements of both hands. When these memories are degraded, as in lesions of the left parietal lobe, the inability to produce or discriminate gestures (i.e. mode/modality consistency) results (see lesion A of Fig. 1) (a functional notion supported more recently by McDonald, Tate, & Rigby, 1994). In contrast, when these gesture memories can no longer interact with the part of the brain responsible for generating "innervatory patterns" (which are discussed later) (see lesion B of Fig. 1) or the innervatory patterns cannot gain access to the motor area for implementation, the result is apraxia without gesture discrimination difficulties (i.e. mode/modality inconsistency).

The Innervatory Pattern. To perform a skilled movement, the arm must traverse a critical set of spatial loci in a specific temporal pattern. Even though the final effectors are muscles activated by motor nerves, these motor nerves are controlled by cortico-spinal neurons found in the precentral gyrus of the frontal cortex (Brodmann's area 4). Because it has been posited that the movement formulae stored in the inferior parietal lobe are coded in a three-dimensional supramodal code before a skilled movement takes place, this space–time representation must be transcoded into an innervatory pattern. Although Geschwind (1965) thought that convexity premotor cortex was important for praxis, apraxia has not been reported from a lesion limited to this frontal cortical area (which did not also involve deep structures), and the function of this area in humans remains uncertain.

Electrical stimulation of the precentral motor cortex induces single simple movements. Stimulation of the medial premotor cortex, called the supplementary motor area (SMA), however, induces complex movements of the fingers, arms, and hands (Penfield & Welch, 1951). The SMA receives projections from the parietal lobe and projects to the neurons in the motor cortex. SMA neurons also discharge before motor neurons (Brinkman & Porter, 1979). Studies of

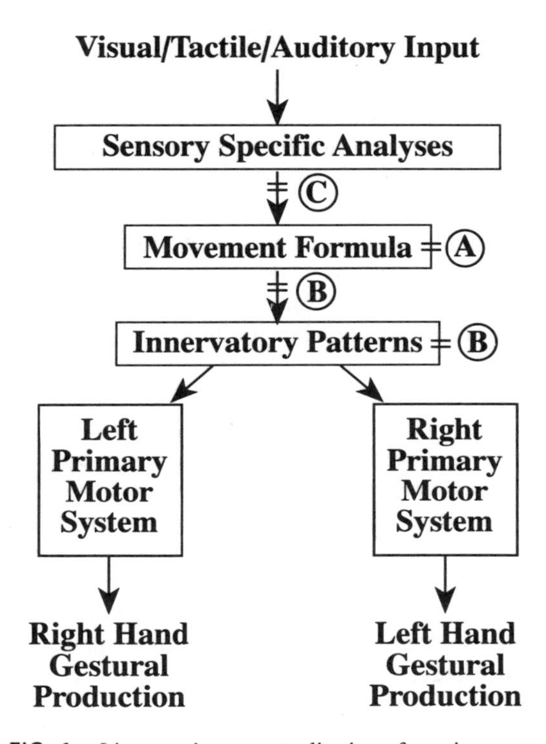

FIG. 1. Liepmann's conceptualisation of praxis processing.

cerebral blood flow, which reflects metabolic activity and hence neuronal activity, reveal that although simple repetitive movements increase activity in the contralateral motor cortex, complex movements increase activity of both the motor cortex and the SMA (Roland, Larsen, Lassen, & Skinhoj, 1980). When participants plan complex movements but do not move, activity is increased in the SMA but not in the motor cortex.

Based on these observations, it would appear that the SMA would be ideally suited to transcode space–time representations of skilled limb movements into the innervatory pattern which would provide information to the motor cortex for implementation, possibly utilising the assistance of the movement "packages" characteristic of the kinetic memories. Liepmann (1980) suggested that these kinetic memories were "properties of the senso-motorium."

Watson, et al. (1986) described several patients with left SMA lesions who had bilateral apraxia for transitive (i.e. involving tool use, such as a hammer) but not intransitive (i.e. not involving tools or objects, such as hitchhiking) movements. Unlike patients with parietal lesions, these patients could comprehend and discriminate pantomimes.

The Action-lexicon. Similarities between how the brain processes language and how the brain processes praxis were posited by Rothi and Heilman (1985, p. 66) when they proposed that "the area of visuokinesthetic motor engrams programs skilled motor movements in much the same way as Wernicke's area programs linguistic acts." In language, the term "lexicon" is defined as that portion of the language system that provides a processing advantage for words that the language user has had prior experience. We propose that the term lexicon, as defined here, might also apply (as a general analogy) to the "movement memories" of Liepmann and the "visuokinesthetic motor engrams" of Heilman and Rothi (1985). We will refer to these movement representations as the information contained in the "action lexicon."

Input versus Output Action Lexicons. The model of praxis processing represented in Fig. 1 was designed to account for the reception (comprehension and/or discrimination) and production (pantomime to command) of skilled movements of the left and of the right hands in response to visual, tactile, or verbal input. The same model can also be used to account for the ability to imitate gestures. The model in Fig. 1 suggests that imitation failure should occur concomitant with a gesture reception deficit, with a gesture production deficit (access or egress disorders), or as the cumulative effect of the simultaneous occurrence of reception and production difficulties (loss or degradation of the representation itself called a "degraded trace" by some investigators). However, Bell (1994) gave 23 patients with single left hemisphere lesions tests of pantomime recognition and imitation and found that performance on these two tasks did not correlate. Mehler (1987) and Ochipa, Rothi, and Heilman (1994) report three patients who imitated limb movements significantly worse than they received or produced limb movements to command. Mehler (1987) described two cases who displayed an inability to "imitate bilateral, intransitive, nonsymbolic hand or arm movements or positions" in the context of no significant sensory or motor difficulties and no impairments with actual implement use, pantomime to command, or gesture reception. Unfortunately, Mehler (1987) did not report testing the imitation of transitive or symbolic gestures. In contrast, Ochipa, Rothi, and Heilman (1994) report a patient who was asked to comprehend, produce to command, and imitate the same transitive gestural stimuli. They report that the patient showed no difficulties with pantomime comprehension but was significantly worse on imitation than on pantomime to command with the same stimuli. This increased difficulty with imitation cannot be accounted for by either a receptive defect alone or as the cumulative effect of receptive and productive difficulties as the patient had no difficulty with gesture reception. To account for this spared gesture reception in the context of these productive difficulties, we separated input and output processing of praxis requiring division of the action lexicon into an input action lexicon (containing "information relative to a code about the physical attributes of a 'perceived-action' "; Rothi & Heilman, 1996) and an output action lexicon (containing "a code about the physical attributes of a 'to-be-performed action' "; Rothi & Heilman, 1996) similar to that found in models of word recognition and production (Fig. 2).

Spared gesture reception in the presence of impaired imitation could be accounted for by dysfunction within this lexical "connection" at some point after egress from the input action lexicon. Because pantomime to command was less severely impaired than imitation, it would suggest that spoken language gains access to the output action lexicon without having to be processed by the input action lexicon (Fig. 3). Therefore, impairment in pantomiming to com-

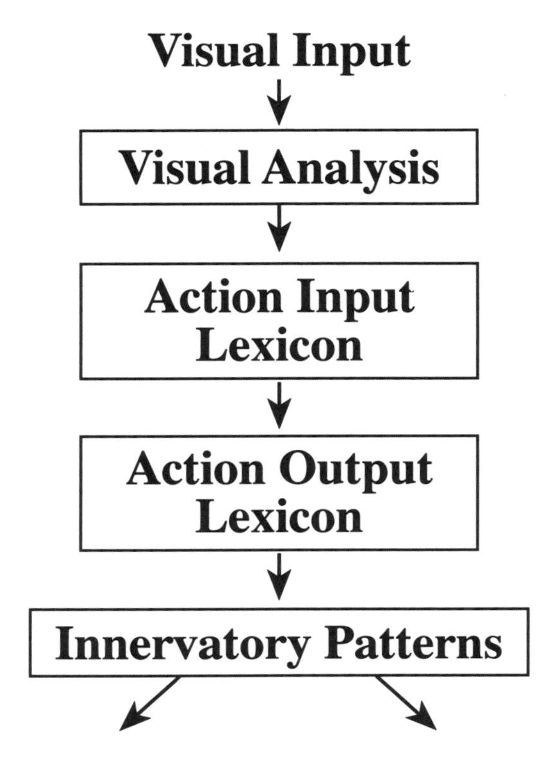

FIG. 2. Modification of Liepmann's model to account for selective impairment of gesture imitation.

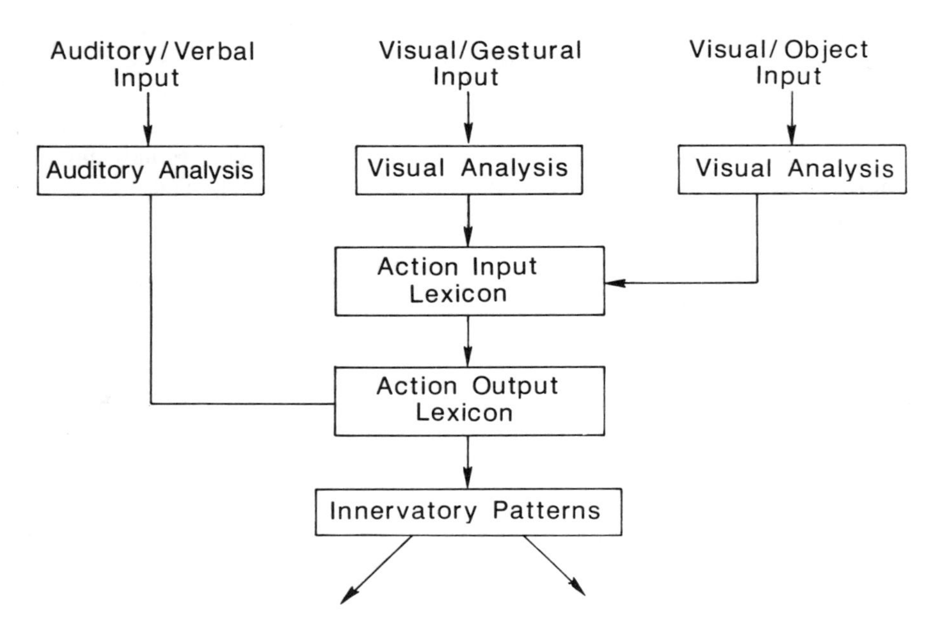

FIG. 3. Modification of praxis model to account for selective modality deficits.

mand together with an impairment of gesture imitation would be the result of dysfunction at or after access to the output action lexicon. Impairment of gesture imitation to a greater degree than gesture to command is not well described by a single level of deficit within this model. It may be that poorer gesture imitation implies more than one level of deficit (for example, a problem at both the level of the output action lexicon as well as between the input and output action lexicons) that in cumulation creates a proportionately larger problem with gesture imitation than gesture to command. A definitive explanation awaits further study.

Nonlexical Action Processing

In this section we review the possibility that, in addition to a lexically based system, there may also be a nonlexical action processing system available for gesture imitation.

Rothi, Mack, and Heilman (1986) reported patients who could imitate gestures that they could not comprehend or discriminate. This discrepancy between gesture comprehension and imitation raises a processing dichotomy that is not accounted for by the model represented in Fig. 2. This dichotomy between comprehension and imitation also has a parallel in disorders of language processing as well. Specifically, patients with transcortical sensory aphasia can imitate (i.e. repeat) but not comprehend spoken language. This fractionation of comprehension and imitation in language processing of course implies that, at least at some levels, they are functions that do not share unitary underlying mechanisms. Beauvois, Derouesne, and Bastard (1980) have suggested that the ability to repeat spoken language may, in fact, involve at least two mechanisms; one that processes the material lexically and one that processes it nonlexically.

In contrast, for the processing of skilled limb movement, the model represented in Fig. 2 implies that the ability to imitate gesture adequately requires the same input processing as that required for gesture comprehension, namely visual analysis and access to the input action lexicon. We propose that, although imitation then requires access to the output action lexicon, comprehension requires access to semantics, and the patients described by Rothi et al. (1986) could not gain access to semantics from their input action lexicon. However, the alternative explanation (parallel to that described by Beauvois et al., 1980; Coslett, Roeltgen, Rothi, & Heilman, 1987; and McCarthy & Warrington, 1984, for imitation of spoken language) is that gesture imitation in our cases could theoretically be performed without using the action lexicon: in other words, a nonlexical route of gesture imitation (see Fig. 4).

It would follow that because normal individuals can imitate meaningless gestures, a "nonlexical" action processor must be available for imitation of this novel material. The nonlexical imitation of action may be mediated by a direct route between visual systems and motor systems (e.g. visual association and premotor cortex). That an independent "nonlexical" action process exists is further supported by Mehler's (1987) patient who had deficits of imitation limited to nonfamiliar limb movements. Because the movements were novel, there were no movement memories or "lexical addresses" in the proposed action lexicons that could be accessed. In addition, patients with ideomotor apraxia often improve on imitation tasks. Is it possible that the patient with ideomotor apraxia who improves on imitation is the one who has selective sparing of this "direct" or nonlexical route of praxis imitation? In turn, is it possible that the patient with ideomotor apraxia who does not improve on imitation is the case who has impairment of both of the routes described in Fig. 4?

Input Modality Selectivity

In this section we review a number of cases that provide evidence that there is selective input into the action lexicons according to modality. Rothi et al., (1986) reported two patients who performed gestures to command but who could not comprehend or discriminate visually presented gestures. These patients had left occipito-temporal lesions that may have prevented visual stimuli from gaining access to the input action lexicon (see lesion C of Fig. 1). The

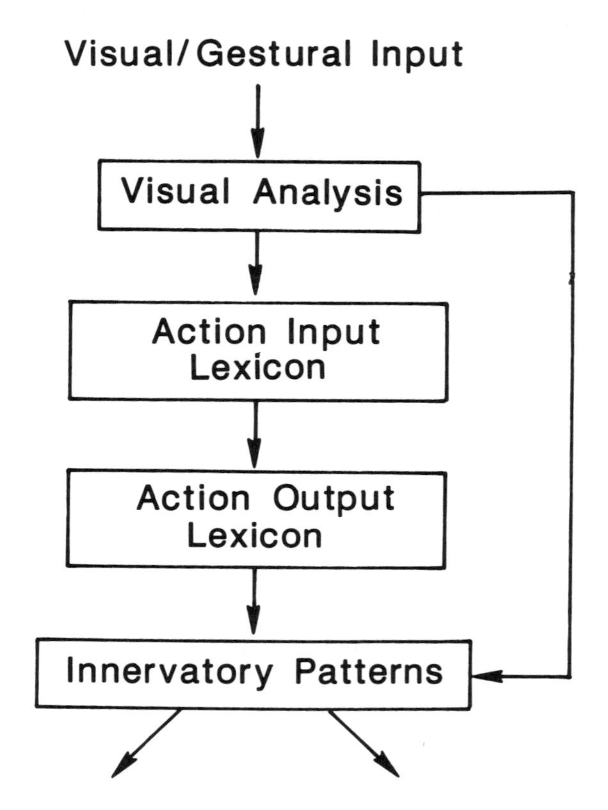

FIG. 4. Modification of praxis model to account for selective sparing of gesture imitation.

pantomime reception deficit in these cases may be similar to the word deafness seen with temporal lobe lesions that prevent processed auditory signals from gaining access to Wernicke's area (the phonological input lexicon). Heilman (1973) described three cases who were impaired when pantomiming to command but who imitated pantomimes flawlessly. One interpretation of these cases is that they suffered an impairment in the ability of spoken language to gain access to the output action lexicon. Unfortunately, these cases were not tested for their ability to comprehend and discriminate pantomimes. However, based on our model, one would expect that their functions would not be impaired.

In addition, patients have been described who have difficulty with pantomiming in response to stimuli presented in specific modalities. For example, Rothi et al. (1986) described cases who pantomimed the function of named tools relatively well; however, when asked to pantomime the function of pictured tools, one patient was 100 % accurate whereas one was never accurate. Coslett and Saffran (1989), Kaplan, Verfaellie, and Caplan (1990), Peña-Casanova,

Roig-Rovira, Bermudez, and Tolosa-Sarro (1985), and Raymer, Greenwald, Richardson, Rothi, and Heilman (1992) also described cases who gestured poorly to visual (picture or seen object) command, though they were able to produce gesture to auditory command. In cases where visual object recognition deficits are ruled out, Peña-Casanova et al. (1985) termed this disorder optic apraxia. De Renzi, et al. (1982) performed a group study and not only found results similar to those of Heilman (1973) and Rothi et al. (1986), but also found patients who had difficulty even in the somaesthetic modality. It would appear that there is a fractionation of input modalities feeding into the action input lexicon, suggesting that the input systems to the action lexicon should be specified for modality as well as the nature of the input material. Again, this is similar to that found in models of word recognition and production. Therefore, we suggest in Fig. 3 the addition of separate input systems for visually presented gestural information, visually presented tools, and auditorily presented verbal information.

Action Semantics

In this section we review the proposal presented by Roy and Square (1985) and supported by others (Benke, 1993; Clark, Merians, Kothari, Poizner, Macauley, Rothi, & Heilman, 1994) that praxis processing is mediated by a two-part system involving both "conceptual" and "production" components. According to these authors, the praxis conceptual system involves three kinds of knowledge; knowledge of the functions that tools and objects may serve, knowledge of actions independent of tools, and knowledge about the organisation of single actions into sequences. We define a tool as that which is used to provide a mechanical advantage in an action, and we define an object as the recipient of an action. As we have described in the preceding sections, the praxis production system involves the sensorimotor component of action knowledge, including the information contained in action programs and the translation of these programs into action. Thus action is dependent upon the interaction of conceptual knowledge related to tools, objects, and actions (what we call action semantics) and the structural information contained in motor programs. Within this framework, the syndrome known as ideomotor apraxia would result with a destruction of the praxis production system. In contrast, a failure of the praxis conceptual system would not produce the production disorder characteristic of ideomotor apraxia. Instead, the failure would involve difficulties in recognising the specific mechanical advantages provided by tools, the specific mechanical requirements to achieve an action goal (possibly implied by an object), and so on. In 1905 Liepmann (1980) described a patient whom he felt suffered a deficit of this conceptual system; he displayed a variety of praxis error types, including tool substitution (using a razor as a comb) and inappropriate action selection (placing his glasses on his outstretched tongue). Liepmann felt that these (as well as other) types of errors reflected a difficulty with the ideation of tool use and in turn labelled it ideational apraxia. That these errors were qualitatively different from those errors produced by cases he termed ideomotor apraxia led him to believe that they reflected distinctly different problems (and in turn were potentially separable systems). The operational definition of ideational apraxia has become rather diluted in more recent history. For example, that the sequencing of multistage actions could be disordered became a predominate dimension of the disorder such that ideational apraxia was tested exclusively using tasks requiring sequential action segments such as writing, addressing, and mailing a letter. Therefore, to focus the issue on the conceptual dimension of Liepmann's description of ideational apraxia, Ochipa, Rothi, and Heilman (1992) used the term conceptual apraxia to describe exclusively the apraxia resulting from a disturbance of the praxis conceptual system and not the sequencing deficit noted by others (Poeck, 1983).

Evidence that conceptual apraxia reflects a specific disruption of the praxis conceptual system was provided in a case described by Ochipa, Rothi, and Heilman (1989). Following a right-hemisphere stroke, this left-handed patient was noted using implements inappropriately in natural settings. For example, he was observed eating with a toothbrush and brushing his teeth with a spoon and a comb. Interestingly, this patient was able to name these same tools and point to them on command, suggesting that he was not agnosic. However, he could not point to tools when their function was described, nor could he verbally describe tool function. His deficit could not be attributed to a language comprehension deficit as he could follow one- and two-step commands not involving tool function. His inability to use tools could not be explained solely by a production deficit (severe ideomotor apraxia which extended to actual tool use) because he could not match tools with the objects on which they were used, suggesting an impairment in the appreciation of the functional relationship between tools and the objects they act upon. Thus, this patient's inability to use tools suggested an impairment of conceptual knowledge related to tool use; that is, an impairment of the conceptual praxis system or action semantics. More recently, Motomura and Yamadori (1994) describe a case with sparing of object pantomime in the context of impaired object use resulting from a focal, ischemic lesion. Heilman, Maher, Greenwald, and Rothi (1995) have also described conceptual apraxia resulting from focal unilateral left-hemisphere, ischemic lesions although most commonly it has been described as the result of more diffuse diseases such as Alzheimer's disease (Ochipa, Rothi & Heilman, 1992; Schwartz, Marin & Saffran, 1979). As these studies have reported impaired praxis conceptual knowledge in the context of a spared action production system, it is interesting to note that Rapcsak, Ochipa, Anderson, and Poizner (1995) also describe the praxis processing of a case with a progressive disease but in this instance it is the praxis production system that is impaired while the praxis conceptual system is relatively spared.

We believe that to account for the representational nature of gesture in general, including not only pantomime but also emblems (e.g. signs such as okay, crazy, etc.), at least some interaction between semantics and the action lexicons must be specified. Therefore, in Fig. 5 we have added the semantic system. We need to acknowledge a controversy that presently exists, however, about the nature of the proposed semantic system. Specifically in relation to action processing, this controversy is raised by reports of optic aphasia where the patients are able to pantomime the function of visually presented tools they are unable to name. This occurs in the context of the spared ability to perform tactile and auditory naming of these same tools. The explanation of some (Beauvois & Saillant, 1985; Shallice, 1987) of optic aphasia as well as the modality-specific aphasias is that there are multiple semantic systems that reflect differences in the modality of input as well as the nature of the material processed. In addition, these multiple semantic systems communicate with one another. The multiple notations of sparing of pantomime in the context of inability to name through vision alone have led some to reflect that at least in the case of optic aphasia, the evidence points only to a fractionation of "action semantics" from all other semantic information. Ochipa et al. (1992) and Raymer (1992) in separate studies each examining praxis knowledge in groups of Alzheimer's patients but using different experimental paradigms both support this notion of an action semantic system that is at some level dissociable from other forms of semantic knowledge. But for others (Riddoch, Humphreys, Coltheart, & Funnell, 1988, p. 9), the interpretation of gestural sparing in optic aphasia has led to the notion that "it seems possible that gestures could be made on the basis of other nonsemantic forms of information (the perceptual attributes of tools or following access to stored structural knowledge)." Riddoch, Humphreys, and Price (1989) and Pilgrim and Humphreys (1991) extended this notion of direct access between structural knowledge about tools and action output in cases of unilateral apraxia as well.

During our initial attempt to develop this praxis model (Rothi, Ochipa, & Heilman, 1991), we attempted to indicate recognition of the difference of opinion expressed in regard to optic aphasia when Riddoch et al. (1988) suggested a direct connection between stored structural knowledge about implements (object recognition system) and the output action lexicon, possibly through the input action lexicon (as indicated in Fig. 5 by dotted lines). The alternative, also indicated by a dotted line in Fig. 5, is the Shallice (1987; 1988) proposal of multiple semantic systems, at least as far as postulating a distinction for action semantics specifically. Subsequently, we have altered this model slightly to solidify the distinction between action semantics and other forms of conceptual knowledge (Ochipa et al., 1992; Raymer, 1992) and believe that this form of conceptual knowledge can be differentially damaged by disease as well as focal brain lesions. In addition, on the basis of a study by Raymer, Rothi, and Heilman (1995) whereby 12 patients with Alzheimer's disease (and 12 normal

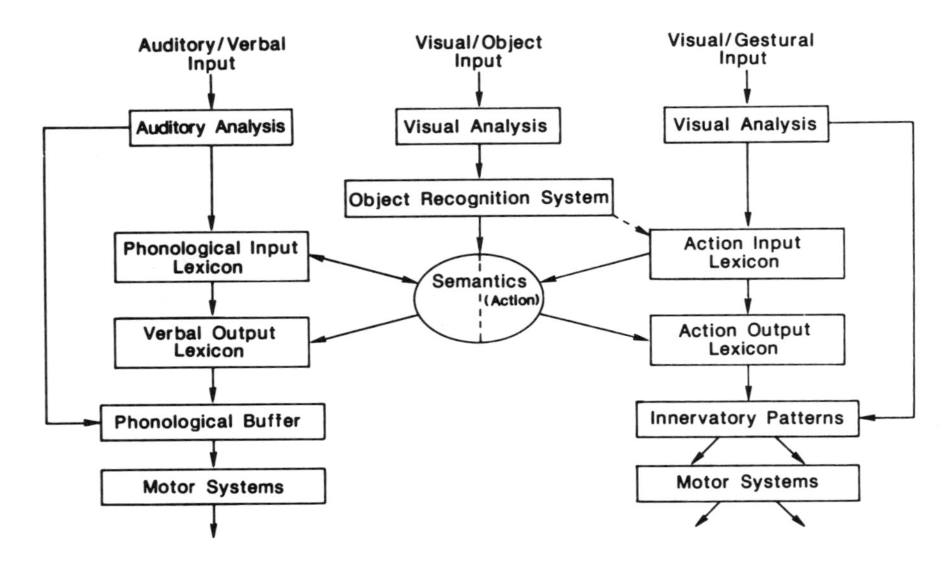

FIG. 5. Rothi, Ochipa, and Heilman's 1991 version of a cognitive neuropsychological model of limb praxis.

controls) were administered 7 experimental semantic tasks contrasting modality of input (verbal name, verbal description of function, and viewed tools) and mode of output (verbal name, verbal description of function, and gestures). In looking at individual data, Raymer and colleagues (1995) noted that some patients were able to gesture to and to name viewed tools while unable to gesture to or to name verbally described functions of the same tools (presumably requiring access to the semantic system). This finding supports the conclusion that, as proposed by Riddoch, Humphreys, and colleagues (Riddoch et al., 1988; Riddoch et al., 1989; Pilgrim & Humphreys, 1991), there is a direct communication between the structural descriptions of viewed tools and stored action production knowledge as shown in Fig. 6. (See Chapter 5 for a more extensive discussion of action semantics.)

CONCLUSION

In summary, brain-impaired individuals may have impaired production, imitation, and/or reception of skilled limb movements. Based on studies and reports of apraxic individuals, we have developed a cognitive neuro–psychological model of the systems that may mediate the performance of these learned, skilled movements. In many respects this model is similar to that proposed for language processing, with specific modalities having input into an input action

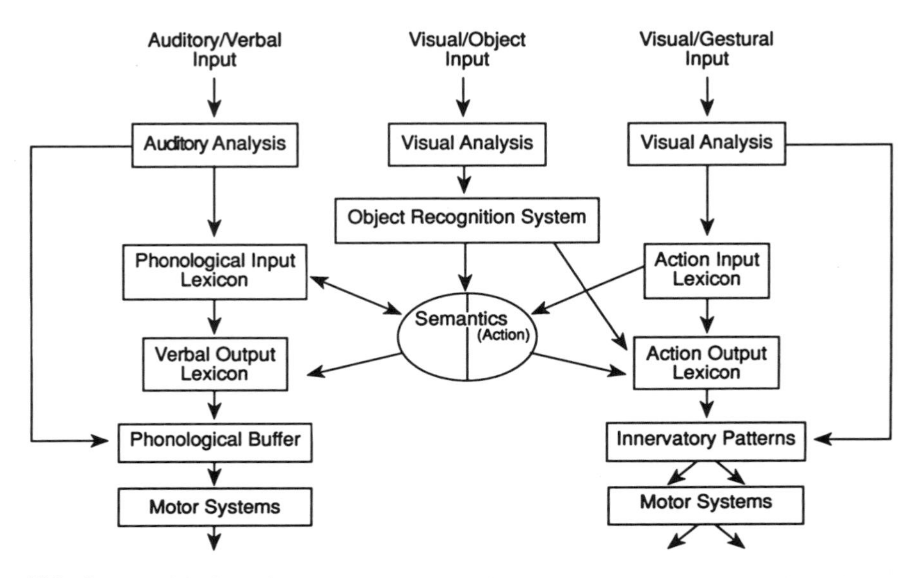

FIG. 6. A model of praxis processing and its relation to semantics, naming, and word and object recognition.

lexicon. The input action lexicon has afferent input into systems that contain knowledge about the results of action and how tools may influence these results (action semantics). This system of action semantic knowledge is at least at some level dissociable from other forms of semantic knowledge. The input action lexicon also has input to an output action lexicon, from the structural descriptions of viewed tools and objects, and from semantics—most specifically from action semantics. The output action lexicon contains "time–space" representation for skilled movements or "movement formulae". These "time–space representations" are subsequently transcoded into innervatory patterns and finally these innervatory patterns are played out by the motor systems.

Although a complex model has been proposed, many questions remain unanswered and many questions remain to be asked. For example, there are cases of apraxia that have been described that cannot be explained by this model (see Pilgrim & Humphreys, 1991, and Riddoch et al., 1989, for examples). It is not at all uncommon in apraxia research to find group designs where an apraxia group is defined by a single test such as pantomime imitation with the suggestion that the groups are equated on all variables influencing praxis processing. At the very least, the complexity of this praxis model underscores the need to expand our testing methods to include more precision in determining the functional locus of praxis failure. In addition, the notion that "apraxia," even terms such as "ideomotor apraxia," fail to recognise the complexity of the system and the potential for multiple explanations for a particular gestural outcome. To answer the many unanswered or unasked questions left in apraxia and skilled movement planning, more elaborate and meticulous methodology than that traditionally used in apraxia research will have to be used. We hope this model will be an impetus for the development of such methodological precision and will allow new opportunities for the development of research that links apraxia and normal praxis performance.

ACKNOWLEDGEMENT

This chapter is a revised version of a paper that originally appeared as Rothi, L. J. G., Ochipa, C., & Heilman, K. M. (1991). A cognitive neuropsychological model of limb praxis. *Cognitive Neuropsychology*, 8, 443–458.

REFERENCES

Beauvois, M. F., Derouesne, J., & Bastard, V. (1980). *Auditory parallel to phonological alexia.* Paper presented at the Third European Conference of the International Neuropsychological Society, Chianciano, Italy.

Beauvois, M.F. & Saillant, B. (1985). Optic aphasia for colours and colour agnosia: A distinction between visual and visuo-verbal impairments in the processing of colours. *Cognitive Neuropsychology, 2,* 1–48.

Bell B. D. (1994). Pantomime recognition impairment in aphasia: An analysis of error types. *Brain and Language, 47,* 269–278.

Benke, T. (1993). Two forms of apraxia in Alzheimer's disease. *Cortex, 29,* 715–725.

Boldrini, P., Zanella, R., Cantagallo, A., & Basaglia, N. (1992). Partial hemispheric disconnection syndrome of traumatic origin. *Cortex, 28,* 135–143.

Brinkman, C. & Porter, R. (1979). Supplementary motor area in the monkey: Activity of neurons during performance of a learned motor task. *Neurophysiology, 42,* 681–709.

Clark, MA., Merians, A. S., Kothari, A., Poizner, H. Macauley, B., Rothi, L. J. G., & Heilman, K. M. (1994). Spatial planning deficits in limb apraxia. *Brain, 117,* 1093–1106.

Coltheart, M. (1984). Editorial. *Cognitive Neuropsychology, 1,* 1–8.

Coslett, H. B., Roeltgen, D. P., Rothi, L. G., & Heilman, K. M. (1987). Transcortical sensory aphasia: Evidence for subtypes. *Brain and Language, 32,* 362–378.

Coslett, H. B., & Saffran, E. M. (1989). Preserved object recognition and reading comprehension in optic aphasia. *Brain, 112,* 1091–1110.

De Renzi, E., Faglioni, P., Scarpa, M., & Crisi, G. (1986). Limb apraxia in patients with damage confined to the left basal ganglia and thalamus. *Journal of Neurology, Neurosurgery and Psychiatry, 49,* 1030–1038.

De Renzi, E., Faglioni, P., & Sorgato, P. (1982). Modality-specific and supramodal mechanisms of apraxia. *Brain, 101,* 301–312.

De Renzi, E. & Lucchelli, F. (1988). Ideational apraxia. *Brain, 111,* 1173–1185.

Duffy, J. R., Watt, F., & Duffy, R. J. (1981). Path analysis: A strategy for investigating multivariate causal relationships in communication disorders. *Journal of Speech and Hearing Research, 24,* 474–490.

Duffy, R. J., Duffy, J., & Pearson, K. (1975). Pantomime recognition in aphasia. *Journal of Speech and Hearing Research, 18,* 115–132.

Duffy, R. J. & Duffy, J. R. (1990). The relationship between pantomime expression and pantomimic recognition in aphasia: The search for causes. In G. R. Hammond (Ed.),

Advances in psychology: Cerebral control of speech and limb movements (417–449). Amsterdam: Elsevier.

Duffy, R. J. & Liles, B. Z. (1979). A translation of Finkelnburg's (1870) lecture on aphasia as asymbolia with commentary. *Journal of Speech and Hearing Disorders, 44*, 156–168.

Geschwind, N. (1965). Disconnexion syndromes in animals and man. *Brain, 88*, 237–294.

Geschwind, N. & Kaplan, E. (1962). A human cerebral disconnection syndrome. A preliminary report. *Neurology, 12*, 675–685.

Head, H. (1926). *Aphasia and kindred disorders of speech.* Cambridge: Cambridge University Press.

Heilman, K. M. (1973). Ideational apraxi—a re-definition. *Brain, 96*, 861–864.

Heilman, K. M., Maher, L. H., Greenwald, L., & Rothi, L. J. G. (1995). Conceptual apraxia from lateralized lesions. *Neurology, 45*, A266.

Heilman, K. M. & Rothi, J. G. (1985). Apraxia. In K. M. Heilman & E. Valenstein (Eds.), *Clinical Neuropsychology* (131–150). New York: Oxford University Press.

Heilman, K. M., Rothi, J., & Valenstein, E. (1982). Two forms of ideomotor apraxia. *Neurology, 32*, 342–346.

Kaplan, R. F., Verfaellie, M., & Caplan, L. R. (1990). Visual anomia and visual apraxia in a patient with a left occipital brain lesion. *Journal of Clinical and Experimental Neuropsychology, 12*, 89.

Kempler, D. (1988). Lexical and pantomime abilities in Alzheimer's disease. *Aphasiology, 2*, 147–159.

Liepmann, H. (1977). The syndrome of apraxia (motor asymboly) based on a case of unilateral apraxia. (A translation from *Monatschrift für Psychiatrie und Neurologie*, 1900, 8, 15–44). In D. Rottenberg & F. H. Hockberg (Eds.), *Neurological classics in modern translation*. New York: Macmillan.

Liepmann, H. (1980). The left hemisphere and action. (A translation from *Münchener MedizinischeWochenschrift*, 1905, 48–49). Translations from Liepmann's essays on apraxia. In *Research Bulletin #506*. Department of Psychology, The University of Western Ontario.

Liepmann, H. & Maas, O. (1907). Fall von linksseitiger Agraphie und Apraxis bei rechstseitiger Lahmung. *Zeitschrift für Psychologie und Neurologie, 10*, 214–227.

McCarthy, R. & Warrington, E. K. (1984). A two-route model of speech production: Evidence from aphasia. *Brain, 107*, 463–485.

McDonald, S., Tate, R. L., & Rigby, J. (1994). Error types in ideomotor apraxia: A qualitative analysis. *Brain and Cognition, 25*, 250–270.

Mehler, M.F. (1987). Visuo-imitative apraxia. *Neurology, 37*, 129.

Motomura, N. & Yamadori, A. (1994). A case of ideational apraxia with impairment of object use and preservation of object pantomime. *Cortex, 30*, 167–170.

Ochipa, C., Rothi, L. J. G., & Heilman, K. M. (1989). Ideational apraxia: A deficit in tool selection and use. *Annals of Neurology, 25*, 190–193.

Ochipa, C., Rothi, L. J. G., & Heilman, K. M. (1992). Conceptual apraxia in Alzheimer's Disease. *Brain, 115*, 1061–1071.

Ochipa, C., Rothi, L. J. G., & Heilman, K. M. (1994). Conduction apraxia. *Journal of Neurology, Neurosurgery and Psychiatry, 57*, 1241–1244.

Paillard, J. (1982). Apraxia and the neurophysiology of motor control. *Philosophical Transactions of the Royal Society of London, B 298*, 111–134.

Peña-Casanova, J., Roig-Rovira, T., Bermudez, A., & Tolosa-Sarro, E. (1985). Optic aphasia, optic apraxia, and loss of dreaming. *Brain and Language, 26*, 63–71.

Penfield, W. & Welch, K. (1951). The supplementary motor area of the cerebral cortex. *Archives of Neurological Psychiatry, 66,* 289–317.
Pilgrim, E. & Humphreys, G. W. (1991). Impairment of action to visual objects in a case of ideomotor apraxia. *Cognitive Neuropsychology, 8,* 459–473.
Poeck, K. (1983). Ideational apraxia. *Journal of Neurology, 230,* 1–5.
Poizner, H., Soechting, J. F., Bracewell, M., Rothi, L. J. G., & Heilman, K. M. (1989). Disruption of hand and joint kinematics in limb apraxia. (Abstract.) *Society for Neuroscience, 15,* 481.
Rapcsak, S. Z., Ochipa, C., Anderson, K. C., & Poizner, H. (1995). Progressive ideomotor apraxia: Evidence for a selective impairment of the action production system. *Brain and Cognition, 27,* 213–236.
Raymer, A. M. (1992). *Dissociations of semantic knowledge: Evidence from Alzheimer's disease.* Doctoral dissertation, University of Florida.
Raymer, A. M., Greenwald, M. L., Richardson, M. E., Rothi, L. J. G., & Heilman, K. M. (1992). Optic aphasia and optic apraxia: Theoretical interpretations. *Journal of Clinical and Experimental Neuropsychology, 14,* 396.
Raymer, A. M., Rothi, L. J. G., & Helman, K. M. (1995). Nonsemantic activation of lexical and praxis output systems in Alzheimer's subjects. *Journal of the International Neuropsychological Society, 1,* 147.
Riddoch, M. J., Humphreys, G. W., Coltheart, M., & Funnell, E. (1988). Semantic systems or system? Neuropsychological evidence re-examined. *Cognitive Neuropsychology, 5,* 3–25.
Riddoch, M. J., Humphreys, G. W. & Price, C. J. (1989). Routes to action: Evidence from apraxia. *Cognitive Neuropsychology, 6,* 437–454.
Roland, P. E., Larsen, B., Lassen, N. A., & Skinhoj, E. (1980). Supplementary motor area and other cortical areas in organisation of voluntary movements in man. *Journal of Neurophysiology, 43,* 118–136.
Rothi, L. J. G., & Heilman, K. M. (1996). Liepmann (1900 and 1905): A definition of apraxia and a model of praxis. In C. Code C. W. Wallesch, A. R. Lecours & Y. Joanette (Eds.), *Classic cases in neuropsychology.* Hove, UK: Psychology Press.
Rothi, L. J. G. & Heilman, K. M. (1985). Ideomotor apraxia: Gestural learning and memory. In E. A. Roy (Ed.), *Neuropsychological studies in apraxia and related disorders* (65–74). New York: North Holland.
Rothi, L. J. G. Heilman, K. M., & Watson, R. T. (1985). Pantomime comprehension and ideomotor apraxia. *Journal of Neurology, Neurosurgery, and Psychiatry, 48,* 207–210.
Rothi, L. J. G., Mack, L., & Heilman, K. M. (1986). Pantomime agnosia. *Journal of Neurology, Neurosurgery, and Psychiatry, 49,* 451–454.
Rothi, L. J. G., Ochipa, C., & Heilman, K. M. (1991). A cognitive neuropsychological model of limb praxis. *Cognitive Neuropsychology, 8,* 443–458.
Roy, E. A. & Square, P. A. (1985). Common considerations in the study of limb, verbal, and oral apraxia. In E. A. Roy (Ed.), *Neuropsychological studies of apraxia and related disorders* (111–161). New York: North Holland.
Schwartz, M. F., Marin, O. S. M., & Saffran, E. M. (1979). Dissociations of language function in dementia: A case study. *Brain and Language, 7,* 277–306.
Shallice, T. (1987). Impairments of semantic processing: Multiple dissociations. In M. Coltheart, G. Sartori, & R. Job (Eds.), *The cognitive neuropsychology of language.* (111–127). Hove, UK: Lawrence Erlbaum Associates Ltd.
Shallice, T. (1988). Specialisation within the semantic system. *Cognitive Neuropsychology, 5,* 133–142.
Square-Storer, P. A., Roy, E. A., & Hogg, S. C. (1990). The dissociation of aphasia from apraxia of speech, ideomotor limb, and buccofacial apraxia. In G. E. Hammond (Ed.), *Cerebral Control of Speech and Limb Movements* (451–476). Amsterdam: Elsevier Science.
Steinthal, P. (1871). *Abriss der Sprachwissenschaft.* Berlin.

Tanaka, Y., Iwasa, H., & Obayashi, T. (1990). Right hand agraphia and left hand apraxia following callosal damage in a right-hander. *Cortex*, *26*, 665–671.
Watson, R. T., Fleet, S., Rothi, L. J. G., & Heilman, K. M. (1986). Apraxia and the supplementary motor area. *Archives of Neurology*, *43*, 787–792.
Watson, R. T. & Heilman, K. M. (1983). Callosal apraxia. *Brain*, *106*, 391–403.
Wernicke, C. (1874). *Der Aphasische Symptomencomplex*. Breslau: Cohn & Weigart.

5 Conceptual Praxis

Anastasia M. Raymer and Cynthia Ochipa

INTRODUCTION

Rothi and her colleagues (Rothi, Ochipa, & Heilman, 1991; Rothi, Ochipa, & Heilman, Chapter 4 of this volume) have proposed a cognitive neuropsychological model of the action processes involved in skilled movement. At the centre of their model is the semantic system, the core conceptual system which is the store of meaningful information for words and objects (tools and all other objects) such as category, functions, and other associated relationships (Caramazza, Hillis, Rapp, & Romani, 1990; Riddoch, Humphreys, Coltheart, & Funnell, 1988).

Allport (1985) proposed that the semantic system is a complex system that consists of multiple subsystems relevant to different input modalities and output modes of processing, including an action-oriented domain. These semantic subsystems are functionally dissociable in that brain damage can affect one subsystem selectively (Funnell & Allport, 1987). Shallice (1988a) has argued, however, that these semantic subsystems may not be discrete modules, but that they represent a distributed network in which regions are specialised for modalities and modes of information, one of which relates to action.

Rothi and colleagues (1991; 1996) incorporated into their model of action processing Allport's (1985) proposal that the semantic system may have a specialised subsystem of action semantic knowledge. In this chapter we will expand upon the notion of an action semantic subsystem and its role in action processing.

As discussed in previous chapters, in focusing specifically on the domain of action knowledge, Roy and Square (1985) proposed that the normal performance of an action is mediated by a two-part praxis processing system involving both "conceptual" and "production" components. According to these authors, the conceptual component of the praxis system involves three kinds of knowledge: knowledge of the functions that tools and objects may serve, knowledge of actions independent of tools, and knowledge about the serial organisation of single actions into a sequence. In contrast, the praxis production system involves the sensorimotor component of this knowledge, including the space–time information contained in action programs, and the translation of these programs into action.

In considering the cognitive neuropsychological model of action processing (see Chapter 4), the praxis production system encompasses action mechanisms beyond the semantic domain and will not be a point of further discussion. Our use of the term praxis conceptual system refers to knowledge subsumed in the action semantic subsystem. This is in contrast to Roy's notion of the praxis conceptual system which some may interpret as including knowledge relevant to the action semantic subsystem and the action output lexicon, mechanisms that Rothi and colleagues dissociate in a manner analogous to models of lexical processing.

The behavioural disturbance arising from the disruption of the praxis conceptual system has been termed ideational appraxia (De Renzi & Lucchelli, 1988; De Renzi, Pieczuro, & Vignolo, 1968; Liepmann, 1905; Ochipa, Rothi, & Heilman, 1989). Unfortunately, the term ideational appraxia has not been uniformly defined in the neuropsychological literature. Some authors, such as Poeck (1983), maintain that ideational apraxia refers only to those patients who are impaired in the sequencing of tool use. Others (De Renzi et al., 1968) describe ideational apraxia as the failure to use single tools appropriately. De Renzi and his colleagues (Barbieri & De Renzi, 1988; De Renzi & Lucchelli, 1988) consider ideational apraxia an amnesia of use in that there is a failure to recall the appropriate action when attempting to use a tool. To avoid confusion arising from the use of the term ideational apraxia, we prefer the term conceptual apraxia to refer to the behaviours which may result from impairment of the praxis conceptual system (Ochipa, Rothi, & Heilman, 1992). Because praxis conceptual knowledge may consist of different types of tool–action knowledge, a variety of action processing disturbances may come under the domain of conceptual apraxia.

In the first part of this chapter we will review neuropsychological literature that supports the notion that the semantic system may be composed of a number of knowledge subsystems, one of which relates specifically to conceptual aspects of praxis or action semantics. Subsequently, we will describe in more detail what the consequences of damage to the praxis conceptual system may be in patients with conceptual apraxia.

EVIDENCE FOR SEMANTIC SUBSYSTEMS

The structure of the semantic system has been the focus of much recent debate. Some researchers have proposed that meaning information is stored in a single unitary semantic system in which a semantic representation contains information relevant to all modes/modalities of processing (Caramazza et al., 1990; Riddoch et al., 1988; Seymour, 1973). One line of evidence for this view comes from the performance of patients who demonstrate quantitatively and qualitatively similar patterns of impairment among lexical tasks of naming, reading, and writing (e.g. Hillis, Rapp, Romani, & Caramazza, 1990). Abilities in gesture-processing tasks seldom have been reported in these patients.

Others propose that semantics can be isolated into subsystems of knowledge whose representations store mode/modality specific conceptual information (Allport, 1985; Beauvois & Saillant, 1985; Paivio, 1971, 1986; Shallice, 1988b). Paivio (1971, 1986) proposed that semantic subsystems are specialised for two different types of information: an imagery system which subserves nonverbal information concerning viewed items and events, and the language system devoted to meaning relevant to language. Allport (1985) elaborated upon this idea and proposed that the semantic system has specialised subsystems for all modalities and modes of processing. Neuropsychological evidence for semantic subsystems of knowledge comes from a number of different sources: optic aphasia, category-specific aphasias, and dissociations of praxis and language.

Optic Aphasia

Patients with optic aphasia, which was first described by Freund (1889) have difficulty naming visually presented pictures and actual objects (tools and other sorts of objects). However, they can name the same items presented through other modalities such as through touch or auditory definitions. Most striking is the observation that patients with optic aphasia are able to demonstrate recognition of unnamed items by either gesturing the appropriate use of the manipulable tools or describing their use. Their ability to correctly recognise viewed items argues against a visual agnosia as the basis for their visual modality-specific naming failure. (For recent reviews of this syndrome see Farah, 1990; or Iorio, Falanga, Fragassi, & Grossi, 1992).

Beauvois and Saillant (1985) noted that the pattern of impairment in optic aphasia is difficult to explain using a neuropsychological framework incorporating a unitary semantic system. Because these patients can name through sensory modalities other than vision, semantic and phonological lexical retrieval processes must be functioning adequately. The presence of a modality-specific deficit may implicate a defect at some stage in presemantic visual processing. However, patients provide gestures and describe the actions of viewed items, suggesting that presemantic visual processing is intact. To account for the dissociations observed in optic aphasia, researchers have advocated a number of interpretations (Farah, 1990; Hillis et al., 1990), including some proposals incorporating modifications in lexical/praxis models (Beauvois & Saillant, 1985; Coslett & Saffran, 1989; Hillis & Caramazza, 1995; Ratcliff & Newcombe, 1982; Raymer, Greenwald, Richardson, Rothi, & Heilman, 1992; Riddoch & Humphreys, 1987). (A review of all the theoretical accounts of optic aphasia is beyond the scope of this chapter.) Beauvois and Saillant (1985) proposed a lexical model incorporating modality-specific semantic systems for visual and verbal semantics, reminiscent of Paivio's (1971, 1986) theory, and suggested that optic aphasia results from an impairment in the interactions of the two semantic systems. Patients can provide gestures for viewed items, because visual semantic information is sufficient to activate action output systems for gesture. Patients cannot name viewed items because both visual and verbal semantic systems must be activated in order for visual information to access output phonological lexical systems for speaking.

If the visual/verbal modality-specific semantic system view is correct (Beauvois & Saillant, 1985), the situation is further complicated by the existence of modality-specific aphasias for auditory input (Denes & Semenza, 1975) and tactile input (Beauvois, Saillant, Meininger, & Lhermitte, 1978). By analogy, one may propose a complex structure for the semantic system, with mechanisms specified for all modalities of input including tactile semantics, auditory semantics, and so forth.

Category-Specific Aphasias

Other patients have been described whose naming and word comprehension abilities were selectively spared or selectively impaired for specific semantic categories such as living and nonliving things (Warrington & McCarthy, 1983), fruits and vegetables (Farah & Wallace, 1992; Hart, Berndt, & Caramazza, 1985), and animals (Hillis & Caramazza, 1991). Such dissociations may suggest that the semantic system is structured according to semantic category relationships. However, some researchers have proposed a more principled account for these divergent category deficits. Warrington and her colleagues (Warrington & McCarthy, 1983; Warrington & Shallice, 1984) initially explained the category-specific deficits for living versus nonliving things in relation to the type of semantic information that is critical for specification of knowledge in the two semantic categories. Function information (verbal semantics) is important for nonliving things and sensory feature information (visual semantics) is important for living things.

Warrington and McCarthy (1987) revised this proposal when they observed another patient whose dissociation was within the category of nonliving things. This patient experienced more difficulty comprehending names of small manipulable things than names of large things. Both categories presumably are related in that function information is crucial for their semantic interpretation. Therefore, Warrington and McCarthy considered the influence of motor information in the specification of the semantics of certain categories, an idea consistent with Allport's (1985) action-oriented semantic domain. Rothi and her colleagues (1991; Chapter 4) have proposed that this motor information could be termed "action semantics." This type of category-specific dissociation suggests that the semantic system may be fractionated in terms of both sensory input modality and output mode specificity.

Dissociations of Praxis and Language

The dissociation of verbal- and action-semantic knowledge has been supported in additional studies. As mentioned in Chapter 4, Ochipa, Rothi, and Heilman (1989) described a left-handed patient who, following a right-hemisphere stroke, could name tools that he did not know how to use. For example, he inappropriately used a toothbrush to eat, but when examined he could name eating utensils and the toothbrush. When presented with mechanical problems (e.g. "what instrument do you use for eating peas?"), he could not name the correct instrument (fork). This pattern may indicate impaired action semantic knowledge in the presence of preserved verbal semantic knowledge allowing for correct naming of tools that he did not know how to use. A patient with progressive language deterioration reported by Schwartz, Marin, and Saffran (1979) presented with the contrasting pattern of preserved action knowledge in the face of impaired verbal semantic knowledge. Specifically, they reported that their patient had "no trouble manipulating actual objects and utilizing them in motor sequences" (p.280), and she could provide accurate mimes for familiar objects she was completely unable to name.

A study of 32 patients with Alzheimer's disease (AD) (Ochipa et al., 1992) provided further evidence for the dissociation of verbal and action semantic knowledge. Ochipa and colleagues reported that a subset of their AD patients had preserved lexical semantic abilities in a word comprehension task, but nevertheless had significant impairments in tasks assessing tool and action knowledge (e.g. object/tool matching, providing gestures associated with incompleted object tasks). They proposed that this pattern reflected a dissociation between impaired action semantic and retained verbal semantic knowledge.

Because of concern that lexical semantic and action semantic knowledge was not tested for the same items in the prior study (Ochipa et al., 1992), a follow-up study was completed. Raymer (1992) tested 12 patients with Alzheimer's disease using a battery of seven naming and gesture tasks which incorporated as stimuli the same 20 tools across tasks. Results indicated no significant differences in accuracy of performance among naming and gesture tasks. However, correlations, controlling for dementia severity, indicated significant relationships among gesture tasks and among verbal tasks, but not between gesture and verbal tasks. One interpretation of these findings is that this pattern reflects differential impairments of action semantics and verbal semantics, although both semantic systems presumably are vulnerable to the effects of the pathology associated with AD, as the AD patients were significantly impaired across all verbal and gesture tasks compared to matched controls.

THE NATURE OF ACTION SEMANTIC KNOWLEDGE

Neuropsychological evidence reviewed thus far supports the notion that the semantic system is composed of a number of specialised subsystems. Earlier proposals suggested modality-specific semantic systems including verbal semantics and visual semantics. In addition, based on a number of neuropsychological studies we have discussed a mode specific action semantic system that can be dissociated from verbal semantics in some patients with stroke and Alzheimer's disease. We now review ideas about types of knowledge that may be subsumed under the action semantic subsystem, and describe deficits in patients that we propose reflect impairments of action semantic knowledge.

As noted earlier, the action semantic subsystem encompasses knowledge related to the praxis conceptual system proposed by Roy and Square (1985) to involve at least three kinds of tool–action knowledge: knowledge of functions of tools, knowledge of tool actions, and knowledge about the serial organisation of actions into a sequence. Hence, several types of action errors may be observed in patients with impairments related to the praxis conceptual system or conceptual apraxia.

Action Errors

Patients with conceptual apraxia may fail to recall the type of actions associated with tools or the objects they act upon (tool–object action knowledge). For example, when asked to pantomime the use of a key, the patient may produce an unrelated hammering movement, or may use an actual key as if it were a hammer. The patient may at other times produce a related movement such as turning a doorknob. We believe these semantically related and unrelated *content errors* result from a loss of knowledge or impaired access to knowledge concerned with tool action.

Patients with conceptual apraxia also may be unable to associate tools with objects that receive their action (tool–object associative knowledge) leading to the inappropriate selection of tools for particular tasks. For example, when shown a nail that has been partially driven into a piece of wood, the patient may be unable to select correctly a hammer from an array of tools for task completion. Patients have particular difficulty in this task when the foil tools are closely related to the correct tool choice, as when a hammer is presented with other tool selections such as a screwdriver, pliers, and wrench.

Finally, patients with conceptual apraxia may be unaware of the mechanical advantages afforded by tools (mechanical knowledge). Our colleagues have investigated two aspects of mechanical knowledge (Heilman, Maher, Greenwald, & Rothi, 1995; Heilman & Rothi, 1993; Ochipa et al., 1992). The first involves the ability to determine what kind of action is needed when presented with a task in which associative tool–object knowledge cannot be used. That is, if the appropriate tool is not available to perform a required action, the patient may be unable to select a tool which shares the crucial attributes of the appropriate tool. For example, if a hammer is not available to complete the task of driving a nail into wood, the patient may choose an inappropriate tool such as a spoon instead of a more appropriate alternative tool that is hard, rigid, and heavy such as pliers. Also, these patients may know what is supposed to happen to an object, but may not be able to imagine a tool action to accomplish the task. So for the nail example, the patient may try to push the nail into the wood by hand, failing to recognise the advantage a hammer would play in the action.

The second aspect of mechanical knowledge involves the ability to devise tools which provide the desired mechanical actions (pushing, lifting, etc.). To perform these tasks of mechanical knowledge correctly, patients need to understand the goal of the task, the type of mechanical action needed for its performance, and what type of tool may assist in performing the task. For example, we use a task in which patients must remove a small block from a container using a metal rod that they can bend into a tool of the proper shape. Some patients are unable to devise a tool and elect to remove the object with their hands. Other patients fashion an improper tool such as bending the rod into a hook when a more effective tool would be a prong.

These types of action errors may be observed in patients with impairments of semantic processing. Therefore, Ochipa and her colleagues (1992), in a study referred to earlier in the chapter, tested a group of individuals with Alzheimer's disease, who are thought to demonstrate semantic deficits, to determine if they demonstrated action errors consistent with conceptual apraxia. Previous studies of apraxia in Alzheimer's disease (Foster, Chase, Mansi, et al., 1984; Rapcsak, Croswell & Rubens, 1989) had emphasised ideomotor apraxia (impairment of the praxis production system) in these individuals. Impairment of the praxis conceptual system (action semantic knowledge) had not been documented in this population. Participants were divided into four subgroups on the basis of presence or absence of ideomotor apraxia and/or semantic language impairment, and were administered tasks of tool–object action knowledge, tool–object associative knowledge, and mechanical knowledge such as those tasks described previously. Results indicated that each of the four Alzheimer's subgroups significantly differed from controls on all tasks of conceptual apraxia, suggesting that Alzheimer's disease patients do have an impairment of conceptual knowledge related to tools.

Ecological Consequences

The manifestation of conceptual apraxia outside the clinical setting has not been well documented. As mentioned earlier, Ochipa and colleagues (1989) described a patient whom they observed using tools inappropriately in natural settings. For example, he ate with a toothbrush and brushed his teeth with a comb. He was not agnosic as he could name these same tools and point to them on command. However, he could not point to a tool when its function was described, nor could he verbally describe tool functions. His ability to follow commands not involving tool function was spared. His failure to use tools correctly could not be explained solely by a production deficit (severe ideomotor apraxia) as he was unable to match tools with their corresponding objects (e.g. hammer–nail), even when actual tool use was not required. His deficit was thought to reflect a loss of knowledge related to tool–object function. That is, the behaviours observed implicated dysfunction of the praxis conceptual system.

We continued to study left-hemisphere stroke patients performing daily living activities in the hospital environment to examine the real-world consequences of apraxia in these patients (Foundas, Macauley, Raymer, et al., 1995). Specifically, we videotaped 10 apraxic stroke patients during mealtime and compared their mealtime behaviours to hospitalised neurologically normal controls. The apraxic stroke patients demonstrated eating patterns that differed substantially from control subjects. Some differences were likely a manifestation of the effects of ideomotor apraxia in the stroke patients, resulting in less efficient implementation of actions during the meal. However, some types of errors observed in the apraxic group may have reflected impairments of praxis conceptual knowledge. Some patients produced errors in which they used tools improperly or misselected tools for an intended action. For example, we observed a nonagnosic patient who used his slice of bread to wipe his mouth, later eating the bread, and another who used his knife to stir his tea.

Therefore, unbeknownst to caregivers, patients may experience difficulties in aspects of tool–object action knowledge in the natural environment. At times the impairment may go undetected because patients ultimately accomplish the desired task, albeit in an inefficient and, at times, dangerous manner. Other times patients simply give up and do not complete the task. These observations suggest that conceptual apraxia may be a relatively common phenomenon in left-hemisphere stroke patients and it may require our attention in clinical management and family counselling. (See Chapter 7, for a discussion of these issues.)

ACKNOWLEDGMENTS

We thank Leslie Gonzalez Rothi and Ken Heilman for their mentorship and contributions to all facets of the research studies described in this chapter, and for their helpful comments on an earlier version of the chapter. Preparation of this chapter was supported by a grant from National Institute of Mental Health (MH48861).

REFERENCES

Allport, D.A. (1985). Distributed memory, modular subsystems and dysphasia. In S. Newman, & R. Epstein, (Eds.), *Current perspectives in dysphasia*. New York: Churchill Livingstone.

Barbieri, C. & De Renzi, E. (1988). The executive and ideational components of apraxia. *Cortex, 24*, 535–453.

Beauvois, M.F. & Saillant, B. (1985). Optic aphasia for colours and colour agnosia: A distinction between visual and visuo–verbal impairments in the processing of colours. *Cognitive Neuropsychology, 2*, 1–48.

Beauvois, M.F., Saillant, B., Meininger, V., & Lhermitte, F. (1978). Bilateral tactile aphasia: A tacto–verbal dysfunction. *Brain, 101*, 381–401.

Caramazza, A., Hillis, A.E., Rapp, B.C., & Romani, C. (1990). The multiple semantics hypothesis: Multiple confusions? *Cognitive Neuropsychology, 7*, 161–189.

Coslett, H.B. & Saffran, E.M. (1989). Preserved object recognition and reading comprehension in optic aphasia. *Brain, 112*, 1091–1110.

Denes, G. & Semenza, C. (1975). Auditory modality-specific anomia: Evidence from a case study of pure word deafness. *Cortex, 11*, 401–411.

De Renzi, E. & Lucchelli, F. (1988). Ideational apraxia. *Brain, 111*, 1173–1183.

De Renzi, E., Pieczuro, O. & Vignolo, L.A. (1968). Ideational apraxia: A quantitative study. *Neuropsychologia, 6*, 41–52.

Farah, M.J. (1990). *Visual agnosia*. Cambridge, Mass: MIT Press.

Farah, M.J. & Wallace, M.A. (1992). Semantically-bounded anomia: Implications for the neural implementation of naming. *Neuropsychologia, 30*, 609–621.

Foster, N.L., Chase, T.N., Mansi, L., Brooks, R., Fedio, P., Patronas, N.J. & Dichiro, G. (1984). Cortical abnormalities in Alzheimer's disease. *Annals of Neurology, 16*, 649–654.

Foundas, A., Macauley, B.L., Raymer, A.M., Maher, L.M., Rothi, L.J.G., & Heilman, K.M. (1995). Ecological implications of limb apraxia: Evidence from mealtime behavior. *Journal of the International Neuropsychological Society, 1*, 62–66.

Freund, C.S. (1889). Über optische Aphasie und Seelenblindheit. *Archiv fur Psychiatrie und Nervenkrankheiten, 20*, 276–297, 371–416.

Funnell, E. & Allport, A. (1987). Non-linguistic cognition and word meanings: Neuropsychological exploration of common mechanisms. In A. Allport, D.G. Mackay, Prinz, W., & E. Scheerer, (Eds.), *Language perception and production*. London: Academic Press.

Hart, J., Berndt, R.S., & Caramazza, A. (1985). Category-specific naming deficit following cerebral infarction. *Nature, 316*, 429–440.

Heilman, K.M., Maher, L.M., Greenwald, M.L., & Rothi, L.J.G. (1995). Conceptual apraxia from lateralized lesions. *Neurology, 45*, A266 (supplement).

Heilman, K.M., & Rothi, L.J.G. (1993). Limb apraxia. In K.M. Heilman, & E. Valenstein, (Eds.), *Clinical Neuropsychology* (3rd ed.). New York: Oxford University Press.

Hillis, A.E. & Caramazza, A. (1991). Category-specific naming and comprehension impairment: A double dissociation. *Brain, 114*, 2081–2094.

Hillis, A.E. & Caramazza, A. (1995). Cognitive and neural mechanisms underlying visual and semantic processing: Implications from "optic aphasia." *Journal of Cognitive Neuroscience, 7*, 457–478.

Hillis, A.E., Rapp, B., Romani, C., & Caramazza, A. (1990). Selective impairment of semantics in lexical processing. *Cognitive Neuropsychology, 7*, 191–243.

Iorio, L., Falanga, A., Fragassi, N.A., & Grossi, D. (1992). Visual associative agnosia and optic aphasia. A single case study and a review of the syndromes. *Cortex*, *28*, 23–37.

Liepmann, H. (1905). Über Storungen des Handelns bei Gehirnkranken. Berlin: Karger.

Ochipa, C., Rothi, L.J.G., & Heilman, K.M. (1989). Ideational apraxia: A deficit in tool selection and use. *Annals of Neurology*, *25*, 190–193.

Ochipa, C., Rothi, L.J.G., & Heilman, K.M. (1992). Conceptual apraxia in Alzheimer's disease. *Brain*, *115*, 1061–1071.

Paivio, A. (1971). *Imagery and verbal processes*. New York: Holt, Rinehart, and Winston.

Paivio, A. (1986). *Mental representations: A dual coding approach*. New York: Oxford University Press.

Poeck, K. (1983). Ideational apraxia. *Journal of Neurology*, *230*, 1–5.

Rapcsak, S.Z., Croswell, S.C., & Rubens, A.B. (1989). Apraxia in Alzheimer's disease. *Neurology*, *39*, 664–669.

Ratcliff, G. & Newcombe, F. (1982). Object recognition: Some deductions from the clinical evidence. In A.W. Ellis. (Ed.), *Normality and pathology in cognitive functions*. London: Academic Press.

Raymer, A.M. (1992). *Dissociations of semantic knowledge: Evidence from Alzheimer's Disease*. Unpublished doctoral dissertation, University of Florida.

Raymer, A.M., Greenwald, M.L., Richardson, M.,Rothi, L.J.G., & Heilman, K.M. (1992). Optica phasia and optic apraxia: Theoretical interpretations. *Journal of Clinical and Experimental Neuropsychology*, *14*, 396 (abstract).

Riddoch, M.J. & Humphreys, G.W. (1987). Visual object processing in optic aphasia: A case of semantic access agnosia. *Cognitive Neuropsychology*, *4*, 131–185.

Riddoch, M.J., Humphreys, G.W., Coltheart, M., & Funnell, E. (1988). Semantic systems or system? Neuropsychological evidence re-examined. *Cognitive Neuropsychology*, *5*, 3–25.

Rothi, L.J.G., Ochipa, C., & Heilman, K.M. (1991). A cognitive neuropsychological model of limb praxis. *Cognitive Neuropsychology*, *8*, 443–458.

Rothi, L.J.G., Ochipa, C., & Heilman, K.M. (1997). A cognitive neuropsychological model of limb praxis and apraxia. In L.J.G. Rothi, & K.M. Heilman, (Eds.), *Apraxia: The neuropsychology of action*. Hove, UK: Psychology Press.

Roy, E.A. & Square, P.A. (1985). Common considerations in the study of limb, verbal and oral apraxia. In E.A. Roy, (Ed.), *Advances in psychology (Vol.* 23). *Neuropsychological studies of apraxia and related disorders* (pp. 111–162). Amsterdam: North-Holland.

Schwartz, M.F., Marin, O.S.M., & Saffran, E.M. (1979). Dissociations of language function in dementia: A case study. *Brain and Language*, *7*, 277–306.

Seymour, P.H.K. (1973). A model for reading, naming and comparison. *British Journal of Psychology*, *64*, 35–49.

Shallice, T. (1988a). *From neuropsychology to mental structure*. New York: Cambridge University Press.

Shallice, T. (1988b). Specialisation within the semantic system. *Cognitive Neuropsychology*, *5*, 133–142.

Warrington, E.K. & McCarthy, R. (1983). Category specific access dysphasia. *Brain*, *106*, 859–878.

Warrington, E.K. & McCarthy, R. (1987). Categories of knowledge: Further fractionations and an attempted integration. *Brain*, *110*, 1273–1296.

Warrington, E.K. & Shallice, T. (1984). Category specific semantic impairments. *Brain*, *107*, 829–854.

6 Limb Praxis Assessment

Leslie J. Gonzalez Rothi, Anastasia M. Raymer, and Kenneth M. Heilman

Liepmann (1900; 1905) raised almost 100 years ago the issue that praxis assessment required more than passing consideration. That is, methods of praxis assessment require careful attention to procedural precision as well as respect for the functional underpinnings of the praxis system. For example, Liepmann (1900) noted that praxis imitation should be tested in addition to praxis production to verbal command and that each of these tasks was uniquely informative. While assessment attributes such as these have been encouraged historically, few remnants have been found in recent history. Assessment methods typically reported encourage gesture to command without mention of the concomitant testing of gesture imitation or vice versa. None have encouraged the testing of gesture reception at the time of gesture production testing where the items are consistent across tasks.

Almost no standardised tests for limb apraxia are presently commercially available (see Helm-Estabrooks, 1992, for one exception) and none are available that test limb praxis in a manner suitable for assessing the psychological underpinnings of the various forms of the disorder discussed previously in our chapter on a cognitive neuropsychological model of limb praxis. We also have no standardised methods of praxis assessment to offer; however, in this chapter we will briefly review the methods we have developed in the course of our research to assess limb praxis in neurologically impaired patients. In addition to their use in research, we have found many of these methods useful in our clinical practices as well. (For another view of praxis assessment issues and controversies, see De Renzi, 1985.)

THE FLORIDA APRAXIA SCREENING TEST (REVISED)

In investigating the incidence of limb apraxia in lateralised brain damage, Liepmann (1905) reported that only the left hemisphere lesioned patients displayed recognisable levels of limb apraxia whereas the right hemisphere lesioned patients did not. While the incidence of limb apraxia within Liepmann's left-hemisphere lesioned group was very high (20 out of 41 right-hemiplegic patients), a number of left-hemisphere lesioned patients did not show observable limb apraxia. When we began to study the problem of limb apraxia and limb praxis some 20 years ago, we had a difficult time identifying a measure sensitive to the presence or absence of limb apraxia in our left-hemisphere damaged patients. While the literature reported various testing methods, we found none of these measures met our specifications. For example, we wanted a measure for assessing praxis to verbal command, but as mentioned earlier, many of the reported methods utilised an imitation paradigm. In addition, available materials contained stimuli that intermixed such praxis-relevant attributes as transitive (i.e. requiring a tool such as "hammering") versus intransitive (i.e. not requiring a tool or object such as "hitchhiking") movements, unimanual versus bimanual movements, or discriminant body part usage (buccofacial versus limb). Therefore we created a short screening instrument called the Florida Apraxia Screening Test (FAST) (Rothi & Heilman, 1984) which was later revised and expanded and called the Florida Apraxia Screening Test–Revised (FAST–R) as part of the Florida Apraxia Battery (Rothi, Raymer, Ochipa, Maher, Greenwald, & Heilman, 1992). The FAST–R is a 30-item gesture to verbal command test using the stimuli listed in Table 1. These stimuli were selected because they could be administered in a verbal command format; they all related to arm/hand movement; and they all could be completed with only one hand. The items include 20 transitive and 10 intransitive pantomimes to allow us to examine differences within patients on these two forms of gesture.

In the FAST–R the patient uses the dominant hand/arm to produce gestural responses, unless motor symptoms prevent effective use of the limb. In that event, the patient performs gestures with the nondominant (ipsilesional in the case of a left-hemisphere lesion) hand. It is not uncommon for non-brain-injured individuals to produce inadequate gestural responses (imprecise production and/or lacking elaboration) on this task possibly due to diminished effort, discomfort with the notion of "performing" in front of someone, or a lack of appreciation for the precision and elaboration expected when performing pantomimes. Therefore we have found it helpful to spend time before testing to practise pantomiming, reviewing each individual's pantomimes with them, and underscoring the need to produce the gesture as though they are "actually holding" the imagined tool. Also we discuss the need to "actually imagine the object that you are acting upon." While some may claim that we are falsely inflating the test scores, we have never found it possible to

TABLE 1
Stimuli used in the FAST–R

Show me:

1. How you salute.
2. How to use scissors to cut a piece of paper out in front of you.
3. How to use a saw to cut a piece of wood out in front of you.
4. How you hitchhike.
5. How to use a bottle opener to remove a cap on a bottle out in front of you.
6. How to use wire cutters to snip a wire out in front of you.
7. Stop.
8. How to use a salt shaker to salt food on a table out in front of you.
9. Go away.
10. How to use a glass to drink water.
11. How to use a spoon to stir coffee on a table out in front of you.
12. How to wave goodbye.
13. How to use a hammer to pound a nail into a wall in front of you.
14. How to use a comb to fix your hair.
15. How to use a knife to carve a turkey on a table in front of you.
16. How to use a brush to paint a wall out in front of you.
17. Come here.
18. How to use a screwdriver to turn a screw into a wall out in front of you.
19. How to use a pencil to write on paper on a table out in front of you.
20. Someone is crazy.
21. How to use a key to unlock a doorknob on a door out in front of you.
22. Be quiet.
23. How to use an iron to press a shirt out in front of you.
24. How to use a razor to shave your face.
25. OK.
26. How to use an eraser to clean a chalkboard out in front of you.
27. How to use a vegetable peeler to shred a carrot.
28. How to make a fist.
29. How to use an ice pick to chop ice out in front of you.
30. How to use a scoop to serve ice cream.

SOURCE: Rothi et al., 1992. Unpublished.

talk our many apraxic patients out of their gestural production difficulties (Raymer, Maher, Foundas, Heilman & Rothi, 1996).

As a matter of course, we recommend that examiners videotape record all patient responses for later scoring. The scoring system we use includes multiple error types as listed in Table 2 (Rothi, Mack, Verfaellie, Brown, & Heilman, 1988). This system is one we have developed based on the systems used by Stokoe (1960) and Klima and Bellugi (1979) for describing the productions of deaf signers in their American Sign Language (ASL) research.

In our laboratory we use a minimum of two judges to score each participants' videotaped gestural productions. The judges are allowed to review each production as often as necessary. For research purposes we score the

TABLE 2
Praxis error types

Content

P = Perseverative—The patient produces a response that includes all or part of a previously produced pantomime.

R = Related—The pantomime is an accurately produced pantomime associated in content to the target. For example, the patient might pantomime playing a trombone for a target of a bugle.

N = Non-related—The pantomime is an accurately produced pantomime *not* associated in content to the target. For example, the patient might pantomime playing a trombone for a target of shaving.

H = Hand—the patient performs the action without benefit of a real or imagined tool. For example, when asked to cut a piece of paper with scissors, they pretend to rip the paper. Another example would be turning a screw by hand rather than with an imagined screwdriver.

Temporal

S = Sequencing—Some pantomimes require multiple positionings that are performed in a characteristic sequence. Sequencing errors involve any perturbation of this sequence including addition, deletion, or transposition of movement elements as long as the overall movement structure remains recognisable.

T = Timing—This error reflects any alteration from the typical timing or speed of a pantomime and may include abnormally increased, decreased, or irregular rate of production.

O = Occurrence—Pantomimes may characteristically involve either single (e.g. unlocking a door with a key) or repetitive (e.g. screwing in a screw with a screwdriver) movement cycles. This error type reflects any multiplication of characteristically single cycles or reduction of a characteristically repetitive cycle to a single event.

Spatial

A = Amplitude—Any amplification, reduction, or irregularity of the characteristic amplitude of a target pantomime.

IC = Internal configuration—When pantomiming, the fingers and hand must be in a specific spatial relation to one another to reflect recognition and respect for the imagined tool. This error type reflects any abnormality of the required finger/hand posture and its relationship to the target tool. For example, when asked to pretend to brush teeth, the patient's hand may close tightly into a fist with no space allowed for the imagined toothbrush handle.

BPT = Body part as tool—The patient uses his/her finger, hand, or arm as the imagined tool of the pantomime. For example, when asked to pretend to smoke a cigarette, the patient might puff on the end of an extended index finger.

EC = External configuration—When pantomiming, the fingers/hand/arm and the imagined tool must be in a specific relationship to the object receiving the

(Continued)

TABLE 2
(Continued)

Content

		action. Errors of this type involve difficulties orienting to the object or in placing the object in space. For example, the patient might pantomime brushing teeth by holding his hand next to his mouth without reflecting the distance necessary to accommodate an imagined toothbrush. Or, when asked to hammer a nail, the patient might hammer in differing locations in space reflecting difficulty placing the imagined nail in a stable orientation.
M	=	Movement—When acting on an object with a tool, a movement characteristic of the action and necessary to accomplishing the goal is required. Any disturbance of the characteristic movement reflects a movement error. For example, when asked to pantomime using a screwdriver, a patient may orient the imagined screwdriver correctly to the imagined screw but instead of stabilising the shoulder and wrist while twisting at the elbow, the patient stabilises the elbow and twists at the wrist or shoulder.

Other

C	=	concretisation—The patient performs a transitive pantomime not on an imagined object but instead on a real object not normally used in the task. For example, when asked to pretend to saw some wood, they pantomime sawing on their leg.
NR	=	No response
UR	=	Unrecognisable response—a response that is not recognisable and shares no temporal or spatial features of the target.

SOURCE: Rothi et al., 1988, with modifications reported in Greenwald, Rothi, Mayer, Ochipa, Chatterjee, and Heilman, 1992; and Ochipa et al., 1992).

productions independently but for clinical purposes we find it helpful for judges to discuss scores item by item and come to a single consensus score. The complexity of this scoring system and the dynamic nature of the pantomimes make this portion of praxis assessment effortful indeed. In addition, we have found that when learning to score praxis productions, judges tend to be very forgiving of errant dimensions of movement production that do not ablate a gesture's intent or obscure a gesture's meaning. The novice judge is cautioned to resist this temptation because, unlike speech production, a gestural production can be significantly aberrant before its meaning is unintelligible. For example, body-part-as-tool errors are ones which do not impair the reception of the meaning of a gesture such as "using a scissor." These errors are commonly produced by normal participants; but, they seem to occur far more frequently in pathology. If the instance of this error is not counted each and every time it occurs, frequency of the error cannot be determined. Thus, if a judge forgives this error because it (1) is an error that normals produce and/or (2) the meaning is not impaired, many praxis problems may go undetected. Therefore, all dimensions of praxis problem listed in Table 2 should be noted and, when present, the item is to be counted as an error.

Finally, a mention about norms for this instrument when using the sensitive scoring system. In our research we have found that groups of left-hemisphere disease patients typically perform in a bimodal distribution (Foundas, Raymer, Maher, Rothi, & Heilman, 1993). That is, some left-hemisphere damaged patients perform very poorly on this test and others perform fairly well on this test, with few patients distributed between. Therefore, it is clear that the poorly performing left hemisphere damaged patients have limb apraxia, but it is not clear whether the well-performing patients are within or outside of the distribution of normal performance. Therefore, the "cut-off" score of "normal" performance is clinically important. Also, it is unclear whether pantomime praxis performance is sensitive to variations in age, geographic distribution, culture, race, or gender. At this point, we use a normal cut-off score for the FAST–R of 15 out of 30 trials correct (Foundas, Macauley, Raymer, Maher, Heilman, & Rothi, 1995) that is based on the particular population we study. Whether this score is applicable to populations elsewhere is unclear. The researcher/clinician interested in using the FAST–R is encouraged to develop a standard of normal performance using a normal population comparable to their brain damaged patient population.

THE FLORIDA APRAXIA BATTERY

Although the FAST–R is effective in screening for limb apraxia, to disambiguate dimensions of limb praxis performance relevant to discrete levels of processing implied by our praxis model, we need a number of tasks that engage different levels of praxis processing. In our research activities we have developed the Florida Apraxia Battery (FAB) (Rothi et al., 1992) to assist us in this process. Each subtest employs items from the same corpus of stimuli used in the FAST–R (and described in Table 1), however readers are encouraged to develop their own corpus if more relevant to the subject population with which they are dealing. It is important, however, to draw all stimuli for this and the remaining tasks from a single corpus so that comparisons across tasks are possible. By analysing patient performance (both accuracy and error patterns) across tasks in which we contrast praxis input, praxis output, and praxis semantic processing demands, we generate converging evidence to support hypotheses about the cognitive mechanism underlying impaired praxis performance for a given patient. Recognising that these methods are purely experimental and not standardised, only a description of each subtest is given below.

Measures of Gesture Reception

Measures of gesture reception are designed to assess the integrity of the praxis mechanisms involved in processing viewed gestures including the action input lexicon, action semantics, and the connection between. For consistency in presentation of gestures across patients, we videotape all gesture reception measures and present them on a video monitor for viewing.

Gesture Naming. In this subtest the examiner performs a gesture and the participant is to "tell me what I am doing." The response is scored correct if the participant names the action or tool. Productions that have phonemic paraphasias but are recognisable should be accepted. Because impairment in this task may relate to gesture input processing (pantomime agnosia or conceptual apraxia) or verbal output processing (aphasia), caution must be used in interpreting results of impaired performance.

Gesture Decision. In this subtest the examiner performs one gesture that is either a real gesture or an unreal gesture. Real gestures are defined as accurately produced, familiar gestures such as hammering or saluting. In contrast unreal gestures are constructed by taking items from the corpus of real motions and altering their performance temporally and/or spatially to create an unreal gesture. For example, for the target of "turn a key" the examiner places their hand in correct spatial orientation but instead of stabilising the elbow and rotating the wrist along a single axis, they destablize the wrist such that there is a large 'arc'-like motion of the hand. The participant's task is to decide whether or not the production is an accurate rendition for a named familiar real gesture. The response is scored as correct if the participant correctly accepts real gestures or correctly rejects unreal gestures. The response is scored as incorrect if the participant incorrectly rejects real gestures or incorrectly accepts unreal gestures.

Gesture Recognition. In this subtest the examiner performs a series of three gestures (one is the target and two are foils) while saying, "which one of these gestures is correct for 'using a target tool and object in a context' or 'intransitive gesture name'? Is it gesture 1, 2 or 3?". The foils are either aberrations of the correct production (nonsense foils created from the target as described above) or are correctly produced gestures representing the use of another tool or intransitive gesture. The response is scored as correct if the participant correctly selects the accurately performed target gesture.

Measures of Gesture Production

These subtests are designed to assess the integrity of the praxis mechanisms involved in producing familiar gestures including the semantic system (possibly action semantics more specifically), the action output lexicon and the connections between. Interpretation of praxis performance must also consider whether praxis impairment reflects deficits related to input processing modalities as specified in each subtest.

Gesture to Verbal Command (FAST–R). We have described this subtest in the previous section.

Gesture to Visual Tool. The examiner shows the patient a tool and says "show me how you use this." The patient is to pantomime the use of the viewed tool and does not touch or hold the tool in hand. The response is scored as correct or incorrect using the scoring system described in Table 2.

Gesture to Tactile Tool. With eyes closed or covered, the participant inspects tools by hand as the examiner says, "show me how you use this tool I am placing in your hand." The examiner places a tool in the patient's hand, then removes it, and the patient pantomimes the tool function. The response is scored in the same manner as the preceding subtest.

Praxis Imitation

These subtests are designed to assess the nonlexical praxis imitation system (gesture as well as nonsense praxis imitation subtests) as well as lexical praxis imitation (gesture imitation subtests only) as described in Chapter 4 of this volume.

Gesture Imitation. The examiner says, "I will produce a gesture and I want you to do it the same way I do it. Don't name the gesture and don't produce your gesture until I am finished." Participants then imitate the familiar gestures. The responses are scored in the same manner as the preceding gesture production subtests.

Nonsense Praxis Imitation. The examiner says, "I will produce a gesture and I want you to do it exactly the way I do it. It is not a real gesture like the other ones you've been doing. It is nonsense. Now don't produce your gesture until I am finished." Participants then imitate the viewed gesture. The response is scored as correct or incorrect using the coding system as it applies to the temporal and spatial dimensions of the intended movement.

Measures of Action Semantics

While we attempt to utilise data converging from a variety of tasks to assess the praxis processing components (and this is especially true when examining the conceptual system), we describe only one subtest that Ochipa, Rothi and Heilman (1992) demonstrated to be particularly sensitive to deficits of action semantics. Also, performance patterns in subtests of gestural reception and gestural production may converge to indicate impairments of action semantics. Raymer and Ochipa (Chapter 5) describe other types of tasks used to detect impairments of action semantics. Also the reader is referred to Ochipa et al., 1992 and Raymer, 1992 for descriptions of other methods.

Tool Selection Task. In this task the participant views an object representing an incomplete action. For example, for the target action of "sawing" the participant sees a piece of lumber that is only partially cut in half. In addition, the participant also views an array of three tool choices (one of which is the target tool, a handsaw) and must "point to the tool that goes with the object." The response is scored as correct if the participant points to the appropriate tool.

Control Tasks

The FAB also includes a number of tasks designed to serve as control measures to contrast with performance in praxis-related tasks where individuals must produce verbal responses.

Tool Naming. The examiner shows the participant a tool and says, "tell me the name of this." The examiner scores the utterances as correct if they produce a recognisable target word. That is, productions that have phonemic paraphasias but are recognisable as the target are accepted.

Gesture Name Repetition. The examiner will say to the patient, "I will say a word and I want you to say it after me." The set of words corresponds to the names of each of the 30 FAST–R gestures. The examiner will score the patient's utterances as correct or incorrect if they recognisably approximate the target word. That is, productions that have phonemic paraphasias but are recognisable as the target word are accepted.

Summary

The FAB consists of a complex set of tasks, varying in praxis processing demands. For a given individual, examiners may find it useful to administer only a limited set of subtests to investigate aspects of impaired praxis processing, Although test administration can be time consuming, we have found the battery useful in understanding unusual dissociations of praxis performance in patients with "conduction apraxia" (Ochipa, Rothi, & Heilman, 1994; Raymer, Adair, Rothi, & Heilman, 1995) who are more impaired at imitating gesture than at producing gesture to verbal command, and "optic apraxia" (Raymer, Greenwald, Richardson, Rothi, & Heilman, 1992) who demonstrate greater impairment at gesturing to viewed tools/objects than at gesturing to verbal command. Careful assessment with the FAB allows us to discover various aberrations of the limb praxis process that we would not detect with the FAST–R alone.

PRAXIS IN NATURAL CONTEXT

To test for the "real-world relevance" of apraxia in the lives of our patients, we have also observed patients operating with tools in natural settings (Foundas et al., 1995; Ochipa, Rothi, & Heilman, 1989). Standardisation of measurement methods for this purpose is obviously problematic because the more the clinician tries to control for extraneous variables, the less "natural" the context becomes. With a desire to balance these two approaches, we developed the following method to study praxis in natural context (Foundas et al., 1995). We focused on one occasion common to all of our hospitalised patients in our clinical/research setting (eating their noontime meal in their rooms) but any condition common to all patients could be studied (e.g. grooming).

At the usual time and place, a lunch tray on a bedside table is placed before the patient and a video camera is turned on. In addition to the standard tools and condiments found on the daily tray, we add foil tools to the tray including a toothbrush, a comb and a pencil. The foils and standard utensils are intermingled to reduce the effect of left to right placement on the lunch tray. The patient receives no assistance in preparing the tools, condiments, or food, and after the camera is turned on, the examiner leaves for 30 or more minutes.

Scoring of these tapes focuses on a number of levels of analysis. First, we have divided the overall infrastructure of meal eating into three basic parts. The preparation phase includes only meal preparation behaviours such as opening the condiment packages and removing food covers. The eating phase begins with the patient's second bite of food, as patients will often complete additional preparations such as seasoning the food after the first bite. Finally we note a phase that includes clean-up behaviours (e.g. putting the napkin on the tray). All of the normal individuals we have studied (Foundas et al., 1995) produced behaviours in each of these stages in succession and they consistently respected the distinctiveness of each of these phases. That is, these stages rarely overlap temporally. In addition, within the eating phase, normal individuals vary their eating such that they are noted to eat in idiosyncratic but consistent patterns as they move through the meal. For example, they might eat a bite of the main course, followed by a bite of bread, followed by a sip of tea and this pattern will be repeated until the meal is completed. The eating patterns of the apraxic patients stand in contrast to normal patterns in that they are disorganised, possibly never developing a consistent pattern, and behaviours characteristic of one meal phase often intrude into inappropriate meal phases. In other words, when compared to other individuals, the apraxics appear to lack or differ in the normal infrastructure of meal eating. Therefore, for assessment of praxis in normal context we examine this task to see if the apraxic patient develops an idiosyncratic eating pattern and whether their phases of eating are temporally distinctive.

In addition to studying the infrastructure of meal eating we also use methods to quantify meal eating actions. An action is operationally defined as a sequence of movements that results in the accomplishment of a definable function. Each patient's videotape is individually evaluated and each action is identified and counted. In addition, total eating time (from picking up the first utensil to pushing the tray away) is measured by a stopwatch. Each action is then categorised as tool or non-tool related. A tool-related action is, of course, an action that uses a tool while a non-tool-related action does not use a tool. One example of a non-tool action during meal eating would be moving a fruit cup to another location on the tray. We then use the middle 5-minute segment of the videotape to determine the proportion of correct actions (number of correct actions divided by number of total actions during those 5 minutes). We calculate this score for tool actions as well as non-tool actions. Finally, we also record the incidence of misuse or mis-selection of tools, sequencing errors, or timing errors during the total mealtime.

Surveying the videotapes of our apraxic patients (Foundas et al., 1995) it was interesting to note that while each patient was disorganised in overall meal plan, inaccurate in selecting tools on the basis of their mechanical attributes, and inefficient and inaccurate in performing the sequences of movements we call tool actions, in every case the meal was eventually eaten and the patient was no longer hungry or thirsty. Nurses returned to find satisfied patients without complaints. If meal preparation and eating were not observed (as is the case in many hospital wards), no one would know that something was amiss. This points to the clinical relevance of assessing the use of tools in natural contexts for some apraxic patients. Again, it is unclear whether praxis performance in this particular context is sensitive to variations in age, geographic distribution, culture, race, or gender. Clearly, performance measures on tasks in the natural context would be influenced by variations in the clinical context. Therefore no norms are available for these measures and we recommend that examiners develop norms within their own clinical setting.

CONCLUSION

The purpose of this chapter was simply to review the attributes of a limb praxis assessment that would be informative relative to the model of praxis performance reviewed in Chapter 4. It is important to note here, however, that others have reported on methods to assess attributes of praxis performance in brain damaged individuals as well. The reader is referred to De Renzi (1985) for examples of other approaches.

Finally, we need to mention that when testing patients with possible limb apraxia, as with all adequate neuropsychological assessment, the possible contribution of comorbid factors must be taken into account. For example, the impact of a coexisting aphasia should surely be considered for any task utilising verbal language. This would most critically influence gesture to verbal command performance in the case of an auditory comprehension failure for named tools. Another would be a gesture reception subtest where the patient is asked to name a verb or tool name for a seen gesture. In these and all of the remaining subtests, the clinician should remain able to distinguish the effects of comorbid factors such as aphasia influencing praxis task performance by, for example, comparing the accuracy across various input modalities. This, of course, remains the challenge for all of clinical neuropsychology and is not specific to measures of limb apraxia.

ACKNOWLEDGEMENTS

The methods described in this chapter represent the collective efforts of many researchers at the University of Florida, including Anne Foundas, Barbara Haws, Margaret Leadon Greenwald, Beth Macauley, Linda Mack, Lynn Maher, Cynthia Ochipa, and Mieke Verfaellie.

REFERENCES

De Renzi, E. (1985). Methods of limb apraxia examination and their bearing on the interpretation of the disorder. In E. A. Roy (Ed.), *Neuropsychological studies of apraxia and related disorders*, (pp. 45–64). Amsterdam: Elsevier Science.

Foundas, A., Macauley, B. L., Raymer, A. M., Maher, L. M., Heilman, K. M., & Rothi, L. J. G. (1995). Ecological implications of limb apraxia: Evidence from mealtime behavior. *Journal of the International Neuropsychological Society*, *1*, 62–66.

Foundas, A. L., Raymer, A., Maher, L. M., Rothi, L. J. G., & Heilman, K. M. (1993). Recovery in ideomotor apraxia. *Journal of Clinical and Experimental Neuropsychology*, *15*, 44.

Greenwald, M. L., Rothi, L. J. G., Maher, L. M., Ochipa, C., Chatterjee, A., & Heilman, K. M. (1992). Concrete object errors in praxis performance. *Journal of Clinical and Experimental Neuropsychology*, *14*, 379.

Helm-Estabrooks, N. (1992). *Test of oral and limb apraxia*. Chicago: Riverside Publishing Company.

Klima, E. & Bellugi, U. (1979). *The signs of language*. Cambridge, MA: Harvard University Press.

Liepmann, H. (1900). Das Krankheitsbild der Apraxie (Motorische Asymbolie), *Monatschrift für Psychiatrie und Neurologie*, *8*, 15–44 (Translated in 1977. The syndrome of apraxia (motor asymboly) based on a case of unilateral apraxia. In D. A. Rottenberg & F. H. Hochberg (Eds.) *Neurological classics in modern translation*. pp. 155–1244. NewYork: Macmillan.

Liepmann, H. (1905). The left hemisphere and action. *Münchener Medizinische Wochenschrift*, 48–49. (Translated in 1980. In *Research Bulletin*, 506. Department of Psychology, The University of Western Ontario).

Ochipa, C., Rothi, L. J. G., & Heilman, K. M. (1989). Ideational apraxia: A deficit in tool selection and use. *Annals of Neurology, 25, 115*, 190–193.

Ochipa, C., Rothi, L. J. G., & Heilman, K. M. (1992). Conceptual apraxia in Alzheimer's Disease. *Brain, 115*, 1061–1071.

Ochipa, C., Rothi, L. J. G., & Heilman, K. M. (1994). Conduction apraxia. *Journal of Neurology, Neurosurgery, and Psychiatry, 57*, 1241–1244.

Raymer, A. M. (1992). *Dissociations of semantic knowledge: Evidence from Alzheimer's disease.* Unpublished doctoral dissertation, University of Florida, Gainesville, Florida.

Raymer, A. M., Adair, J. C., Rothi, L. J. G., & Heilman, K. M. (1995). Another case of conduction apraxia and its theoretical implications. *Journal of the International Neuropsychological Society, 1*, 149.

Raymer, A., Greenwald, M., Richardson, M., Rothi, L. J. G., & Heilman, K. M. (1992). Optic aphasia and optic apraxia: Theoretical interpretations. *Journal of Clinical and Experimental Neuropsychology, 14*, 396.

Raymer, A. M., Maher, L. M., Foundas, A., Heilman, K. M., & Rothi, L. J. G. (1996). The significance of body part as tool errors in limb apraxia. *Journal of the International Neuropsychological Society, 2*, 27.

Rothi, L. J. G. & Heilman, K. M. (1984). Acquisition and retention of gesture in apraxic patients. *Brain and Language, 3*, 426–432.

Rothi, L. J. G., Mack, L., Verfaellie, M., Brown, P., & Heilman, K. M. (1988). Ideomotor apraxia: Error pattern analysis. *Aphasiology, 2*, 381–388.

Rothi, L. J. G., Raymer, A. M., Ochipa, C., Maher, L. M., Greenwald, M. L., & Heilman, K. M. (1992). *Florida Apraxia Battery* (Experimental ed). Unpublished.

Stokoe, W. C. (1960). Sign language structure: An outline of the visual communication of the American deaf. *Studies in Linguistics*, (Occasional Papers, 8).

7 Management and Treatment of Limb Apraxia

Lynn M. Maher and Cynthia Ochipa

IMPACT OF LIMB APRAXIA

Limb apraxia is defined operationally as a disorder of learned skilled movement that cannot be explained by an elemental motor deficit such as weakness, akinesia, abnormalities of tone or posture, ataxia, movement disorders, or by intellectual deterioration, lack of understanding or uncooperativeness (Heilman & Rothi, 1993). The production and/or conceptual impairments observed in limb apraxia represent a disruption in the praxis system which normally is responsible for storing skilled motor information for future use. This complex system facilitates our interaction with the environment by providing a processing advantage for previously constructed programs rather than recreating the programs *de novo* each time they are required (Rothi, Ochipa, & Heilman, 1991). Localisation and information-processing models of apraxia suggest that the neural networks that mediate praxis are lateralised to the left hemisphere in most right-handers (Liepmann, 1977; Heilman, Rothi, & Valenstein, 1982). Impairment to the complex, multicomponant praxis system results in a variety of deficits, presumably dependent upon which components are compromised (Geschwind, 1965, 1975; Ochipa and Heilman, 1991, Roy & Square, 1985).

Although the incidence of limb apraxia in left-hemisphere brain damaged patients is high (De Renzi, Faglioni, Lodesani, & Vecchi, 1983; Kertesz & Ferro, 1984), little has been written about the management or treatment of limb apraxia. There are several reasons why a disorder that has been described for over 100 years has had so little attention paid to its remediation or management. First, apraxic patients rarely complain of the disorder and often appear to be unaware of their praxis deficits (Rothi, Mack, & Heilman, 1990). Since apraxic patients frequently present with hemiparesis of their dominant hand, they may dismiss whatever deficits they do recognise as being related to using their nondominant hand (Heilman & Rothi, 1993). Furthermore, limb apraxia frequently co-occurs with aphasia (De Renzi, Motti, & Nichelli, 1980; Goodglass & Kaplan, 1963; Kertesz, Ferro, & Shewan, 1984) which may limit the patients' ability to express concern over their deficits.

Second, even though the incidence of limb apraxia is high, it generally has been believed that patients recover spontaneously (Basso, Capitani, Sala, Laiacona, & Spinnler, 1987). However, others have observed the disorder to be enduring (Poeck, 1985). Recent studies suggest that while improvement in some aspects of limb praxis performance occurs over time, other aspects of the disorder are persistent (Foundas, Raymer, Maher, Rothi, & Heilman, 1993; Maher, Raymer, Foundas, Rothi, & Heilman, 1994).

The third possible reason that treatment/management of limb apraxia has been neglected is that limb apraxia is best identified in the pantomimed use of tools and objects out of the context of their natural use. Also, praxis performance in apraxic patients typically improves with actual tool use. Therefore, many believe apraxia has little negative impact on patients' lives. As Poeck (1985) states, "Apraxia shows only under testing conditions. It does not prevent the patient from using his limbs spontaneously . . . This makes therapy in most instances unnecessary . . ." (p.104). Thus limb apraxia has been considered to be of theoretical and localising value rather than practical significance (Basso et al., 1987; Poeck, 1985).

Several investigators have challenged the idea that patients with limb apraxia are not impaired with actual tool use (Heilman & Rothi, 1993; McDonald, Tate, & Rigby, 1994; Poizner, Soechting, Bracewell, Rothi, & Heilman, 1989). Poizner et al. (1989, and Chapter 8 of this volume) described deficits in the timing and trajectory of movements in apraxic patients during actual tool manipulation, suggesting that while the movements may improve with tool use, they are still aberrant. In addition, a recent study by McDonald, Tate and Rigby (1994) found that of 17 individuals with apraxia, all but two of them made the same types of errors in actual tool use as they did on pantomime to verbal command or imitation. Whereas the results of these studies call to question the assumption that apraxic individuals are not impaired with actual tool use, they do not address the issue of the relevance of limb apraxia outside the artificial context of clinical evaluation. Recently, however, there have been several investigations which suggest that there are practical, real-life consequences for patients with limb apraxia.

A study by Sundet, Finset and Reinvang (1988) evaluated the dependency of patients with left- or right-hemisphere strokes at 6 months post onset on caregiver assistance in activities of daily living. Caregivers assessed the level of assistance required by the patient for common activities such as grooming, dressing etc. Sundet et al. (1988) found that measures of apraxia were highly correlated with levels of dependency estimated by caregivers, and that patients with limb apraxia required more assistance with tasks of daily living than patients with other neuropsychological deficits.

Ochipa, Rothi and Heilman (1989) described a left-handed, right-hemisphere stroke patient who, despite being able to name and recognise tools (ruling out object agnosia), made errors in tool selection and use. The patient was noted to use tools inappropriately in natural settings. For example, he was observed to eat with a toothbrush and brush his teeth with a comb. The authors attributed this deficit to an impairment of the conceptual praxis system or action semantics (Ochipa et al., 1989; Rothi et al., 1991).

Schwartz and her colleagues described action errors during activities of daily living in a population of brain-damaged patients (Mayer, Reed, Schwartz, Montgomery, & Palmer, 1990). Patients were observed making errors such as attempting to butter a hot cup of coffee and attempting to eat cereal with a fork. Furthermore, in a detailed case study of a man with callosal apraxia from a bifrontal injury, Schwartz and her colleagues observed errors in action planning and/or efficiency, tool selection and use, and action timing and sequencing (Schwartz, Reed, Montgomery, Palmer, & Mayer, 1991). They proposed that a disorder of executive control, which they termed "frontal apraxia", was responsible for these errors but acknowledged that deficits in other neural networks, such as the praxis system, may have contributed to these errors of everyday actions.

The question of the ecological impact of limb apraxia on patients' lives prompted Rothi and her colleagues to study a group of acute left-hemisphere brain-damaged (LBD) patients during mealtime (Foundas, Macauley, Raymer, Maher, Heilman, & Rothi, 1995) and compare their performance to age- and education-matched, hospitalised controls. The participants were videotaped while eating their hospital lunch from a standard hospital tray (as described in the previous chapter). In addition to the appropriate eating utensils, tool foils (toothbrush, comb, and pencil) were also placed on the lunch tray. The videotapes of both the LBD and the control groups were analysed with respect to mealtime organisation and adequacy of action components. The participants were also assessed for limb apraxia and of the 10 LBD patients, eight were found to be apraxic. The results of the study indicated that the LBD group was less organised in their meal performance, used fewer tools, and produced fewer tool actions than the control group. Despite this difference in the planning and performance of tool-related actions, there was no difference between the two groups for non-tool-related actions, ruling out the possible influence of an elemental motor dysfunction on performance. In addition, there was a significant positive correlation between the degree of apraxia severity and the number of action errors. Therefore, since limb apraxia can effect actual tool use and may have a negative effect on activities of daily living, it becomes more important to consider the management and treatment of limb apraxia in this population.

There is yet another reason why the clinician may consider treatment of limb apraxia. Aphasia and limb apraxia frequently co-occur and are relatively common in patients with left-hemisphere brain damage (Duffy, & Duffy, 1981; Duffy, Duffy, & Pearson, 1975; Kertesz et al., 1984; Kertesz & Hooper, 1982). LeMay, David, and Thomas (1988) report that aphasic patients, especially those with Broca's aphasia, use more spontaneously generated hand gestures during speech than nonaphasics. Hermann, Reichle, Lucius-Hoene, Wallesch, and Johannsen-Horbach (1988) report that aphasic patients seem to use descriptive and codified gestures as a means of compensating for their aphasia. Glosser, Wiener, and Kaplan (1986) support the idea that aphasics' use of gesture is primarily for communication. Conversely, Borod, Fitzpatrick, Helm-Estabrooks, and Goodglass (1989) report that the presence of limb apraxia negatively affects the use of spontaneous communicative gesture. They suggest that training of praxis skills may positively effect spontaneous use of gestures in conversation.

In addition to spontaneous use of gesture in conversation, gesture is utilised as a facilitator of spoken language (Christopoulou & Bonvillian, Hanlon, Brown, & Gerstman, 1990; Kearns, Simmons, & Sisterhen, 1982; Skelly, Schinsky, Smith, & Fust, 1974), or as an alternative or augmentative communication system in aphasic patients (Baratz, 1985, Bonvillian, & Friedman, 1978; Kirshner, & Webb, 1981). Amer-ind gesture training is a common alternative communication choice in aphasia therapy for patients who have difficulty producing speech. Helm-Estabrooks, Fitzpatrick, and Barresi (1982) promote the use of pantomime training to circumvent the linguistic deficits of global aphasics. While the training of gesture is frequently attempted in aphasia therapy, the influence of limb apraxia on the acquisition and use of gestural communication is controversial. Several reports have suggested that despite the presence of severe apraxia, patients can learn gestures (Code & Gaunt, 1986; Coelho & Duffy, 1990; Cubelli, Trentini, & Montagna, 1991; Feyereisen, Barter, Goossens, & Clerebaut, 1988). However it has also been reported that the presence of limb apraxia negatively affects the quality of referential gestures and the ease with which gestures are acquired (Feyereisen et al., 1988; Rothi & Heilman, 1984). Rothi and Heilman (1984) examined the influence of limb apraxia from parietal lesions on learning lists of gestures and suggested that apraxic–aphasics demonstrated a slower rate of acquisition of gestures, acquired fewer number of gestures and produced fewer total numbers of gestures when compared to nonapraxic aphasics and normal controls. This difficulty in acquiring new gestures, they felt, was reflective of a deficit in consolidating the information into memory.

While many studies have demonstrated the ability of left-hemisphere brain-damaged patients to acquire a gestural repertoire, the success with which these patients have been able to generalise the use of gesture outside the therapeutic context has been variable (Bellaire, Georges, & Thompson, 1988; Kirshner & Webb, 1981; Coelho, 1990). In some cases where generalisation to other contexts did not occur, the presence or absence of limb apraxia was not directly discussed. Notably, however, those patients who demonstrated the most successful generalisation of gesture were able to pantomime tool use relatively well (Kirshner & Webb, 1981, Skelly et al., 1974), allowing for the speculation that the presence of limb apraxia may negatively affect the generalisation of learned gesture for communication purposes.

In summary, limb apraxia has been demonstrated to have a potentially negative impact on independence in activities of daily living, tool selection and use, ease of gesture acquisition and quality of gesture production in patients with left-hemisphere brain damage. For these reasons the clinician may want to address the management and treatment of limb apraxia as part of the overall patient treatment plan.

MANAGEMENT VS. TREATMENT OF APRAXIC DEFICITS

The term management refers to the practice of modifying the apraxic patient's interactions with the environment to accommodate their neurobehavioural deficit (Rothi, 1992). Apraxic patients are typically anosognosic for their praxic deficits (Rothi et al., 1990) or attribute their disabilities to their right hemiparesis and inexperience using their left arm (Heilman & Rothi, 1993). Furthermore, apraxic deficits are often overlooked in hospital settings where the patients' needs are anticipated or where opportunities for independent tool use are limited. Therefore help is needed in identifying tool use deficits and determining what restructuring of the home environment may be necessary to make the transition from hospital to home less traumatic for both patient and family. If tool use problems cannot be ruled out, hospital staff and family members must be warned to limit the patient's access to those tools which, if misused, may be dangerous. Some management strategies would include replacing tasks that may require tools with those that can be performed without tools wherever possible, limiting the selection of tools or objects to be worked on, avoiding series of tasks, having the patient perform tasks with which they are most familiar, and utilising cues when possible. Deficits in the accurate production of praxis may be managed by utilising proximal movement whenever possible and reducing the complexity of the required movement.

In contrast to management of apraxia, we use the term treatment in reference to the direct application of aid to the deficient behaviours (Rothi, 1992). The deficient behaviours seen in neurologically impaired patients may be the result of their reliance upon performance strategies that are no longer effective. In contrast, the deficient performance may be the result of the lack of reliance upon strategies that may be more effective. In the former case, treatment would best be directed at stopping the patient's use of ineffective behaviours, while in the latter case, treatment would best be directed at instituting effective strategies (Rothi, 1992). With respect to limb apraxia, the use of ineffective strategies when more efficient ones are available was described by Liepmann in the early 1900s (Liepmann, 1907, 1977). He described a patient with unilateral apraxia who made no attempt to utilise his relatively spared left hand, but rather chose to perform inaccurately with his right hand (despite indicating that he was aware his performance was deficient). It was only after his defective limb was restricted that he used his spared hand, a good example of how patients do not exploit their preserved abilities and why treatment should begin with an attempt to limit the patient's use of ineffective strategies.

TREATMENT OF LIMB APRAXIA

There are few reports to date on the direct treatment of limb apraxia. How to proceed will depend largely on one's philosophy of rehabilitation. Rothi (1995) conceptualises treatment approaches as being either restitutive or substitutive in nature. Treatments predicated on restitution of function are designed to speed up or maximise the recovery ultimately governed by the nervous system. Restitutive treatments are designed to address the underlying deficit and encourage maximum return of function within those limits set by the system. That is, a restitutive treatment addresses the impaired function and attempts to reconstruct the way the function was performed premorbidly. In contrast, substitutive treatments aim to achieve the behavioural goal in a new way. Substitutive treatments may be divided into those which are vicariative and those which are compensatory (Rothi, 1995). Vicariative treatments attempt to facilitate a functional reorganisation of the nervous system by having unimpaired functions assume responsibility for impaired functions. A vicariative treatment would ultimately remove the substitutive cue in the hopes that the behavioural goal may still be supported. A compensatory approach would involve altering the task strategy so that the behavioural goal thereafter would be achieved using new functions in a manner different from normal performance. Rothi (1995) suggests that treatments in the earliest post onset phase should be aimed at both restitution and substitution of function, with restitutive treatments directed at those behaviours with the most potential to recover. Substitutive strategies should be directed at those behaviours that have the most potential to remain impaired and may be appropriate beyond the acute phase of recovery. Although admittedly controversial, Rothi (1995) advocates for the use of treatment approaches (restitution, vicariation, compensation) which are compatible with physiological mechanisms of recovery and the natural course of evolution from brain damage.

With respect to limb apraxia, little is known about the natural evolution of the syndrome. A study by Maher et al. (1994) looked at the pattern of recovery in limb apraxia over a 6-month period of time. Using a qualitative error system (Rothi, Mack, Verfaellie, Brown, & Heilman, 1988, and see Chapter 6), they found differential patterns of recovery for intransitive gestures (i.e. gestures not requiring a tool or object, such as "wave goodbye") and transitive gestures (i.e. gestures requiring tool use). For intransitive gestures there was a spontaneous decrease in the number of content errors (i.e. errors where the meaning representation of the gesture was incorrect), whereas for transitive gestures there was a spontaneous decrease in the number of unidentifiable production errors (which included perseverations, unrecognisable responses and failures to respond). Spatial errors (including body part as tool, movement, external configuration, and internal configuration errors) and temporal errors were found to be persistent. The natural course of recovery in apraxia suggests that if performance improves, it is in the areas of meaningfulness and recognisability of their gestures. Therefore, early treatment efforts might focus on improving these aspects of praxis in order to maximise the recovery potential. However, more studies are needed to determine if apraxia is amenable to treatment and, if so, which treatment procedures would be most efficacious. General suggestions for intervention such as the use of tactile, kinaesthetic, and proprioceptive information, physical manipulation of the apraxic limb, backward chaining, and the use of real tools/objects (Quintana, 1989) lack empirical evidence to support or refute their efficacy in treatment.

To determine if praxic production of apraxic patients could be improved by direct treatment, Maher, Rothi, and Greenwald (1991) studied the effects of apraxia treatment on a 55-year-old man, C.R., with a 22-month history of ideomotor apraxia and preserved gesture recognition. Twenty gestures of common tools and household items were chosen for the study (10 of which were used in treatment and 10 of which were untreated). A within-subject ABA withdrawal design was used (McReynolds & Kearns, 1983). Following the establishment of a stable baseline, daily one-hour therapy sessions were initiated. The treatment sessions consisted of presenting the patient with multiple cues (tool, object, visual model, and feedback) and the patient was to demonstrate the use of the target tool. The cues were then systematically withdrawn until the criterion of 90% accuracy was reached at the final step (gesture in response to visual presentation of the tool). Treatment of error responses throughout the hierarchy involved immediate accuracy feedback and correction of the error responses. The patient was required to produce the gesture correctly three more times before proceeding to the next item. Correction of errors involved both imitation of the movement and physical manipulation of the limb when necessary. Treatment lasted for two weeks.

The patient's praxis performance to verbal command on the 20 items was videotaped for three trials immediately prior to treatment, three trials immediately post treatment, and one trial 2 weeks post treatment. These tapes were randomly ordered and scored by two independent judges using the Florida Apraxia Screening Test scoring system (Rothi et al., 1988, and described in Chapter 6,). A probe measure of 10 meaningless gesture sequences was also taken daily. Comparisons of pretest and post test measures reflect a substantial improvement in the quality of gesture production of both treated and untreated gestures. Prior to the initiation of treatment C.R.'s mean performance for both the treated and untreated gestures was 30% accurate, while post treatment mean performance for treated and untreated gestures was 76% and 80%, respectively. The probe measure remained stable throughout the treatment phase, indicating that experimental control was maintained.

This lack of improvement for meaningless gestures on the probe measure suggests that the gains made on the production of meaningful gestures were related to the treatment rather than to a generalised improvement in motor function. Repeated testing 2 weeks later resulted in a decreased performance on both treated and untreated items, though some gains on the treated items were maintained. Untreated items returned to baseline. Curiously, performance on the probe measure improved over the 2-week withdrawal period, during which the patient acknowledged he had "practised" this task. The decline in praxis performance with the withdrawal of treatment was considered evidence that the observed gains following the treatment phase were related to the treatment. These findings suggested that at least for some patients, training in praxis performance can improve the quality of gesture production.

Since the treatments of the Maher et al. (1991) study were somewhat global in nature, no conclusions could be drawn as to which feature of the procedures was most effective. We attempted to clarify some of these issues in subsequent studies (Ochipa, Maher, & Rothi, 1995). Specifically, we developed a treatment program to study the effects of treating particular error types (e.g. movement errors, external configuration errors) on praxis performance in two chronically apraxic patients.

G.R. is a 44-year-old, right-handed man with a 3-year history of Broca's aphasia, ideomotor apraxia, and right hemiplegia subsequent to a left-hemisphere stroke. Initial praxis test results (Florida Apraxia Battery; Rothi, Raymer, Ochipa, Maher, Greenwald, & Heilman, 1992) in the pantomime to verbal command subtest revealed a preponderance of external configuration errors (EC) where the hand movements were directed incorrectly, movement errors (M) where the incorrect joints were activated to produce the gesture, and internal configuration errors (IC) where the hand was not positioned correctly. Gesture recognition was relatively spared.

R.P. is a 66-year-old, right-handed man with a 4-year history of Broca's aphasia and ideomotor apraxia. His stroke was in the frontoparietal area of the left hemisphere. R.P.'s praxis performance was characterised primarily by errors of movement (M) and internal configuration (IC) when initially tested. Like G.R., his gesture recognition was relatively spared.

Twenty gestures of common tool/household item use were chosen for the study and were divided into two treatment stimulus sets (10 items each). A within-subject multiple baseline design across behaviours was used (McReynolds & Kearns, 1983). Separate treatments were designed to address the dominant error types seen in the pantomime to command performance on initial testing. For treatment of external configuration errors (represented by an open circle in Figs. 7.1 and 7.2), the patient was trained to orient his hand correctly to the objects being acted upon (e.g. correctly and consistently directing hand movement to an imagined nail in space when asked to demonstrate the use of a hammer). For movement treatment (represented by a black square in Figs. 7.1 and 7.2) the patient was given verbal descriptions to guide correct joint movement while gesturing (e.g. can opener: "Rotate your fingers, keeping your wrist fixed"). Internal configuration treatment (represented by a black triangle in Fig. 7.2) focused on the appropriate positioning of the hand and fingers to accommodate the tool being pantomimed (e.g. holding the hand in a curved, open position for pretending to hold a glass).

Each treatment was applied to one set of treatment items until the patient achieved two consecutive trials with 10% or less occurrence of the targeted error type on treated items. For example, if external configuration treatment was applied (as indicated by EC Tx at the top of Fig. 7.1) and the criterion performance was achieved, the same treatment was then applied to the second treatment stimulus set (as indicated by Set 2 EC Tx and charted on the bottom graph of Fig. 7.1). When criterion was achieved for the second set of stimulus items, another error type was targeted (e.g. treatment of movement errors as indicated by M Tx at the top of Fig. 7.1). For each response the patient gave, multiple error types were possible; however, feedback was given only for the error type targeted. For example, during movement treatment, errors other than movement errors were ignored and only movement errors were corrected.

G.R. received treatment once a day, four times weekly for a total of 44 sessions. R.P. received treatment twice a day, 2 days per week for a total of 24 sessions. Probes of half the treatment items (Sets 1 and 2) pantomimed to auditory command were taken daily, such that a complete probe of all 20 items (treated and untreated) was obtained every two sessions. These probes were videotaped and scored independently by one of the examiners (C.O.) using the error typing system of Rothi et al. (1988). A subset of the probes was scored by a second examiner (L.M.M.) for reliability. Interjudge reliability for error types was greater than 85%.

Treatment results are summarised in Figs. 7.1 and 7.2. These results suggest that both patients achieved considerable improvement in praxis performance. Post treatment scores obtained at 2 weeks after treatment was terminated

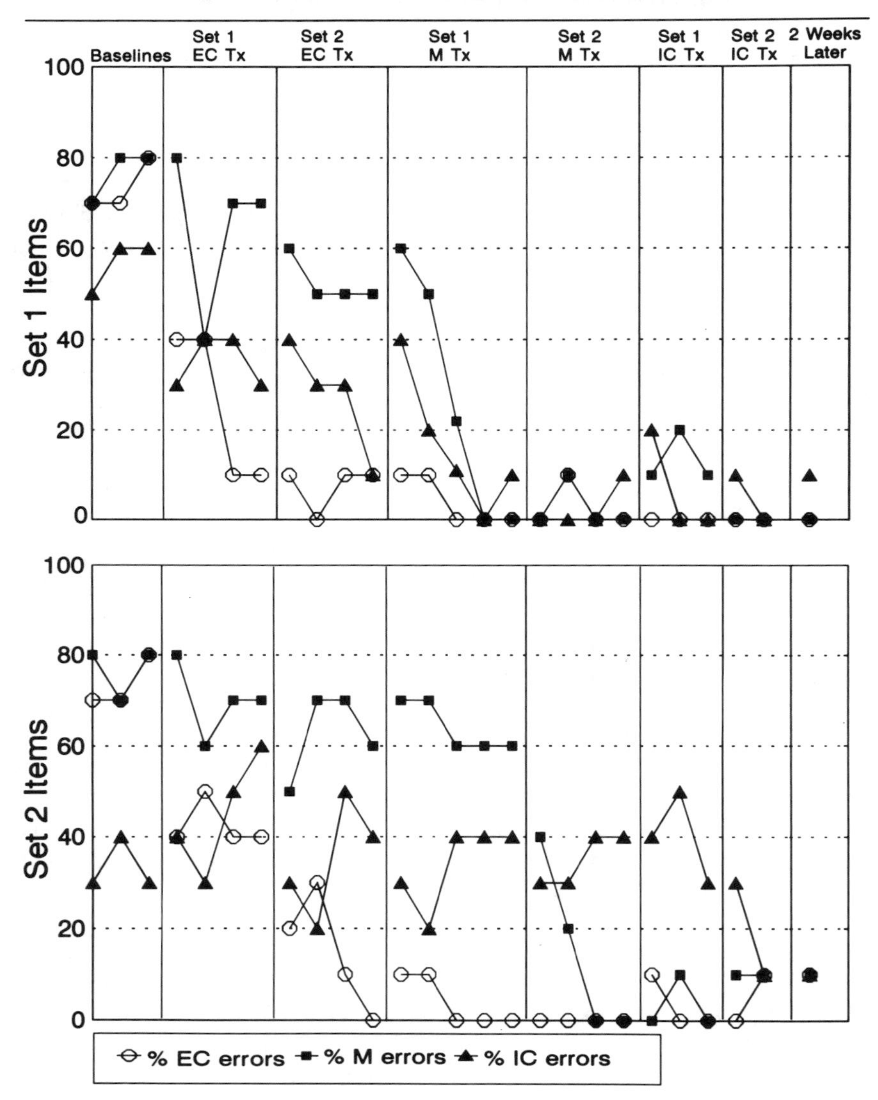

FIG. 1. Treatment data for G.R. for external configuration treatment, movement treatment, and internal configuration treatment. The stimulus set being treated and the type of treatment being conducted over time are indicated horizontally along the top of the graph. The dark vertical lines indicate when a change was made in the items being treated (Set 1 vs. Set 2) or in the type of treatment being administered once criterion was achieved (e.g. external configuration treatment EC Tx vs. movement treatment M Tx). The numbers on the vertical axis indicate the percentages of error type occurrence on the treatment stimulus Set 1 items (above) and treatment stimulus Set 2 items (below).

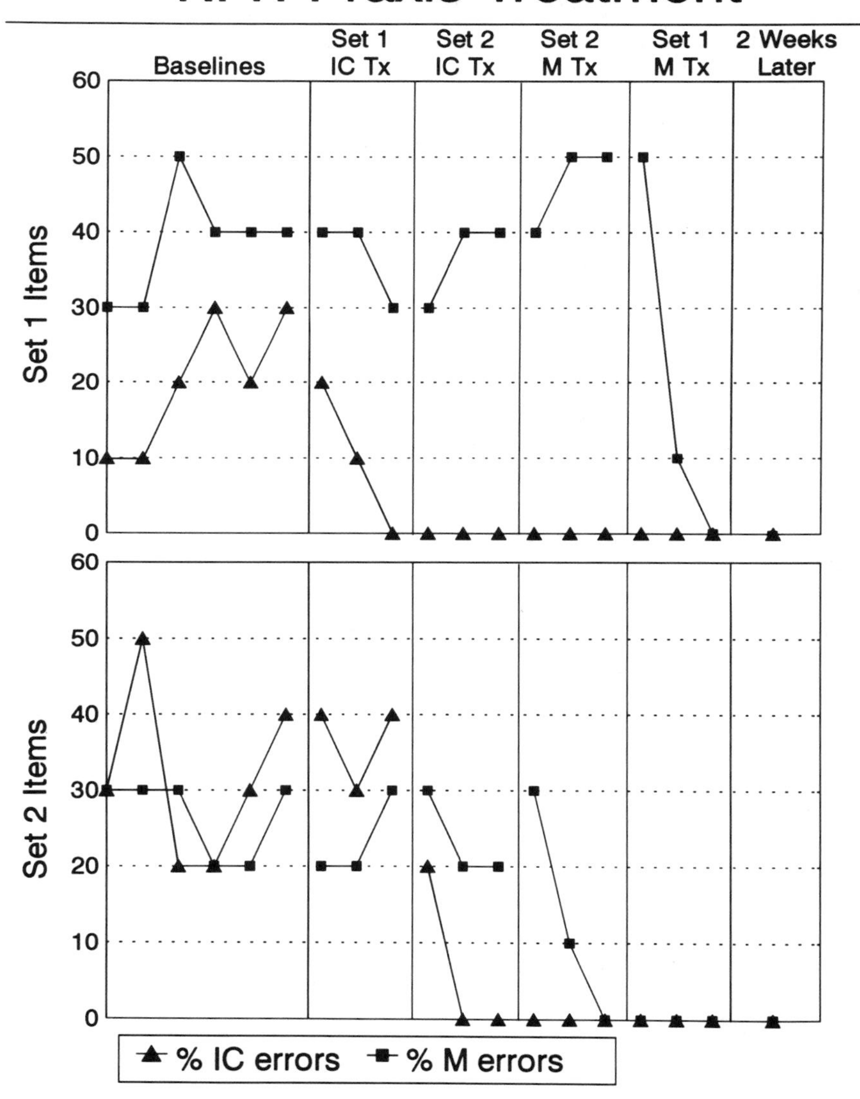

FIG. 2. Treatment data for R. P. for internal configuration treatment and movement treatment. The stimulus set being treated and the type of treatment being conducted over time are indicated horizontally along the top of the graph. The dark vertical lines indicate when a change was made in the items being treated (Set 1 vs. Set 2) or in the type of treatment being administered once criterion was achieved (internal configuration treatment IC Tx vs. movement treatment M Tx). The numbers on the vertical axis indicate the percentages of error type occurrence on the treatment stimulus Set 1 items (above) and treatment stimulus Set 2 items (below).

revealed that treatment gains were maintained after treatment was discontinued. However, the observed effects were treatment specific in that the incidence of a given error type did not reduce to criterion until it was targeted in treatment. The one exception to this was in the final stage of treatment for G.R., when the incidence of internal configuration errors dropped to criterion prior to treatment. It should be noted that for this patient, internal configuration errors were the least prevalent of his error types prior to the initiation of the study. For the most part however, treating one error type did not generalise to improvement for any other error type. Furthermore, treatment of a specific error type did not improve across untreated gestures. With that one exception, the patients praxis performance on Set 2 items did not improve to criterion until they were directly treated.

Finally, pretreatment and post treatment Florida Apraxia Battery (Rothi et al., 1992) scores revealed no change in the incidence of external configuration, movement or internal configuration errors for either patient. The failure to improve on Florida Apraxia Battery items further indicates that generalisation to untreated items did not occur. Thus, there was no generalisation within tasks to untreated items, nor was there genaralisation across praxis tasks. Despite the restitutive quality of this treatment paradigm, the lack of generalisation suggests that we were retraining specific gesture performance only and not improving praxis processing in general. Failure to generalise might be a result of the nature or stability of the apraxia in these two patients. However, a lack of generalisation is consistent with other reported attempts in apraxia treatment (Code & Gaunt, 1986; Cubelli et al., 1991; Pilgrim & Humphreys, 1994).

Code and Gaunt (1986) reported the use of a modified eight-step task hierarchy in the treatment of a patient with severe chronic aphasia, ideomotor apraxia and ideational apraxia. While the patient was able to acquire some hand signs, there was little change in post treatment measures of limb apraxia. This observation was supported in another case report by Cubelli et al. (1991) where referential gesture was trained using various pragmatic techniques such as PACE (Davis & Wilcox, 1985). Their patient was also successful in improving gesture production such that his gestures were more transparent post treatment. However, post treatment measures of ideational apraxia and ideomotor apraxia did not improve.

The lack of generalisation in apraxia treatment studies speaks to the specificity of the treatment effects. Despite the variety of treatment approaches and patient profiles in these studies, there are a few elements they all have in common. First, to control for the possible confounding effects of "spontaneous recovery" all of these studies involved chronic patients. Perhaps we would find a more generalising influence of apraxia treatment if it were applied in the acute post onset period. Second, all of the apraxic patients described were also aphasic. While these two disorders have been demonstrated to be dissociable (De Renzi et al., 1980; Goodglass & Kaplan, 1963) the influence of aphasia on praxis rehabilitation is not clear based on these studies. Third, the treatment approach in all of these studies was consistent with a restitutive approach to rehabilitation. Perhaps a more substitutive approach would have generalised more readily. In contrast to these studies, Pilgrim and Humphreys (1994) report the results of praxis remediation in a patient without aphasia using a more substitutive/vicariative treatment design.

Pilgrim and Humphreys' patient, G.F., is a left-handed, head injured patient, 23 months post injury, with right temporal and frontal lobe lesions who presented with a unilateral left limb apraxia. The patient was mute initially but regained speech fluency with only minor perseverations. There was some left-sided neglect, left-sided extinction to simultaneous stimuli, and visual memory impairment. G.F.'s praxis performance was worse when asked to gesture the use of visually presented objects, as opposed to the more typical profile of poorest performance to auditory verbal command. Using a cognitive neuropsychological model for action (Riddoch, Humphreys, & Price, 1989) these authors propose that G.F.'s difficulties result from an inability to suppress competing action plans which are generated by the visual presentation of the tools/objects.

Pilgrim and Humphreys (1994) used a modified form of "conductive education" (CE) with G.F. which consisted of "a task-analysis of the movements involved in using common objects and verbalised articulation of the goal-directed task" (p.280). A multiple baseline procedure was used with two sets of five objects: a treatment set and a control set. Treatment was conducted daily for three weeks. The treatment approach could be considered vicariative in nature. Initially the treatment consisted of physical manipulation plus verbalisation of the task elements. Physical manipulation was faded and as the actions were learned the verbalisation cues were also faded. The patient's wife was also trained and conducted the 15-minute practice sessions with the patient daily. Pre- and post treatment measures included gesture to visual presentation, gesture to visual plus tactile presentation and gesture to auditory command. The results of this study suggest that performance on the treated items improved with training, but there was no effect of the training on the untreated items. The authors suggest that verbalisation acted either to inhibit unwanted action plans or generate the correct action plan. They state that "The basic principle of CE is the restructuring of the functional system by incorporating the role of speech in the regulation of motor acts" (p. 283). Despite this vicariative strategy, the treatment effects in this study were still item specific. Furthermore, Pilgrim and Humphreys note that their patient required frequent prompting to use this strategy.

In conclusion, it appears that praxic deficits are amenable to the treatments used thus far; however treatment effects are item specific. Therefore, selection of stimuli for treatment should be based on the functional relevance of the treatment items. Treatment may also be more efficacious if it takes place within the natural context. Bellaire et al. (1988) suggested that communicative gestures were more easily generalised when trained in a natural setting. It may be that treatment may have a more lasting impact and/or ecological impact if it is conducted in an applied fashion.

SUMMARY

Despite the relatively frequent occurrence of limb apraxia in patients with left-hemisphere brain damage, how to proceed once limb apraxia has been identified has not been well studied. In light of recent studies demonstrating the potential negative effects of the disorder in everyday life, the clinician may wish to address the apraxic deficits in order to reduce the negative impact. Basically, the clinician is faced with two alternatives: management of the deficits or direct treatment (though these options are certainly not mutually exclusive).

Management of limb apraxia would involve altering the environment such that the negative impact of limb apraxia on activities of daily living is minimised and the risk of injury is eliminated (e.g. removing dangerous tools from the patient's reach). In addition to management, the clinician may chose to treat the apraxic deficits directly. Preliminary studies suggest that limb apraxia is amenable to treatment in some patients, but the gains made seem to be item specific. To date there has been poor generalisation of treatment gains to untreated gestures or to other settings. This leads to the conclusion that treatment tasks should be selected carefully with the practical application of the target behaviours kept in mind.

The efficacy of limb apraxia treatment requires further study. More studies designed to evaluate the effectiveness of substitutive strategies for motor planning and execution may assist in determining how to treat this disorder. Conversely, future attempts at intervention might be directed toward a more applied, functional and pragmatic treatment approach in the natural environment. Clearly there is much more to be learned about the management and treatment of limb apraxia.

REFERENCES

Baratz, R. (1985). A case study: Manual communication training for a global aphasic. *Aphasia–Apraxia–Agnosia*, 20–27.

Basso, A., Capitani, E., Sala, S.D., Laiacona, M., & Spinnler, H. (1987). Recovery from ideomotor apraxia. *Brain*, *110*, 747–760.

Bellaire, K., Georges, J.B., & Thompson, C.K. (1988). *Acquisition and generalization of gestures in aphasia*. Paper presented at the annual convention of the American Speech, Language and Hearing Association, Boston.

Bonvillian, J.D. & Friedman, R.J. (1978). Language development in another mode: The acquisition of signs by a brain damaged adult. *Sign Language Studies*, *19*, 111–120.

Borod J.C., Fitzpatrick, P.M., Helm-Estabrooks, N., & Goodglass, H. (1989). The relationship between limb apraxia and the spontaneous use of communicative gesture in aphasia. *Brain and Cognition, 10*, 121–131.

Christopoulou, C. & Bonvillian, J.D. (1985). Sign language, pantomime and gestural processing in aphasic persons: A review. *Journal of Communication Disorders, 18*, 1–20.

Code, C. & Gaunt, C. (1986). Treating severe speech and limb apraxia in a case of aphasia. *British Journal of Disorders of Communication, 21*, 11–20.

Coelho, C.A. (1990). Acquisition and generalization of simple manual sign grammars by aphasic subjects. *Journal of Communication Disorders, 23*, 383–400.

Coelho, C.A. & Duffy, R.J. (1990). Sign acquisition in two aphasic subjects with limb apraxia. *Aphasiology, 4*, 1–8.

Cubelli, R., Trentini, P., & Montagna, C.G. (1991). Re-education of gestural communication in a case of chronic global aphasia and limb apraxia. *Cognitive Neuropsychology, 8*, 369–380.

Davis, G.A. & Wilcox, M.J. (1985). *Adult aphasia rehabilitation: Applied pragmatics*. San Diego, CA: College-Hill Press, Inc.

De Renzi, E., Faglioni, P., Lodesani, M., & Vecchi, A. (1983). Performance of left brain-damaged patients on imitation of single movements and motor sequences: Frontal and parietal-injured patients compared. *Cortex, 19*, 333–343.

De Renzi, E., Motti, F., & Nichelli, P. (1980). Imitating gestures: A quantitative approach to ideomotor apraxia. *Archives of Neurology, 37*, 6–10.

Duffy, R.J. & Duffy, J.R. (1981). Three studies of deficits in pantomime expression and pantomime recognition in aphasia. *Journal of Speech and Hearing Research, 46*, 70–84.

Duffy R., Duffy, J., & Pearson, K. (1975). Pantomime recognition in aphasia. *Journal of Speech and Hearing Research, 18*, 115–132.

Feyereisen, P., Barter, D., Goossens, M., & Clerebaut, N. (1988). Gestures and speech in referential communication by aphasic subjects: Channel use and efficiency. *Aphasiology, 2*, 21–32.

Foundas, A.L., Macauley, B.C., Raymer, A.M., Maher, L.M., Heilman, K.M., & Rothi, L.J.G. (1995). Ecological implications of limb apraxia: Evidence from mealtime behavior. *Journal of the International Neuropsychological Society, 1*, 62–66.

Foundas, A., Raymer, A.M., Maher, L.M., Rothi, L.J.G., & Heilman, K.M. (1993). Recovery in ideomotor apraxia. *Journal of Clinical and Experimental Neuropsychology, 15*, 44.

Geschwind, N. (1965). Disconnexion syndromes in animals and man. *Brain, 88*, 237–294, 585–644.

Geschwind, N. (1975). The apraxias: Neural mechanisms of disorders of learned movement. *American Scientist, 63*, 188–195.

Glosser, G., Wiener, M., & Kaplan, E. (1986). Communicative gestures in aphasia. *Brain and Language, 27*, 345–359.

Goodglass, H. & Kaplan, E. (1963). Disturbance of gesture and pantomime in aphasia. *Brain, 86*, 703–720.

Hanlon, R.E., Brown, J.W., & Gerstman, L.J. (1990). Enhancement of naming in nonfluent aphasia through gesture. *Brain and Language, 38*, 298–314.

Heilman, K.M. & Rothi, L.J.G. (1993). Apraxia. In K. M. Heilman & E. Valenstein (Eds.), *Clinical neuropsychology* (3rd ed.). New York: Oxford University Press.

Heilman, K.M., Rothi, L.J., & Valenstein, E. (1982). Two forms of ideomotor apraxia. *Neurology, 32*, 342–346.

Helm-Estabrooks, N., Fitzpatrick, P.M., & Barresi, B. (1982). Visual action therapy for global aphasia. *Journal of Speech and Hearing Disorders, 47*, 385–389.

Hermann, M., Reichle, T., Lucius-Hoene, G., Wallesch, C.W., & Johannsen-Horbach (1988). Nonverbal communication as a compensatory strategy for severely non-fluent aphasics? A quantitative approach. *Brain and Language, 33*, 41–54.

Kearns, K.P., Simmons, N.N., & Sisterhen, C. (1982). Gestural sign (Amer-Ind) as a facilitator of verbalization in patients with aphasia. In R. Brookshire (Ed.), *Clinical aphasiology conference proceedings*. Minneapolis: BRK Publishers.

Kertesz, A. & Ferro, J.M. (1984). Lesion size and location in ideomotor apraxia. *Brain, 107*, 921–933.

Kertesz, A., Ferro, J.M., & Shewan, C.M. (1984). Apraxia and aphasia: The functional-anatomical basis for their dissociation. *Neurology, 34*, 40–47.

Kertesz, A. & Hooper, P. (1982). Praxis and language: The extent and variety of apraxia in aphasia. *Neuropsychologia, 20*, 275–286.

Kirshner, H.S. & Webb, W.G. (1981). Selective involvement of the auditory-verbal modality in an acquired communication disorder: Benefit from sign language therapy. *Brain and Language, 13*, 161–170.

LeMay, A., David, R. & Thomas, A.P. (1988). The use of spontaneous gesture by aphasic patients. *Aphasiology, 2*, 137–145.

Liepmann, H. (1977). The syndrome of apraxia (motor asymboly) based on a case of unilateral apraxia. (Translation from *Monatschrift für Psychiatrie und Neurologie*, 1907.) In D. A. Rottenberg & F.H. Hochberg (Eds.), *Neurological classics in modern translation*. New York: Macmillan Publishing Co.

Maher, L.M., Raymer, A.M., Foundas, A., Rothi, L.J.G., & Heilman, K.M. (1994). *Patterns of recovery in ideomotor apraxia*. Paper presented at the annual meeting of the International Neuropsychological Society, Cincinnati, OH.

Maher, L.M., Rothi, L.J.G., & Greenwald, M.L. (1991). Treatment of gesture impairment: A single case. *ASHA, 33*, 195.

Mayer, N.H., Reed, E., Schwartz, M.F., Montgomery, M., & Palmer, C. (1990). Buttering a hot cup of coffee: An approach to the study of errors of action in patients with brain damage. In D.E. Tupper & K.D. Cicerone (Eds.), *The neuropsychology of everyday life,* (Vol. 2). *Assessment and basic competencies*. Norwell, MA: Kluwer Academic Publishing.

McDonald, S., Tate, R.C., & Rigby, J. (1994). Error types in ideomotor apraxia: A qualitative analysis. *Brain and Cognition, 25*, 250–270.

McReynolds, L.V. & Kearns, K.P. (1983). *Single-subject experimental designs in communicative disorders*. Austin, TE: Pro-ed.

Ochipa, C., Maher, L.M., & Rothi, L.J.G. (1995). Treatment of ideomotor apraxia. *Journal of the International Neuropsychological Society, 2*, 149.

Ochipa, C., Rothi, L.J.G., & Heilman, K.M. (1989). Ideational apraxia: A deficit in tool selection and use. *Annals of Neurology, 25*, 190–193.

Pilgrim, E. & Humphreys, G.W. (1994). Rehabilitation of a case of ideomotor apraxia. In M.J. Riddoch & G.W. Humphreys (Eds.), *Cognitive neuropsychology and cognitive rehabilitation*. Hove, UK: Lawrence Erlbaum Associates Ltd.

Poeck, K. (1985). Clues to the nature of disruptions to limb praxis. In E.A. Roy (Ed.), *Neuropsychological studies of apraxia and related disorders*. New York: North-Holland.

Poizner, H., Soechting, J.F., Bracewell, M. Rothi, L.J.G., & Heilman, K.M. (1989). Disruption of hand and joint kinematics in limb apraxia. *Society for Neuroscience, 15*, 481.

Quintana, L.A. (1989). Cognitive and perceptual evaluation and treatment. In C.A. Trombly (Ed.), *Occupational therapy for physical dysfunction*. Baltimore, MD: Williams and Wilkins.

Riddoch, M.J., Humphreys, G.W., & Price, C.J. (1989). Routes to action: Evidence from aphasia. *Cognitive Neuropsychology, 6*, 437–454.

Rothi, L.J.G. (1992). Theory and clinical intervention: One clinician's view. In NIDCD monograph, *Aphasia treatment: Current approaches and research opportunities*. NIH Publication No. 93-3424, 91–98.

Rothi, L.J.G. (1995). Behavioral compensation in the case of treatment of acquired language disorders resulting from brain damage. In R.A. Dixon & L. Backman (Eds.), *Psychological compensation: Managing losses and promoting gains*. Hove, UK: Lawrence Erlbaum Associates Ltd.

Rothi, L.J.G. & Heilman, K.M. (1984). Acquisition and retention of gestures by apraxic patients. *Brain and Cognition, 3*, 426–437.

Rothi, L.J.G. & Horner, J. (1983). Restitution and substitution: Two theories of recovery with application to neurobehavioral treatment. *Journal of Clinical Neuropsychology, 5*, 73–81.

Rothi, L.J.G., Mack, L., & Heilman, K.M. (1990). Unawareness of apraxic errors. *Neurology, 40* (Suppl.1), 202.

Rothi, L.J.G., Mack. L., Verfaellie, M., Brown, P., & Heilman, K.M. (1988). Ideomotor apraxia: Error pattern analysis. *Aphasiology, 2*, 381–388.

Rothi, L.J.G., Ochipa, C., & Heilman, K.M. (1991). A cognitive neuropsychological model of limb praxis. *Cognitive Neuropsychology, 8*, 443–458.

Rothi, L.J.G., Raymer, A.M., Ochipa, C., Maher, L.M., Greenwald, M.L., & Heilman, K.M. (1992). *Florida Apraxia Battery* (Experimental ed.).

Roy, E.A. & Square, P.A. (1985). Common considerations in apraxia. In E.A. Roy (Ed.), *Neurological studies of apraxia and related disorders*. New York: North-Holland.

Schwartz, M.F., Reed, E.S., Montgomery, M., Palmer, C., & Mayer, N.H. (1991). The quantitative description of action disorganization after brain damage: A case study. *Cognitive Neuropsychology, 8*, 381–414.

Skelly, M., Schinsky, L., Smith, R., & Fust, R. (1974). American Indian Sign (Amer-Ind) as a facilitator of verbalization for the oral-verbal apraxic. *Journal of Speech and Hearing Disorders, 39*, 445–456.

Sundet, K., Finset, A., & Reinvang, I. (1988). Neuropsychological predictors in stroke rehabilitation. *Journal of Clinical and Experimental Neuropsychology, 10*, 363–379.

8 Kinematic Approaches to the Study of Apraxic Disorders

Howard Poizner, Alma S. Merians, Mary Ann Clark,
Leslie J. Gonzalez Rothi, and Kenneth M. Heilman

INTRODUCTION

Apraxia provides a special window into complex motor function. The movement impairment of apraxic patients cannot be explained by weakness, incoordination, sensory deficits, impaired comprehension, or lack of attention to commands. Furthermore, since the definition of ideomotor apraxia implies that the movements are learned, apraxia is a disorder of the control of the execution of certain classes of movement (Geschwind & Damasio, 1985; Heilman & Rothi, 1993). Although neurologists have noted the problems that apraxic patients have in controlling limb movement, there have been relatively few quantitative analyses of the nature of their movement disorder. Given that it may well be aspects of movement timing, joint coordination, and three-dimensional spatial patterning that have been disrupted, just such quantitative analyses are critical to understanding the nature of the limb control deficits and, ultimately, to better understand the neural control of limb movement.

This chapter presents some new strategies for investigating the nature of the movement breakdown in patients with apraxia based on recent technological advances in movement analysis. These techniques involve the creation and analysis of digital records of the spatiotemporal properties of upper limb movement. We present these techniques, and illustrate their use through three-dimensional analyses that we have been carrying out on patients with ideomotor limb apraxia (Clark, Merians, Kothari, Poizner, Macauley, Rothi, & Heilman, 1994; Poizner, Clark, Merians, Macauley, Rothi, & Heilman, 1995; Poizner,

Mack, Verfaellie, Rothi, & Heilman, 1990; Poizner & Soechting, 1992; Rapcsak, Ochipa, Anderson, & Poizner, 1995).

THREE-DIMENSIONAL MOTION ANALYSIS

Three-Dimensional Data Acquisition

Traditional methods of movement measurement have utilised video recording or high speed cinematography. In these methods, the positions of multiple markers on the image from a camera are digitised frame by frame by a human operator. This is a notoriously laborious and error-prone procedure (Woltring, 1984). Furthermore, the use of a single camera provides only two-dimensional information; the use of multiple cameras presents problems of camera synchronisation. We and others have developed systems for automated movement measurement that offer high spatial and temporal resolution for tracking multiple moving body segments in three-dimensional space (see Poizner, Wooten, & Salot, 1986).[1]

Figure 1 presents the main hardware components of the system we are currently using. Two commercially available optoelectronic cameras (Northern Digital, Inc.'s OPTOTRAK system) are used to record arm movements. These cameras directly sense the positions of infrared emitting diodes which are secured to the participant's left arm at the shoulder, elbow, wrist and hand. A microcomputer synchronises the sequential activation of the diodes with the digitising of the camera signals. The position of each of the four joints is sampled at 100Hz. and low-pass filtered with a modified Butterworth filter using a cut-off frequency of 8Hz. Limb trajectories are recorded from neighbouring views so that three-dimensional coordinates could be reconstructed. In some experiments, targets are presented via a programmable robot arm to study three-dimensional reaching movements (Adamovich, Berkinblit, Smetanin, Fookson, & Poizner, 1994; Fookson, Smetanin, Berkinblit, Adamovich, Feldman, & Poizner, 1994; Kothari, Poizner, & Figel, 1992). In these latter experiments, movement of the robot as well as the person's arm is recorded.

Computergraphic Analysis

The analysis of human motion can be advanced by analysing motion not only numerically, but also graphically. We have developed software for the interactive manipulation and dynamic display of the three-dimensional trajectories reconstructed from the data acquired from the optoelectronic cameras (Jennings & Poizner, 1988; Kothari et al., 1992; Poizner et al., 1986). Using a Silicon Graphics IRIS 4D/80GT work station, the trajectory paths of

[1]Examples of three-dimensional movement measurement systems include Northern Digital, Inc.'s WATSMART and OPTOTRAK systems and Selcom's SELSPOT system.

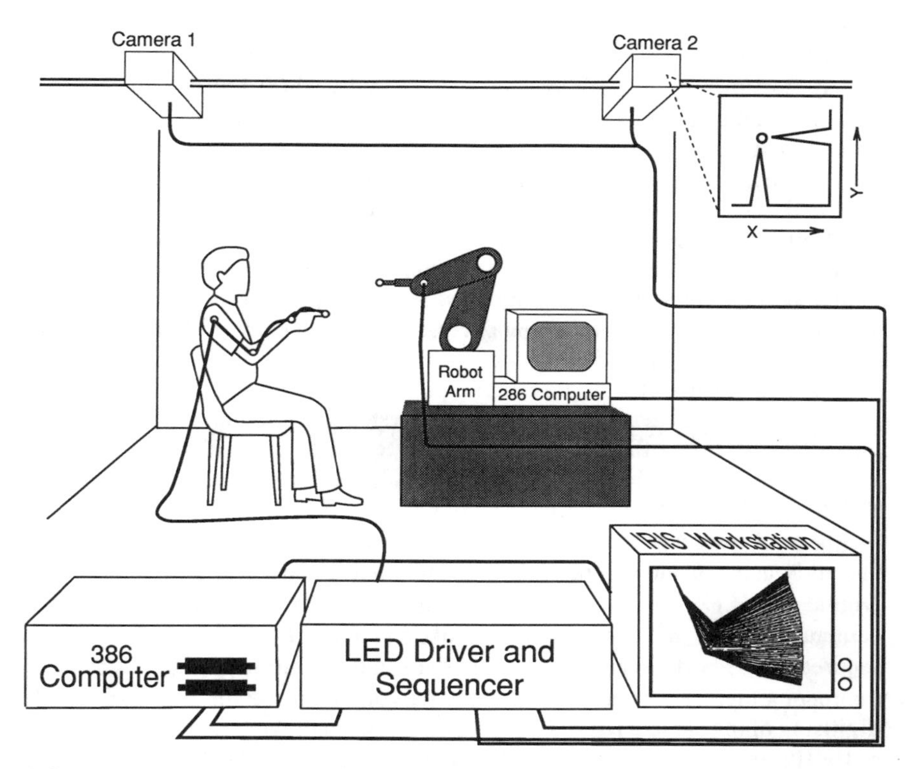

FIG. 1. Three-dimensional movement monitoring system. The main hardware components and the positioning of the LEDs on a participant is shown.

individual limb segments can be displayed simultaneously, and dynamically, with numeric and graphic displays of a variety of trajectory components. The simultaneous graphic and numeric display of data allow us to synchronise data around specific events. The user can interact with the graphic image in real time to rotate, translate, scale, and graphically edit the trajectory.

We have been using these techniques to address the nature of the movement breakdown in limb apraxia (see Clark et al., 1994; Poizner et al., 1990; Poizner et al., 1995, Poizner & Soechting, 1992; Rapcsak et al., 1995). In this chapter, we will describe impairments that apraxic patients showed in three aspects of movement control: spatial orientation, space–time relations, and joint coordination. Patients were classified as apraxic on the basis of their performance on the Florida Apraxia Screening Test Revised (Rothi & Heilman, 1984). All apraxic subjects had unilateral lesions in the left-hemisphere; detailed case histories and brain lesion analyses of these individuals are presented in Poizner et al. (1990) and in Clark et al. (1994). Motion analyses were performed on a number of learned, skilled movements including slice bread

(SLICE), roll up a car window (WIND), erase a blackboard (ERASE), and unlock a door (KEY). These movements are drawn from a particular subclass of learned, skilled movements, namely, transitive movements, which are movements that require the limb to work in concert with an external instrument, such as a tool. In some experiments, participants were provided cues in a graded fashion, from verbal command (no cues), to tool present, to object of the tool's action present, to actual tool and object manipulation. To obtain measures of movement variability, each gesture is replicated four times. All participants, apraxic and control, performed the movements with their nondominant left hand.Their left hand was used since it was ipsilateral to the side of the lesion for the apraxic participants, and thus would not be affected by weakness or incoordination due to primary motor deficit.

IMPAIRED SPATIAL ORIENTATION OF MOVEMENT

Limb Trajectories

Figure 2 presents the reconstructed trajectories of the limb segments during the SLICE gesture when one control individual and one apraxic individual (MR) produced the gesture to verbal command (left) and when the gesture was produced while manipulating both the tool and object (right). Both the left and right panels show three-dimensional views of successive stick figure positions of the upper arm, forearm and hand drawn every 20ms (every 10ms for the lower left panel). Figure 2 shows that the trajectories of the control individual consist of a series of forward and backward strokes with sharp reversals between the outward and inward components of the movement. The trajectory paths along which the wrist travelled were linear and tightly organised, producing closely overlapping planes of movement. These planes of movement were consistently located in the sagittal plane perpendicular to the goal object. Although the spatial organisation of the trajectory remained relatively unchanged between cue conditions, when the individual actually manipulated the tool and object (right), there was a decrease in the amplitude of the movement. The linear shape of the trajectory path, the tight planar organisation of successive strokes, and the sharp reversal in space at the end of the forward stroke characterised the controls' movements in both conditions.

In contrast, the lower left panel of Fig. 2 shows that the apraxic patient produced a gesture to verbal command which resembled a chopping motion. This motion was oriented predominately in the frontal plane, parallel to the object. The lower right panel of Fig. 2 shows that when the apraxic patient was required to manipulate the tool and object to actually slice the bread, the plane of the motion was modified somewhat, shifting to an off-sagittal plane. This motion resulted in the patient's slicing a piece of bread that was small and triangular in shape. The lower panels of Fig. 2 indicate that the apraxic patient

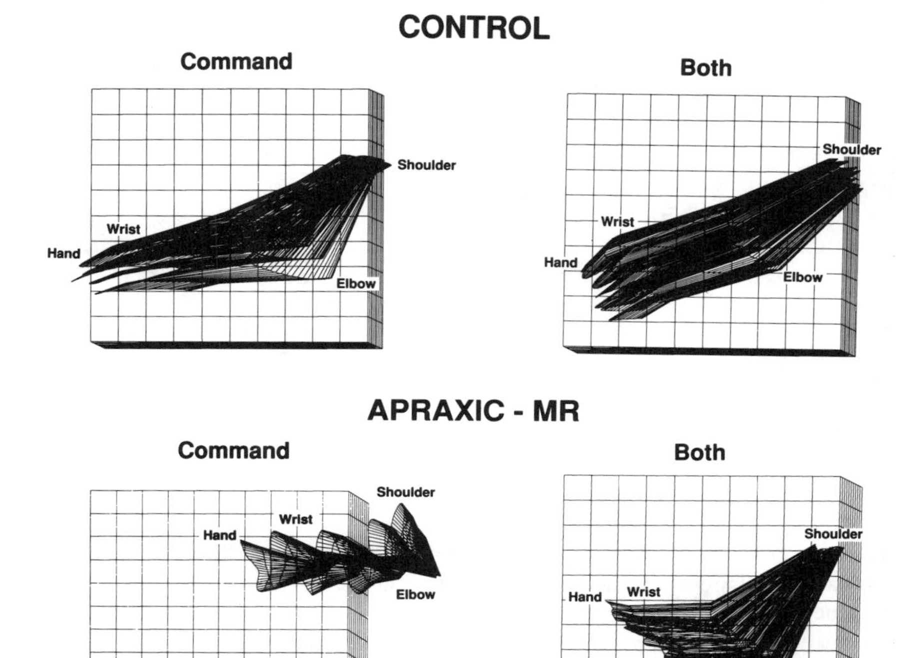

FIG. 2. Reconstructed three-dimensional limb trajectories of a control (top) and apraxic individual (bottom) for movement to verbal command and manipulation of tool and object (both). Stick figures represent successive limb positions at 20ms. intervals for both top and lower right panels, and at 10ms. intervals for lower left panel. Calibration grid lines are spaced .05 meters apart. The trajectory paths of the control are linear and tightly organised, with closely overlapping planes of movement. In contrast, the apraxic patient produced wrist paths which were curved and executed in variable planes of movement both when gesturing to verbal command and when actually manipulating the tool and object.

produced wrist paths which were curved and executed in variable planes of movement both when gesturing to verbal command and when actually manipulating the tool and object.

Plane of Motion

To capture the spatial orientation of the plane of motion and changes in that orientation during the movement, the best-fitting plane of motion of the hand trajectory for the SLICE gesture is presented for three apraxic patients and four controls. This best-fitting plane of motion was computed for each replication of the gesture and for each movement repetition cycle within a given gesture. To

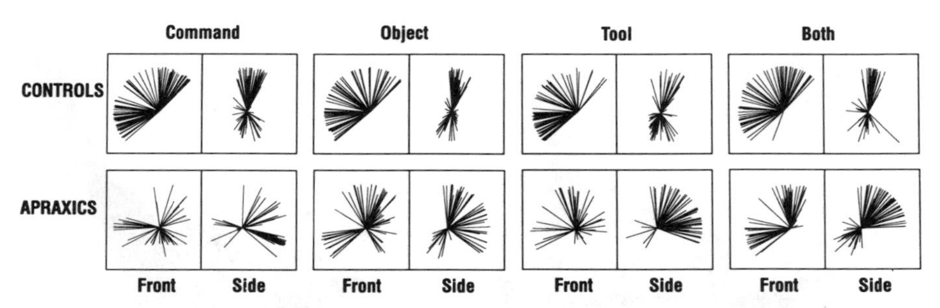

FIG. 3. Directional vectors perpendicular to the best-fitting plane of motion for the wrist trajectories of all apraxic and all controls. Each vector represents the plane normal for one cycle of movement and is shown from front and side views for each cue condition. All vectors have been converted to a unit length of 1. These vectors, representing the orientation of the plane of motion of the trajectories, show the consistent disruption in plane of motion across cue conditions in the apraxic patients. (from Clark, 1994. Reprinted by permission).

specify quantitatively the plane's orientation in three-dimensional space, the direction perpendicular to each plane was calculated. These directional vectors for each movement replication and repetition cycle are presented in Fig. 3.

Vectors along the anterior/posterior axis reflect movement occurring in the coronal (frontal) plane, whereas vectors along the lateral axis reflect movement occurring in the sagittal plane. Fig. 3 shows that vectors for the controls lie predominantly along the lateral axis in all cue conditions. Large displacements along this axis and small displacements along the anterior/posterior axis reflect motion in the sagittal plane. In contrast, the vectors for the apraxic patients lie predominantly along the anterior/posterior axis with minimal displacement along the lateral axis for conditions of verbal command, object, and tool. For the both condition (tool/object manipulation), Fig. 3 shows that the apraxic patients had vectors along both the lateral axis and the anterior/posterior axis. The sagittal component of the movements statistically differed between the apraxic and control individuals for the cue conditions of command, object, and tool. The frontal component of the movements statistically differed between the apraxics and controls for all cue conditions. When both the object and the tool were present, there was no statistically significant difference in the sagittal component. Thus, the apraxic patients oriented their movements improperly in space across all cue conditions. They were able to modify their movement pattern when given full contextual cues, but were able to do so only by increasing the sagittal component of the movement; they were unable to correct the displacement in the frontal plane creating a movement, even with full contextual cues, which was not restricted to a single plane in space. Although with full contextual cues the apraxic patients were able to shift to a more sagittal direction, a predominant component of the gesture remained in the frontal plane.

Thus, control over the direction of the movement axis is impaired in apraxic

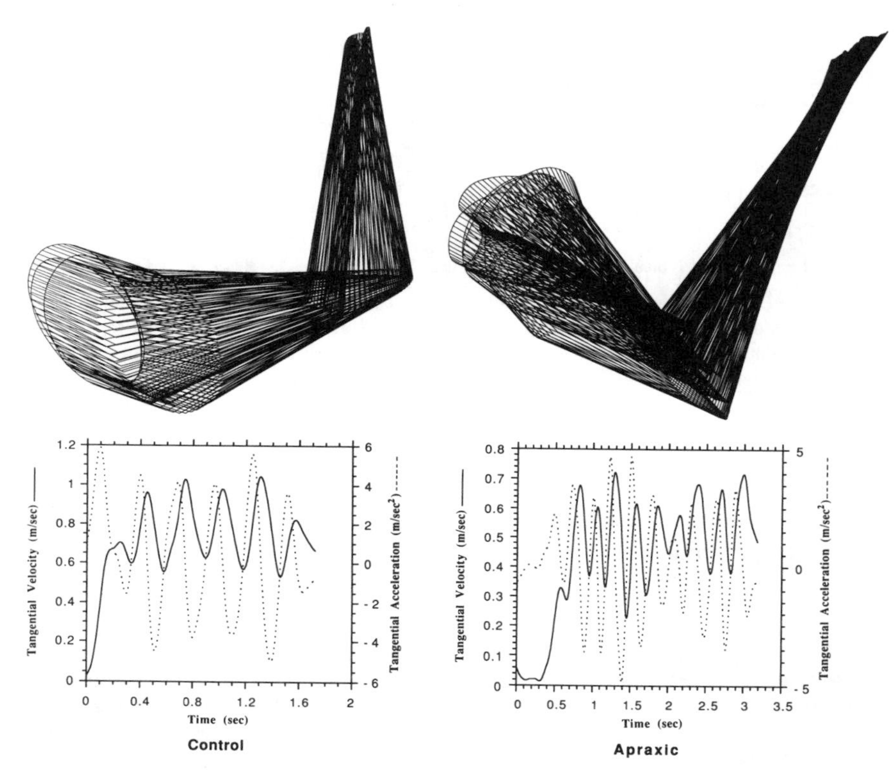

FIG. 4. Three-dimensional reconstructions of the motions of the hand and arm of a control subject and apraxic B.H. performing the gesture WIND. Tangential velocities and accelerations are also presented. Note searching behaviour of apraxic B.H., and the irregular velocity profile. (From Poizner & Soechting, 1992, p. 447. Reprinted with permission.

patients. This impaired movement control is also reflected in the inappropriate addition of movement axes by some apraxic patients. For example, Fig. 4 presents the reconstructed movement of a control and of apraxic B.H. for the gesture WIND. The control shows smooth circular movement repeated about a well-defined centre point. In contrast to the control subject, apraxic B.H. produces repeated circular movement paths, but constantly changes their amplitudes and spatial orientations.

DECOUPLING OF NORMAL SPACE–TIME RELATIONS

Movement trajectories may be described in terms of their spatial path and the time sequence along that path. A strict relationship exists between these two movement attributes: hand velocity is tightly coupled to a spatial aspect of the trajectory, namely, its curvature, or degree of bending (Morasso, 1983; Viviani

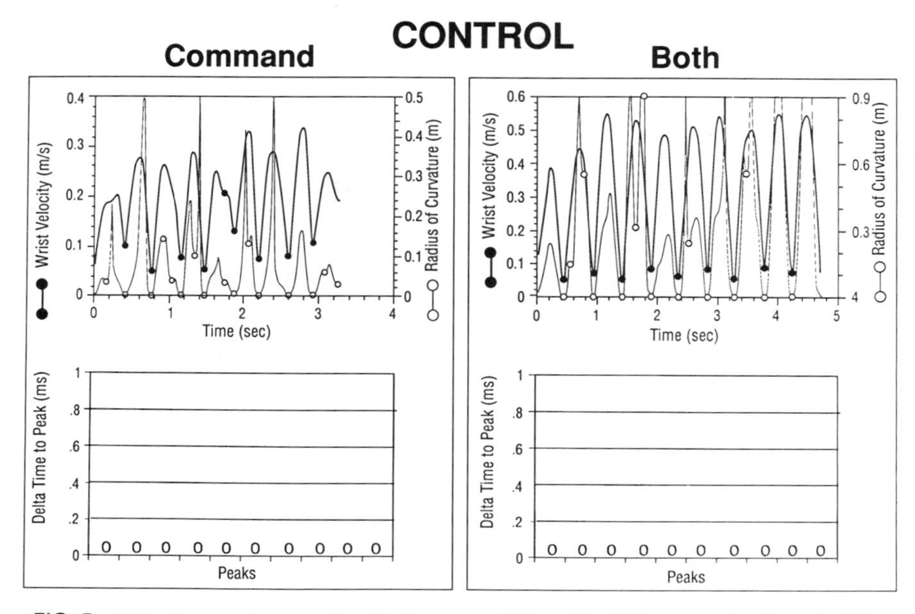

FIG. 5. Velocity–curvature coupling for the cue conditions of verbal command and manipulation of tool and object (both) in a control subject. Tangential wrist velocity (solid circles) and radius of curvature (open circles) are plotted in the upper panels. Time correspondences between minima and radius of curvature minima are plotted in the lower panels. Note the tight correspondence between velocity and radius of curvature in the control subject (from Clark et al., 1994. Reprinted with permission).

& Terzuolo, 1982). This relationship is not due to any physical law, but rather has been related to a control process by which complex trajectories are planned through proportional control of velocity and curvature (Viviani & Terzuolo, 1982). This space–time coupling breaks down in apraxia. Figures 5 and 6 present the temporal course of hand velocity and radius of curvature for a control and for an apraxic patient performing the gesture SLICE. (Radius of curvature, the inverse of curvature, is plotted rather than curvature, so that it may be plotted on the same axis as velocity.)

Figure 5 shows that for the control, there is a strict correlation between velocity and curvature: as velocity increases, curvature decreases. To capture the linkage between the curves, a computer algorithm was written to locate and mark velocity minima and radius of curvature minima. The filled circles in the figure mark these minima. The times of occurrence of the velocity minima were noted, and the difference in time between each velocity minimum and the nearest radius of curvature minimum was computed. These time differences are plotted in the lower panel. Whereas the time correspondence between minimum velocity and peak curvature was almost perfect for controls, there was substantial lack of such

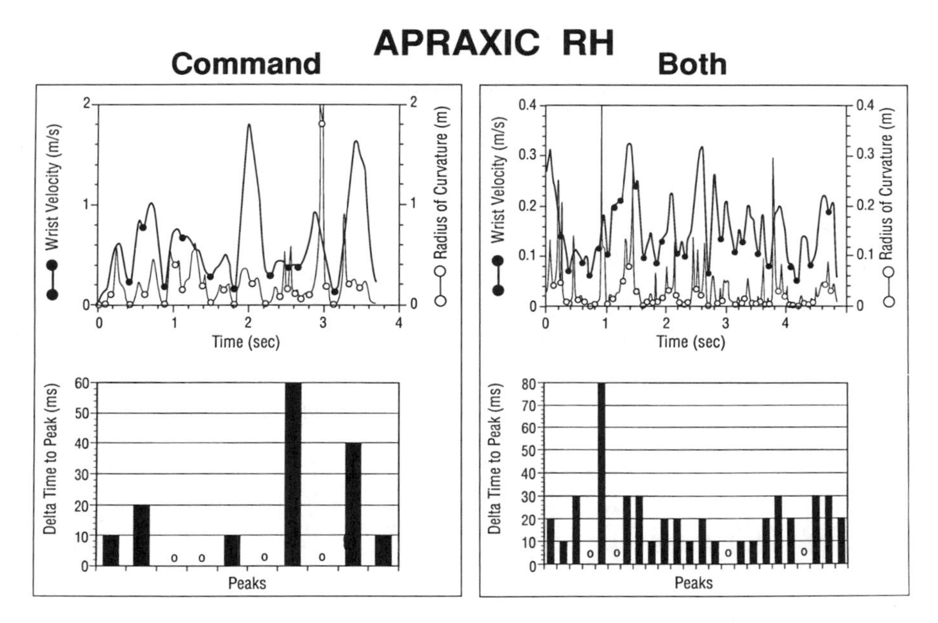

FIG. 6. Velocity–curvature decoupling for the cue conditions of verbal command and manipulation of tool and object (both) in an apraxic patient. Tangential wrist velocity (solid circles) and radius of curvature (open circles) are plotted in the upper panels. Time correspondences between minima and radius of curvature minima are plotted in the lower panels. Note the decoupling between velocity and radius of curvature in the apraxic subject (from Clark, et al., 1994. Reprinted with permission).

correspondence in the apraxic patients. Fig. 6 shows that this velocity–curvature relationship is decoupled in the apraxic patients. Indeed, there was a significant disruption in the linkage between the velocity and curvature of the wrist trajectory across cue conditions in the apraxic patients.

IMPAIRED JOINT COORDINATION

In addition to making spatial orientation errors, the spatial trajectory of an apraxic's limb when making a pantomime is often incorrect (Rothi, Mack, Verfaellie, Brown, & Heilman, 1988). This error in spatial path may be related to incorrect joint use or poor joint coordination. To examine the apraxic patients' relative use of proximal versus distal joints in their movements, we ask whether the apraxic patients may be attempting to control their distal musculature via their proximal. The gesture "unlock a door" provides a good starting point for this analysis. Controls generate the twisting portion of the movement distally from the elbow, and consequently show a small displacement of the elbow itself. In contrast, apraxic patients often generate the motion proximally from the shoulder, and thus exhibit a very large elbow displacement.

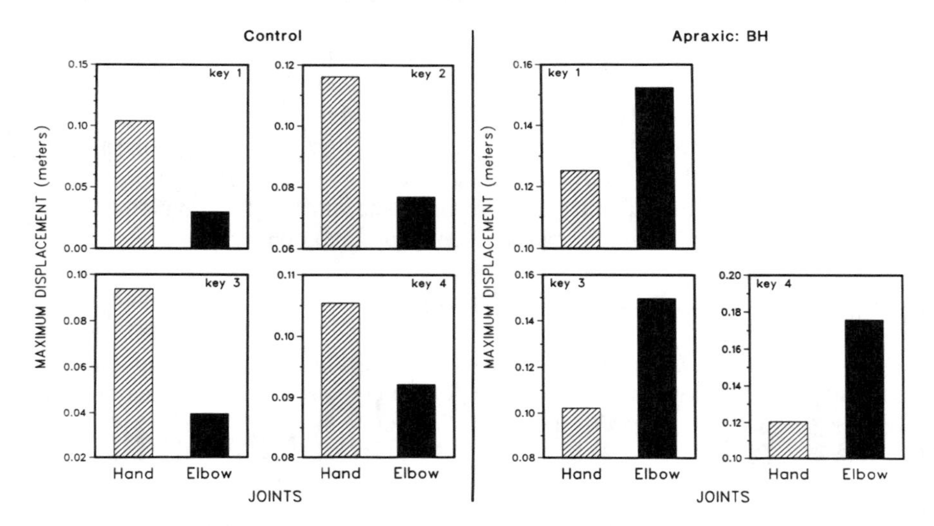

FIG. 7. Maximum displacements of the hand and elbow across trials for the twist portion of the gesture KEY of a control subject and apraxic B.H. Apraxic B.H. shows larger elbow than hand displacements in contrast to the control (from Poizner et al., 1990, pp. 85–101. Reprinted with permission).

The twist portion of each movement was graphically edited out so that it could be analysed directly. To determine the degree to which the movement was generated at the shoulder as opposed to the elbow, the maximum displacement of the hand and elbow was measured. If the movement is distally generated, there should be larger hand than elbow movement. If the movement is proximally generated at the shoulder, there should be an opposite pattern. Figure 7 presents the maximum displacement of the hand and the elbow, across trials, for a control subject and for apraxic B.H.

Figure 7 shows that the controls have substantially larger hand than elbow displacements across movement replications. In contrast, apraxic B.H. shows exactly the opposite pattern. To quantify the relative displacements of hand and elbow movement, the ratio of hand-to-elbow displacement for the twist portion of the movement was calculated for each trial, for each of five controls and for apraxic B.H. The mean hand-to-elbow displacement ratio for the control subjects was 1.86, indicating almost twice as much hand as elbow movement. The mean ratio for apraxic B.H., however, was .73, reflecting greater elbow than hand motion. Apraxic B.H.'s mean ratio was significantly different from that of the controls, with no overlap in the two ratio distributions. These data indicate that whereas the controls were generating the movement distally from a point axis at the elbow, apraxic B.H. was generating it proximally at the shoulder.

In order to analyse more thoroughly patterns of joint coordination, we use an arm coordinate system that has been identified psychophysically as the preferred coordinate system for the recognition of the orientation of the arm in space (Soechting & Ross, 1984; Soechting & Terzuolo, 1986). The parameters chosen are theta, the angular elevation of the upper arm measured in a vertical plane relative to vertical; beta, the angular elevation of the forearm measured in a vertical plane relative to vertical; eta, the angular motion of the upper arm measured in a horizontal plane (yaw) relative to the anterior direction; alpha, the angular motion of the forearm measured in a horizontal plane (yaw) relative to the anterior direction; and phi, elbow flexion/extension. Although the elevation and yaw angles are measured relative to spatial axes, theta may be taken as an indirect measure of shoulder flexion/extension and eta may be taken as an indirect measure of horizontal shoulder abduction/adduction. We computed the angular orientations of the upper arm and forearm limb segments and flexion and extension at the elbow over the course of each movement.

Apraxics Apportion Arm Angles Differently than Controls

Figure 8 presents the variation in the five arm angles over time for one control (top panels) and apraxic R.H. (lower panels) for movement performed to verbal command and performed when actually using the tool and the object. Figure 8 shows that the control produced smooth, sinusoidal variation in all arm angles for both cue conditions. When the control actually manipulated the tool and object, the amplitudes of the angular motions were diminished. However, the patterns of arm angles remained comparable. The largest angular variations for the control in both conditions were in elbow flexion and extension (phi) and in movement of the upper arm in the horizontal plane (eta). Moreover, the control subject showed similar patterns of phase relations in both conditions. For example, in both conditions, as the elbow extended (i.e. as phi increased), the upper arm moved toward the midline of the body (i.e. eta increased), while the forearm moved laterally in the horizontal plane (i.e. alpha decreased). These phase relations among arm angles kept the wrist path straight and oriented in the sagittal plane.

In contrast, the bottom panels of Fig. 8 show that the variation in the arm angles for the apraxic patient was less sinusoidal and more variable than that for the control. Additionally, the frequency of these oscillations were diminished in the apraxic patient as compared to the control. When presented with full contextual cues, the apraxic patient produced angular changes of diminished amplitude. This accommodation to the manipulation of the tool and object is similar to that of the control. However, even with the addition of full contextual cues, the changes in arm angles over time in the apraxic patient remained irregular and poorly coordinated.

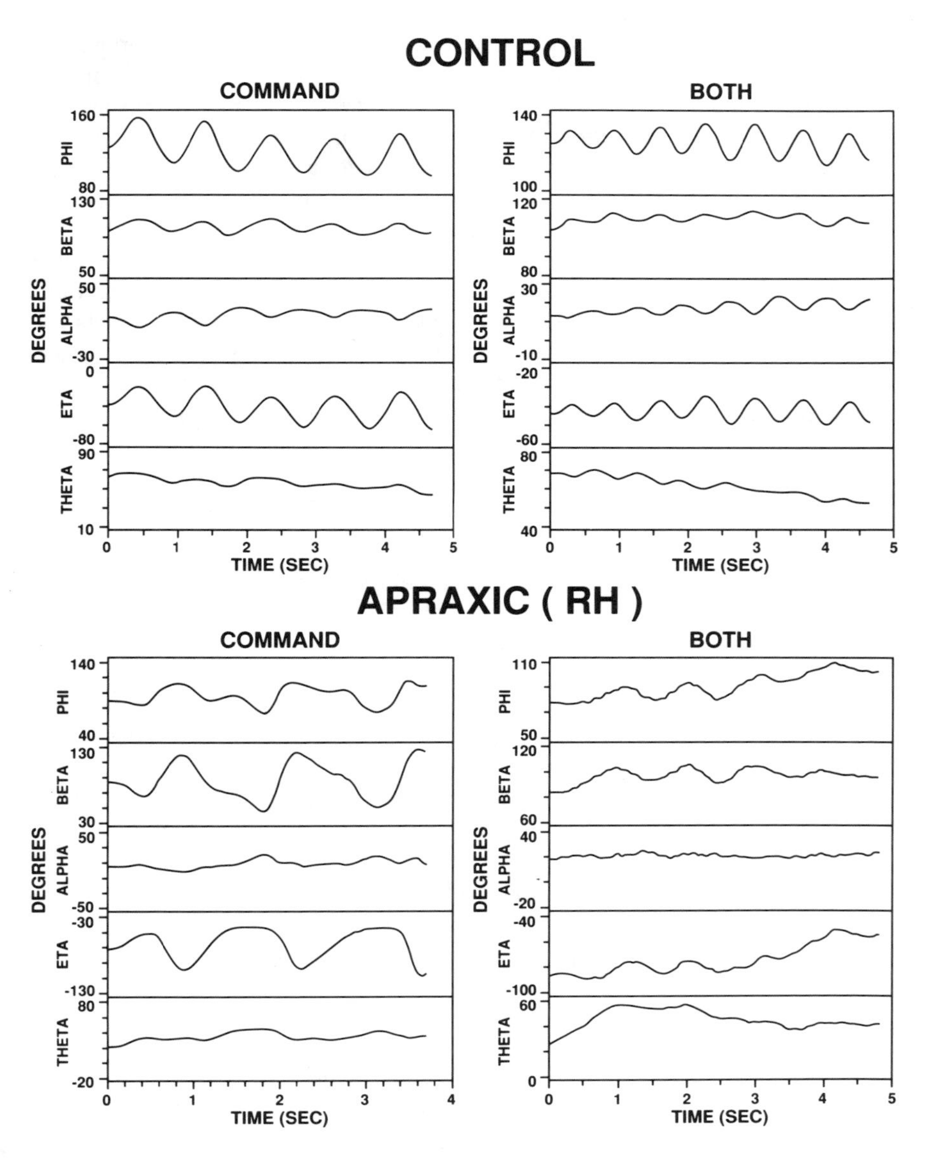

FIG. 8. Variations in arm angles over time for control (top) and apraxic (bottom) for movement to verbal command and to manipulation of tool and object (both). Note that the variation in arm angles for the apraxic patient is distorted across cue conditions (from Poizner et al., 1995. Reprinted with permission.

Figure 9 presents histograms of the relative amplitudes of pairs of arm angles across segments for all replications of the slice gesture for the three apraxic patients (white bars) and the four controls (black bars) for movements performed to verbal command. Ratios of arm angles are taken to normalise for differences in arm size and in absolute size of movements. Arrows in Fig. 9 indicate median ratios. The upper left panel presents amplitude ratios across movement segments for forearm yaw to upper arm yaw (i.e. amplitude of motion of the forearm in the horizontal plane to motion of the upper arm in the horizontal plane). Figure 9 shows that the controls had a median ratio of .42, indicating that they exhibited approximately twice the horizontal displacement of the upper arm than horizontal displacement of the forearm. In comparison, the apraxic patients produced the slicing gesture with approximately equal amounts of horizontal movement of the upper arm and forearm. The distributions of amplitude ratios of forearm to upper arm yaw differed significantly between controls and apraxics. Likewise, the apraxic patients apportioned upper arm elevation and forearm elevation differently from the controls. The upper right panel of Fig. 9 shows that the controls had a median amplitude ratio of theta/beta of 1.04, reflecting movement that used approximately equal amounts of upper arm elevation and forearm elevation; indeed, in some cases there was more upper arm than forearm elevation. In contrast, the median ratio for the apraxic patients was .27, indicating that they used three times as much forearm elevation as upper arm elevation in the slicing movement. These distributions of amplitude ratios differed significantly between control and apraxics.

The amplitude ratios for forearm yaw to elbow flexion/extension and forearm elevation to elbow flexion/extension are shown in the lower panels of Fig. 9. Both lower panels show that the controls produced a slicing gesture using significantly more elbow flexion/extension than either forearm yaw or forearm elevation. In contrast, the motion executed by the apraxic patients consisted of close to equal amplitudes of forearm yaw and elbow flexion/extension, and even greater amplitude of forearm elevation than elbow flexion/extension. Finally, Fig. 9 shows that the distributions of arm angles for the controls are, in general, much tighter than those for the apraxics, reflecting less variability across segments and controls than apraxics.

The combinations of arm angles presented in Fig. 9 demonstrate a marked difference in the manner in which the slicing gesture was produced to verbal command by both groups. Whereas, the control group apportioned the motions of the upper arm and the elbow angle in greater amplitude than motion of the forearm, the apraxic group showed much the reverse pattern. Moreover, during actual manipulation of the tool and object, the apraxic group was able to modify the relative amplitude of upper arm to forearm elevation, bringing it closer to the control value, but nonetheless, this amplitude relation remained significantly different from that of the control group.

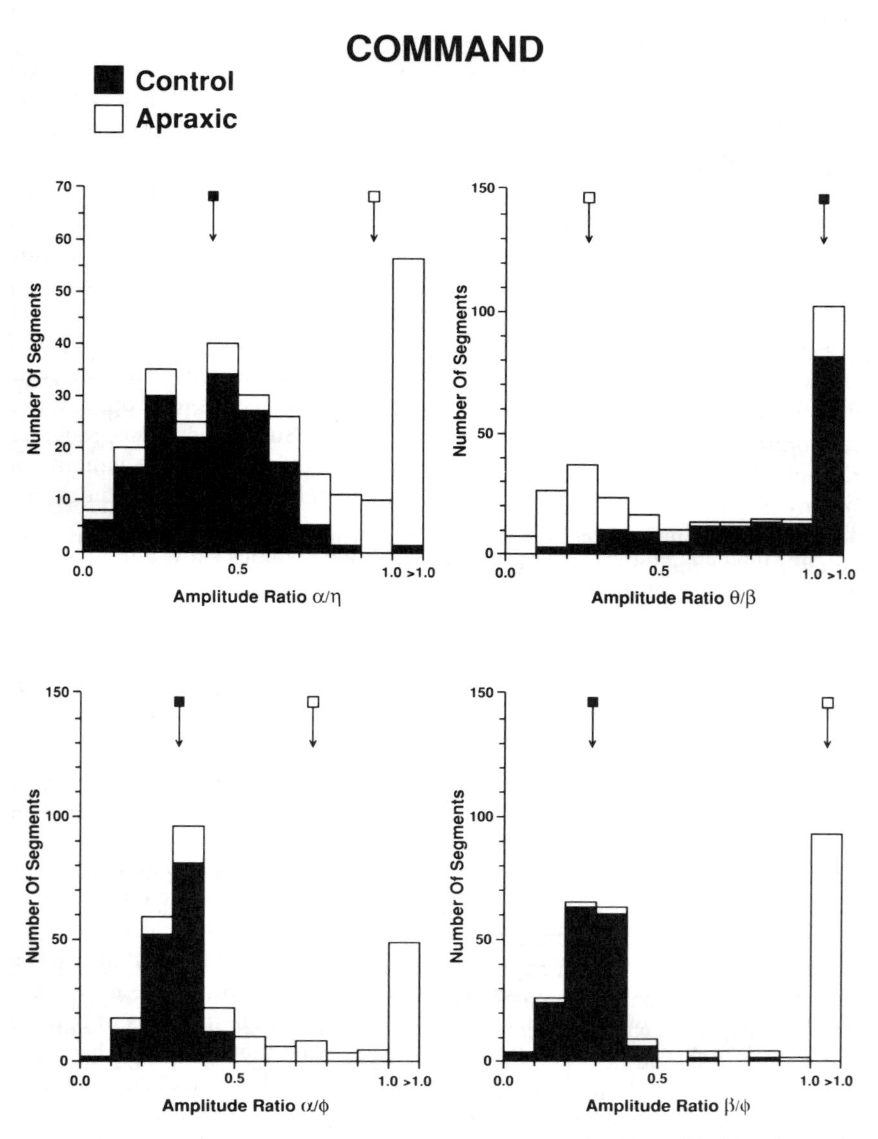

FIG. 9. Relative amplitudes of selected arm angles for control subjects (black) and apraxics (white) for all segments and replications of the slicing gesture performed to verbal command. Arrows represent median amplitude ratios. Ratios between 0 and 1.0 are plotted at intervals of .1 and all ratios above 1.0 are plotted together. Note that the apraxics apportion arm angles differently than the controls (from Poizner et al., 1995. Reprinted with permission).

SUMMARY AND CONCLUSIONS

We have presented some three-dimensional kinematic analyses of the breakdown in motor control in subjects with limb apraxia to illustrate the possibilities afforded by objective, quantitative measurement of the movement disorder. One simply cannot uncover precise patterns of timing and spatial relations in three-dimensional trajectories from observation or from viewing of videotaped recordings. However, just such precise information on the temporal and spatial trajectory variables is essential for uncovering the nature of the loss of motor control in apraxia and, as well, can prove important for the development of therapies for remediation of the disorder. As the noted neurologist Gordon Holmes recognised long ago, "One of the great disadvantages which clinical investigations suffer is difficulty in recording and measuring observations accurately by such methods as physiologists employ" (Holmes, 1939, p. 10). The study of motor function through the quantitative, three-dimensional computer-graphic analysis of patients with motor disorders allows issues to be approached in a quantitative rather than qualitative manner. Such study also provides models of the disorder with potentially important information like patterns of wrist displacement and velocity and patterns of relative joint amplitudes and timing.

Our kinematic analyses have demonstrated that apraxics show deficits in the spatio-temporal attributes of the wrist trajectory and in the coordination of the joint motions. A prominent spatial error was the planar orientation of the movement. When one learns to use a tool or instrument, one learns to use it in a specific spatial relationship to the body. For example, when using a knife to slice bread, one usually moves it in a sagittal plane. The apraxic patients, however, often produced movements improperly oriented with respect to the body. Moreover, the velocity profiles of the wrist trajectories of the apraxic group were often irregular and non-sinusoidal, and the apraxics showed a lack of coupling between spatial and temporal properties (speed and path curvature) of the trajectories. Furthermore, the apraxic patients improperly coordinated their joint motions, often relying more on proximal rather than distal joint control.

Importantly, we found that these movement abnormalities occurred not only when the apraxic patients made movements to verbal command, but also when they were actually engaged in tool/object manipulation. Since the movement deficits were found during actual tool/object manipulation, the basis of apraxia cannot rest on a disconnection between language and motor areas as has been proposed (Geschwind, 1975). In Geschwind's (1975) disconnection model, an apraxic individual is unable to carry out movements to command because the left hemisphere that comprehended the verbal command is disconnected from the right premotor and motor areas which controls the left hand. The disconnection model predicts, therefore, that apraxics should be able to correctly use actual tools since these tasks do not require language. However, we did not find this to be the case. An alternate model, however, can account for these data. If apraxia results from either the destruction of visuokinaesthetic motor representations of learned movement, stored in posterior association cortex of the left hemisphere, or from a separation of these representations from premotor or motor areas (Heilman, 1979; Heilman & Rothi, 1985), then actual tool use also should be impaired in apraxics. Our motion analyses provide compelling evidence that this latter model of the basis of apraxia is correct. In so doing, they indicate the important role that kinematic data can play in resolving competing theories of the basis of motor disorders.

The methodology we have presented for studying apraxia provides a quantitative rather than only a qualitative analysis, and can help us begin to measure quantitatively the breakdown in motor programming in apraxia and in other motor disorders. Such analyses should not only advance our understanding of the neural basis of apraxia and other motor disorders, but also may serve as a useful clinical tool in evaluating motor disorders. For example, these procedures may allow the detection of subtle forms of limb apraxia, and allow discovery of new subclassifications of the disorder that might otherwise remain unrecognised. Moreover, these methods should allow development of a reliable method for measuring the effects of treatment of apraxia and its course of recovery. Finally, the better we understand the nature of a disorder, the better we can design therapies to remediate it. By applying these methods to apraxia and other neural disorders of movement, we can begin to provide a new perspective on neural processes underlying higher level motor control.

ACKNOWLEDGMENTS

This research was supported in part by NIH Grant NS28665, NSF Grant SBR92-13110, and NIH Grant DC01664 to Rutgers University. We thank Gresha Feldman for his help with the computer graphics and John Soechting for his very generous assistance with the software for analysing arm angles.

REFERENCES

Adamovich, S., Berkinblit, M., Smetanin, B., Fookson, O., & Poizner, H. (1994). Influence of movement speed on accuracy of pointing to memorized targets in 3-D space. *Neuroscience Letters, 172*, 171-174.

Clark, M. A., Merians, A. S., Kothari, A., Poizner, H., Macauley, B., Rothi, L.J.G., & Heilman, K. M. (1994). Spatial planning deficits in limb apraxia. *Brain, 117,* 1093–1106.

Fookson, O., Smetanin, B., Berkinblit, M., Adamovich, S., Feldman, G., & Poizner, H. (1994). Azimuth errors in pointing to remembered targets under extreme head rotations. *Neuroreport, 5*, 885–888.

Geshwind, N. (1975). The apraxias: Neural mechanisms of disorders of learned movement. *American Scientist, 63,* 188–195.

Geshwind, N. & Damasio, A. R. (1985). Apraxia. In J. A. M. Frederiks (Eds.), *Handbook of clinical neurology* (pp. 423–432). New York: Elsevier Science.

Heilman, K. M. (1979). Apraxia. In K. M. Heilman & E. Valenstein (Eds.), *Clinical neuropsychology* (pp. 159–185). New York: Oxford University Press.

Heilman, K. M. & Rothi, L. G. (1985). Apraxia. In K. M. Heilman & E. Valenstein (Eds.), *Clinical neuropsychology* (pp. 131–150). New York: Oxford University Press.

Heilman, K. & Rothi, L. J. G. (1993). Apraxia. In K. Heilman & E. Valenstein (Eds.), *Clinical neuropsychology* (pp. 141–164). New York: Oxford University Press.

Holmes, G. (1939). The cerebellum of man. *Brain, 62*, 1–31.

Jennings, P. J. & Poizner, H. (1988). Computergraphic modeling and analysis II: three-dimensional reconstruction and interactive analysis. *Journal of Neuroscience Methods, 24*, 45–55.

Kothari, A. Poizner, H., & Figel, T. (1992). Three-dimensional graphic analysis for studies of neural disorders of movement. *SPIE Visual Data Interpretation, 1668*, 82–92.

Morasso, P. (1983). Three-dimensional arm trajectories. *Biological Cybernetics, 48*, 187–194.

Poizner, H. Clark, M., Merians, A. S., Macauley, B., Rothi, L., & Heilman, K. (1995). Joint coordination deficits in limb apraxia. *Brain, 118*, 227–242.

Poizner, H., Mack, L., Verfaellie, M., Rothi, L. G., & Heilman, K. M. (1990). Three-dimensional computergraphic analysis of apraxia. *Brain, 113*, 85–101.

Poizner, H. & Soechting, J. (1992). New strategies for studying higher level motor disorders. In D. Margolin (Eds.), *Cognitive neuropsychology in clinical practice* (pp. 435–464). New York: Oxford University Press.

Poizner, H., Wooten, E., & Salot, D. (1986). Computergraphic modeling and analysis: A system for tracking arm movement in three dimensional space. *Behavior Research Methods Instruments and Computers, 18(5)*, 427–433.

Rapcsak, S., Ochipa, C., Anderson, K., & Poizner, H. (1995). Progressive ideomotor apraxia: Evidence for a selective impairment of the action production system. *Brain and Cognition, 27*, 213–236.

Rothi, L. G. & Heilman, K. (1984). Acquisition and retention of gestures by apraxic patients. *Brain and Cognition, 3*, 426–437.

Rothi, L. G. J., Mack, L., Verfaellie, M., Brown, P., & Heilman, K. (1988). Ideomotor Apraxia: Error pattern analysis. *Aphasiology, 2*, 381–388.

Soechting, J. & Ross, B. (1984). Psychophysical determination of coordinate representation of human arm orientation. *Neuroscience, 13*, 595–604.

Soechting, J. F. & Terzuolo, C. A. (1986). An algorithm for the generation of curvilinear wrist motion in an arbitrary plane in three-dimensional space. *Neuroscience, 19*, 1393–1405.

Viviani, P. & Terzuolo, C. (1982). Trajectory determines movement dynamics. *Neuroscience, 7*, 431–437.

Woltring, H. J. (1984). On methodology in the study of human movement. In H. Whiting (Ed.), *Human motor actions: Bernstein reassessed* Amsterdam: Elsevier Science.

9 Representations of Actions in Ideomotor Limb Apraxia: Clues from Motor Programming and Control

Deborah L. Harrington and Kathleen Y. Haaland

INTRODUCTION

Ideomotor limb apraxia is a disorder in skilled movement that some clinicians have loosely attributed to a dysfunction in motor programming. Although we will present some evidence that supports this general view, the chapter will focus on the different levels of description at which the motor representation can be studied to disclose specific cognitive mechanisms of motor programming disorders. Our goal is to provide the reader with a greater appreciation for how investigations into motor programming and control can advance empirical findings and refine our theoretical conceptualisations about ideomotor limb apraxia. The theoretical perspective that guides our discussion is derived from both motor control theory and cognitive psychology, and has broad implications for the study of other cognitive–motor disorders. This perspective provides a framework for inquiries into the properties and the organisation of representations of actions through the study of the cognitive operations involved in motor programming and control.

The chapter begins with a summary and appraisal of some early studies that investigated the mechanisms of limb apraxia using methods directly linked to the clinical assessment of the disorder. This methodology continues to dominate research into apraxia so that the level of inquiry into the underlying mechanisms is constrained unnecessarily. This unidimensional approach also runs the risk of biasing theoretical conceptualisations of the disorder. To remedy this problem, we contend that other levels of inquiry should also be embraced, especially those that investigate specific aspects of motor programming. A discussion of motor programming will follow to illustrate how the concept has been useful in studying the motor representation in neurologically intact individuals. We will describe some of the many levels at which motor representations can be studied and show how greater specificity in our theorising about motor programming can provide some clues as to the cognitive mechanisms underlying limb apraxia. This, in turn, should help delineate more clearly the neural systems that give rise to these cognitive–motor operations. We will then proceed to the focus of the chapter, which is to describe and integrate two lines of research that have explored specific aspects of motor programming in individuals with left-hemisphere damage (LHD). We will report on research that directly compares apraxic and nonapraxic individuals with LHD. However, since few studies of motor programming exist that directly examine individuals with limb apraxia, we will also report on investigations that indirectly implicate mechanisms of limb apraxia by studying individuals who have sustained LHD, but may or may not have ideomotor limb apraxia.

THE CLINICAL ASSESSMENT OF IDEOMOTOR LIMB APRAXIA

Ideomotor limb apraxia is a cognitive–motor disorder that is intriguing because the symptoms underscore the intimate and complex functional connections among internal cognitive operations and the physical motor event. Apraxia is more common in individuals with damage to the left hemisphere, and is characterised clinically by a disruption in the performance of familiar gestures. Apraxia is considered a cognitive–motor disorder because it is found in the limb ipsilateral as well as contralateral to the damaged hemisphere (Geschwind, 1965; Haaland & Flaherty, 1984) which rules out primary sensory or motor deficits as potential mediators of the disorder. In addition, language does not appear to be a mediating factor because the degree of gestural disruption can be dissociated from aphasia severity or type (Goodglass & Kaplan, 1972; Lehmkuhl, Poeck, & Willmes, 1983; Rothi & Heilman, 1984), and severe apraxia has been reported in patients without language disturbances (Heilman, Coyle, Gonyea, & Geschwind, 1973). While ideomotor limb apraxia is commonly referred to as a motor programming disorder, the specific cognitive–motor operations that are disrupted remain elusive. This is conspicuous both in the experimental literature where the mechanisms of the disorder have been disputed, and in the definition of apraxia which places more emphasis on the causes to which it cannot be attributed (i.e. weakness, sensory loss, language comprehension deficits, general intellectual deterioration) than the reasons for the disorder (i.e. a disruption in skilled movement). Ideomotor apraxia contrasts with ideational apraxia which is considered a disorder in the conceptualisation of how to perform a movement or use an object (De Renzi, 1985; De Renzi & Lucchelli, 1988), or sequence a series of movements involving the use of objects (Liepmann, 1908; Poeck & Lehmkuhl, 1980).

Most investigations into the mechanisms of ideomotor limb apraxia have employed methods that are closely tied to clinical assessments of the disorder. Before reviewing some of these studies, let us first describe how ideomotor limb apraxia is evaluated and distinguished from other neurological disorders. The clinical assessment of limb apraxia is commonly based on an examination of gesture performance to verbal command (e.g. "Show me how you brush your teeth") and/or imitation (e.g. the examiner pantomimes combing his/her hair for the patient to imitate), the latter of which is less confounded by auditory comprehension deficits. Although performance to verbal command can be worse, most apraxic patients are also impaired on imitation (e.g. Roy, Square-Storer, Hogg, & Adams, 1991). Clinical assessments often contain three types of items to assess the importance of the symbolic value of movements and of imitating how to use objects. The movement types consist of nonsymbolic movements (e.g. "Put your hand on your chin"), intransitive symbolic gestures (e.g. "Salute", "Wave goodbye"), and transitive symbolic gestures which involve object[1] use ("Comb your hair"). Deficits reflective of ideomotor limb apraxia are characterised, in part, by an abnormal spatial-temporal or postural patterning of the arm, hand, or fingers. Examples of such errors include movements that use the wrong finger(s), inaccuracies in the target of the movement, distortions in the orientation or amplitude of the arm, hand, or fingers, and/or inappropriate ordering of movement components. Apraxic individuals also commit "body part as object" errors when imitating transitive gestures such that their finger, hand, and/or arm is used as if it were the imagined implement. Imitation responses are sometimes unrecognisable or are perseverations of a previously produced imitation. This is not an exhaustive list of errors, but it underscores the complex pattern of gesture performance on clinical assessments.

MECHANISMS OF GESTURE PERFORMANCE: AN APPRAISAL OF THE LITERATURE

Several studies have tried to isolate the mechanisms underlying limb apraxia through an analysis of the error patterns, particularly in the performance of familiar gestures (Haaland & Flaherty, 1984; Lehmkuhl et al., 1983; Poeck, 1985; Rothi, Mack, Verfaellie, Brown, & Heilman, 1988; Roy, Square, Adams, & Friesen, 1985), but also unfamiliar sequences of hand and finger movements (Kimura, 1977; Roy, 1981). These studies have contributed immensely to describing the types of breakdowns in the performance of gestures. However, there is ongoing controversy about whether sequencing, perseverative, body part as object, and/or spatial errors are most pathognomonic of apraxia (Duffy &

[1]Historically, gestures that incorporate objects include movements that act upon an object (e.g. turn the doorknob) and those that use tools (e.g. brush your teeth). Our chapter preserves this convention, but the reader is advised that other chapters of this book generally equate object use only with tool use.

Duffy, 1989; Haaland & Flaherty, 1984; Mozaz, Pena, Barraquer, Marti, & Goldstein, 1993; Rothi et al., 1988).

An implicit assumption underlying the delineation of error types in gesture performance is that deficits in certain cognitive–motor operations are manifested by specific kinds of errors. For example, body part as object errors in limb apraxia have been attributed to deficits in integrating memory representations of intrapersonal space with those of extrapersonal space (Haaland & Flaherty, 1984) and a dependence upon contextual information (Roy & Square, 1985). Similarly, some suggest that perseverative errors reflect impairments in transitioning between different motor programs for movements (Kimura, 1977), and sequencing errors signify deficient serial ordering mechanisms (Roy, 1981). A conceptual difficulty with this assumption of a one-to-one mapping between a specific cognitive–motor operation and an error type is that most studies infer mechanisms of apraxia from error data alone. Experimental manipulations designed to affect some cognitive–motor operations but not others have not been done to determine whether error patterns are affected differently depending upon the cognitive requirements of a task. In other words, there has not been a direct examination of whether the types of apraxic errors vary depending upon the cognitive demands of the task. In the absence of a more analytical treatment of error profiles, numerous alternative mechanisms can be put forth to interpret virtually any of the error types observed in apraxia.

The emphasis in the apraxia literature on analysing error types to disclose the mechanism(s) of diminished gesture performance originates from clinical descriptions of the disorder. This is probably the basis for the preponderance of investigations into the performance of highly familiar movements, especially those that involve interactions with objects. In fact, the few studies that are available suggest that transitive movements which involve object use are more impaired in apraxia than intransitive symbolic movements (Goodglass & Kaplan, 1963; Haaland & Flaherty, 1984; Roy et al., 1991). While this may argue that only familiar movements should be studied because apraxia is principally a disorder of skilled movement, deficits are also seen when performing nonsymbolic or meaningless movements (Dee, Benton, & Van Allen, 1970; De Renzi, Motti, & Nichelli, 1980; Haaland & Flaherty, 1984; Lehmkuhl et al., 1983; Rothi & Heilman, 1984). In fact, the apparent greater sensitivity of transitive movements in the assessment of apraxia may have little to do with their familiarity or perhaps even the role that objects play in pantomime or imitation. This is because there are many potential differences among the movement types that could affect the probability of performance deteriorations including the number of movement components, the range of possible error types, the required degree of movement precision, and the use of limb, hand, and/or finger movements. These aspects of movement, as well as others, are typically mixed together in the clinical assessment of apraxia so that it is impossible to pinpoint unequivocally the reasons for the disorder.

It should be apparent that the descriptive approach to disclosing mechanisms of apraxia considers only the overt movement. The other part of the action event, the covert representation, is ignored. This is a problem when the goal is to uncover the central mechanisms for the disorder because an isomorphic relationship does not necessarily exist between a description of an overt movement and its central representation. For example, the duration of an overt movement or some of its biomechanical constraints are probably not directly represented in memory, at least for some types of movements (Marteniuk, MacKenzi, Jeannerod, Athenes, & Dugas, 1987).

Clearly, most conceptualisations of apraxia are based largely on methods closely linked to clinical assessments of the disorder. The problem with this is that despite the distinction between theories and methods, the two are often inextricably linked. Scientists have become increasingly aware that experimental findings and their theoretical interpretations can be biased when the level of inquiry is restricted to a single methodological approach. Unbiased and comprehensive investigations should embrace many levels of inquiry in the pursuit of a unified theory of any behavioural phenomenon.

What implications does this appraisal have for the study of apraxia? First, if we acknowledge that ideomotor limb apraxia may be due to a disruption in some aspect of the central motor representation (Heilman, Rothi, & Valenstein, 1982), then it is reasonable to investigate the functional status of specific levels of the motor program. This is the rationale for our experiments and those of others which we will describe later. Second, the reliance upon error measures in most studies of apraxia unnecessarily restricts the level of analysis to overt movements in which abnormal performance is apparent to the human eye. This practice ignores the bulk of the data in which apraxic individuals "correctly" carry out movements. To assess fully the integrity of the central representation of actions, correct movements also should be studied by manipulating factors that affect specific cognitive–motor processes involved in planning actions. This approach augments an error analysis by examining whether the covert cognitive–motor operations that are utilised to carry out a functionally correct movement are completely normal. The logic of this approach will be developed further in the next section. Finally, it is important to mention briefly that some have dismissed the possibility that research in nonhuman species will contribute to our understanding of apraxia because it is a syndrome that is unique to humans (Poeck, 1985). This position hinges on the fact that clinical assessments of apraxia are impossible to implement in nonhuman species. However, if the level of inquiry is expanded to include methods that systematically examine the areas of the brain that represent specific aspects of the central motor program, then consideration of the cognitive and the neuroanatomical mechanisms of apraxia in nonhuman species (Kolb & Whishaw, 1985) will be important because the behavioural consequences of discrete damage to different areas of the brain can be examined more easily.

REPRESENTATIONS OF MOVEMENTS: THE MOTOR PROGRAM CONCEPT

It is not a simple undertaking to delineate the mechanisms of cognitive–motor disturbances when one contemplates the complexity of the action system. For example, consider the fact that humans have the impressive ability to learn intricate sequences of actions in many different skill domains. This is especially remarkable because there are multiple ways in which movements can be carried out to achieve a particular goal. Although some movement solutions are more preferable than others and may be constrained by the environmental context, many questions remain about how choices are narrowed down from a multitude of possibilities in the action system. Bernstein (1967) had a keen appreciation for the complexity of this dilemma, which he designated the degrees-of-freedom problem. This dilemma is just one of the many challenges confronting theorists. Research into how actions are coded and represented in the brain has a long tradition in psychology and physiology. The complexity of the motor system was appreciated very early on by Lashley (1951) whose insights into the problem of sequencing actions led to the formulation of the concept of a motor program, which is central to theories concerned with how movements are represented in the brain. Not surprisingly, the specification of the motor program is far from complete. Many key problems remain in terms of explaining how motor programs are represented so they can engineer the correction of movement errors (Jeannerod, 1986a), integrate perceptual representations of the environment with codes for movement (Goodale & Milner, 1992), organise components or sequences of movements (Cohen, Ivry & Keele, 1990; Jeannerod, 1986b; Keele, 1987), and control the velocity, accuracy, and timing of movements (Keele & Ivry, 1987; Marteniuk et al., 1987; Meyer, Abrams, Kornblum, Wright, & Smith, 1988). The problems listed are by no means exhaustive, but they help depict the complexity of the action system and hence, the different levels at which breakdowns can occur in movement disorders.

The concept of a motor program has been used in motor control theory and cognitive psychology to study the properties and the organisation of the motor representation by exploring how different exogenous and endogenous variables affect the planning and the implementation of voluntary movements. Motor programs are abstract representations of intended movements that describe the relationships among the goal of an action, the external event or object it is directed toward, and the organism's interactions with the external event. The content of the motor program has been somewhat elusive and it has been criticised for its inherent emphasis on serial hierarchical processing (Wickens, Hyland, & Anson, 1994). Many also have argued that the motor program construct is constrained inappropriately by the computer metaphor which holds that memory representations consist of static, stored contents or subroutines that are called up by programs. Still, a case can be made for the construct's utility when it is defined in broader terms as a functional state which allows an organism to carry out a specific movement or skilled actions (Rosenbaum, 1991). In this vein, the motor program concept can be valuable in explaining how the preparation and control of movements is affected, for example, by the goal(s) of the action or by the physical interactions with the environment. Answers to these kinds of questions lend insight into the nature of the motor representation.

How does one identify experimentally the properties of actions that are represented in motor programs? Many experimental paradigms have been used. One common approach derived from motor control theory has been to examine the kinematics of simple movements. In these investigations, the nature of the motor representation is revealed by the form and temporal patterning of a movement trajectory. For example, kinematic analyses show that grasping movements consist of two components, each of which is affected by different properties of the action and the object. The first component anticipates the size of the object while reaching for it and the second adjusts the final size of the grip before contacting the object (Marteniuk, Leavitt, MacKenzi, & Athenes, 1990). Similarly, the velocity profile and the relative scaling of the acceleration and deceleration phases of a simple aiming movement are dependent upon its distance and precision requirements (Haaland, Harrington, & Grice, 1993; Marteniuk et al., 1987; Meyer et al., 1988).

Concepts derived from cognitive psychology have also been influential for investigations into the properties of the motor representation. Many studies have considered the problem of what characteristics of movements are important for preparing a plan of action. In these studies, the preparatory reaction time (RT) interval, which is the time between a signal to perform an action and the initiation of the movement, is examined to determine how various properties of the movement affect its duration. For example, the duration of preparatory RT is affected by movement precision, the location of a target, movement amplitude, the force and velocity of the movement, and temporal accuracy (Franks & Van Donkelaar, 1990; Haaland et al., 1993; Kerr, 1978). Inquiries into the higher level structure of the motor representation have also been most influenced by concepts central to cognitive psychology. At this level of the motor representation, serial events appear to be organised by rules, schema or other mechanisms that determine the selection, activation and inhibition of responses as well as their serial order and spatio-temporal unfolding (Keele, 1987; MacKay, 1985; Rosenbaum, Inhoff, & Gordon, 1984). There is ample evidence showing that the performance of a specific movement is determined, in part, by the actions that follow the movement. For instance, preparatory RT increases as the number of movements contained within a sequence increases (Harrington & Haaland, 1987; Sternberg, Monsell, Knoll, & Wright, 1978). This finding shows that neurologically intact individuals prepare a plan based on all serial events in a sequence consisting of five to six responses, not simply the first response in the sequence. Similarly, RT is fastest when two sequences in working memory are structurally similar (Gordon & Meyer, 1987; Rosenbaum et al., 1984), which demonstrates that individuals organise serial events by using natural rhythmic, spatial, or motoric properties to chunk or parse responses into smaller units.

This condensed overview of a few selected methods for mapping the content and the organisation of the motor representation portrays the complexity of the issues confronting scientists studying neurologically intact humans. The idea that motor programs consist of many different layers is compatible with the observation that some key functional components of the motor representation appear to be distributed in different areas of the brain (Georgopoulos, Ashe, Smyrnis, & Taira, 1992; Goodale & Milner, 1992; Haaland & Harrington, 1990; Harrington & Haaland, 1991a, 1991b; Jeannerod, Michel, & Prablanc, 1984; Keele & Ivry, 1990). This raises the distinct possibility that there could be more than one mechanism of ideomotor limb apraxia, inasmuch as different levels of the motor representation would be expected to support the performance of the complex movements that are tested in clinical assessments of gesture performance.

In the remaining sections of this chapter we will discuss some preliminary investigations into the specific levels of the motor representation that may be disrupted in limb apraxia. We will first describe some experiments in the tradition of motor control theory that employ kinematic analyses to examine the form and temporal patterning of simple movements in patients with LHD. Then we will turn to a study in our laboratory that is a preliminary investigation into the programming and implementation of hand–posture sequences in individuals with and without limb apraxia. The rationale for the latter study was guided by concepts from cognitive psychology in which mental operations are inferred from changes in performance that are associated with systematic variations in specific characteristics of serial events.

KINEMATIC ANALYSES OF MOVEMENT WITH LEFT HEMISPHERE DAMAGE

Properties and Organisation of the Motor Representation

Kinematic analyses of movement can reveal important information about how motor representations are organised in the central nervous system and, consequently, are promising in terms of revealing some of the reasons for breakdowns in the action system. Kinematic analyses use a variety of mechanical and optical methods to record the velocity, accuracy, and shape of movement trajectories. The simple aiming task is one widely used paradigm in which the kinematic properties of single joint movements have been studied. In this task, individuals move a stylus to a target or point their finger at a target. Figure 1 shows that the movement trajectory can be separated into an initial and a secondary component. The initial component constitutes most of the movement and consists of an acceleration (IMAT) and a deceleration phase (IMDT) which transports the effector to the vicinity of a target. The secondary component (SMT) typically involves lower velocity movements to contact a target, and is usually more consequential for movements to smaller targets (see Fig. 1).

An assumption underlying analyses of movement trajectories is that organisational principles are reflected by factors that either have an invariant effect or a systematic effect on the planning and the online control of a movement across a range of task demands. One invariant characteristic of simple aiming movements is the classic bell-shaped acceleration and deceleration velocity profiles which are found despite changes in movement amplitude or target width. Figure 1 depicts these bell-shaped profiles for two simple aiming movements, one to a 3mm and the other to a 20mm diameter target. This figure illustrates that at one level there is a general or an abstract central representation that accommodates variations in movement context, perhaps by scaling movement velocity to a base form.

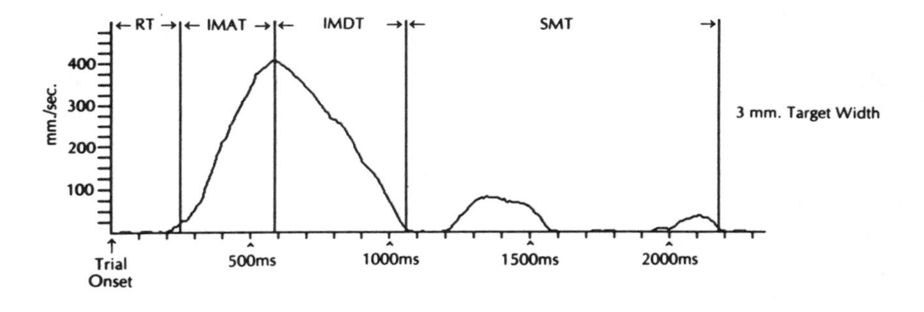

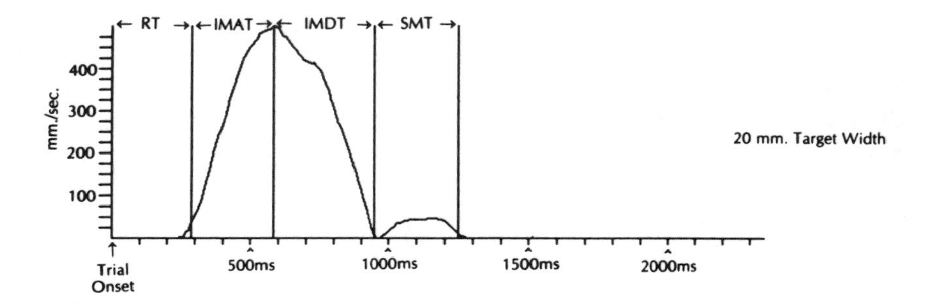

FIG. 1. Velocity graphs of simple aiming movements. The top graph shows a velocity profile for a movement to a 3mm diameter target circle and the bottom graph for a movement to a 20mm diameter target. The ordinate represents the movement velocity in mm/s. The abscissa depicts the time couse of the movement from the onset of the trial to when the stylus remains in the target circle for 400 ms. Measures that are calculated on each trial include: (1) preparatory reaction time (RT), (2) initial movement acceleration time (IMAT), (3) initial movement deceleration time (IMDT), and (4) secondary movement time (SMT).

However, at another level of the motor representation the relative scaling of the acceleration and deceleration phases is systematically modified by changes in movement amplitude and by alterations in the goal or the target of the movement. For example, Fig. 1 shows that the percentage of the movement devoted to the initial acceleration phase is less for a movement to a small target than to a large target (Haaland et al., 1993). Similarly, the acceleration phase of a movement is shorter when subjects grasp a disk than when they point to the disk (Marteniuk et al., 1987). In both examples, the movement trajectory for a goal that requires more precision (i.e. smaller targets, grasping a disk) has a shorter acceleration phase, but a longer deceleration phase, the latter of which better accommodates the greater spatial accuracy requirements of the task. These findings demonstrate that the planning and the control of movements is also partly task specific. This is an adaptive feature of the action system because our prior learning experiences, which help build the central motor representation, allow us to constrain the movement options that are available under certain task demands (Kelso, 1995; Marteniuk, et al., 1987). Hence, the temporal relationships among the different phases of a movement are adjusted optimally for a particular goal.

Inquiries into the role that prior learning plays in building the motor representation have also been concerned with the problem of explaining how sensory feedback is incorporated into the planning and the control of movement. One early theoretical formulation (Keele, 1968) proposed that the initial and secondary components of the movement trajectory differ in their reliance on preprogrammed and sensory-dependent processes. The initial ballistic component was regarded as preprogrammed or open loop because its kinematics were not influenced by the removal of visual information (Jeannerod, 1988). The secondary component was considered closed loop because programming occurs online, and the movements are slow and sensory dependent. Recent studies have shown, however, that the dichotomy between open- and closed-loop components of aiming is not absolute because the initial component sometimes can be modified by the removal of certain types of visual information during the movement (Goodale, Pelisson, & Prablanc, 1986; Haaland et al., 1993; Prablanc, Pelisson, & Goodale, 1986). For example, the kinematics of the acceleration phase are altered when visual information about the target location is eliminated at movement onset but not when individuals are deprived of visual feedback about their hand position (Haaland et al., 1993). By contrast, the kinematics of the deceleration phase are altered when visual information is absent for either the target location or the hand position. These findings show that motor programs can be quickly modified during movement, perhaps through a comparison of efferent signals from the motor program with afferent signals from eye and/or arm movements. For example, knowledge of the target location prior to movement can provide extraretinal target location information (i.e. derived from the efferent signals which control the extraocular muscles) that can be used during early phases of the movement if the target is turned off. Similarly, target location and arm position input appear to be integrated online with the initial motor program in later phases to update and improve aiming movements. Importantly, the finding that distinct aspects of the movement context affect different phases of the movement trajectory is compatible with the view that motor programs are organised hierarchically or heterarchically. In early phases of simple movements, an abstract representation of the goal (e.g. movement direction, target location) is planned because this is essential for efficiently moving an effector to the vicinity of the target. However, this plan can be altered if the goal of the movement is degraded or changed in some way. As the movement approaches the target, information about the spatial location of the effector also unfolds as this becomes important for accurately reaching the target.

To summarise, kinematic analyses have revealed high level abstract properties of the organisation of the central motor representation which are invariant across a wide range of movement contexts. At other levels, however, it is clear that the central representation imposes certain constraints on the temporal characteristics of a movement. Constraints imposed on movement options are task specific in one sense, but in another sense can be viewed as general rules or schema for selecting movements that are developed through our prior learning experiences and also are determined, in part, by biomechanical factors. Schema allow us to recognise readily action goals and in turn, automatically activate the motor programs that will implement them optimally. Movement constraints appear to have some type of organisation in that some are important in planning the initial reaching phases of a movement whereas others come into play as the movement approximates the goal.

Disruptions in the Motor Representation With Left-Hemisphere Damage

These insights into the properties and the organisation of the motor representation in neurologically intact humans have important implications for uncovering the mechanisms underlying spatio-temporal abnormalities and other movement distortions that are commonly observed in clinical assessments of limb apraxia. Specifically, one might ask whether these impairments are due to a disruption in the high level organisation of the movement kinematics, the lower levels that are responsible for selecting and implementing movement constraints, or the ability to utilise sensory feedback to modify ongoing movements and incorporate this information into the central representation. The few existing studies that are relevant to these issues are those that have investigated movement kinematics in patients who have LHD. Although ideomotor limb apraxia was not assessed, the high incidence of the disorder with LHD suggests that patterns of spared and impaired performance may generalise to apraxia. Nonetheless, the implications of this research for understanding the mechanisms of apraxia should be considered preliminary as it clearly is of interest to compare directly the performance of apraxic and nonapraxic individuals.

Fisk and Goodale (1988) were the first to use kinematic methods to examine the production of visually guided movements in 17 patients who had sustained LHD and 11 with right-hemisphere damage (RHD). All patients performed the task using the hand ipsilateral to the damaged hemisphere to control for any primary sensory or motor deficits. Age-matched controls were tested on both hands so that their left-hand performance was compared with that of the LHD group and their right-hand performance was compared to the RHD group. Participants were seated in front of a monitor with their head position secured by a chin and a head rest. At the beginning of a trial they fixated on a point in front of them, after which a target light appeared 10 or 20 degrees horizontally to either side of the fixation point. Video recordings were made of the limb as participants pointed to a target, which remained visible throughout the reach. The results showed that velocity profiles were normal for patients with RHD, but not for those with LHD. Patients with LHD showed no disturbance in the form of the velocity profiles (i.e. the bell-shaped acceleration and deceleration phases), suggesting that the higher level organisation of the motor representation was intact. By contrast, there was a prolonged period of low velocity movements at the end of the pointing movement in the LHD group. This finding was interpreted as a deficiency either in monitoring the motor program so that it could be updated as the movement unfolded, or in utilising sensory information to monitor the position of the moving limb. Both of these interpretations implied that LHD results in closed-loop processing deficits. However, the LHD group also attained a lower peak velocity at the end of the acceleration phase of the movement. Recall that some work shows that this movement component relies more on open-loop processing. This might suggest then that lower levels of the motor program were degraded, diminishing the LHD patients' ability to select an optimal force to propel the arm to the target. This interpretation raised the possibility that some aspect of open-loop processing may also be disrupted with LHD.

These hypotheses were tested in our laboratory by recording the kinematics of a movement trajectory using a simple aiming task (Haaland & Harrington, 1989). Fifteen individuals with LHD due to stroke and 14 individuals with RHD due to stroke were tested using the arm ipsilateral to the lesioned hemisphere. Because of the small sample size, participants were not separated into apraxic and nonapraxic groups. Fifteen of the controls performed the task using their right arm and 14 using their left arm. Participants were seated facing a monitor and a vertical rod attached to a stylus that was mounted on a track on top of a digitising tablet to allow for horizontal movements. The position of the stylus on the tablet could be viewed on the monitor. The digitising tablet was interfaced with a computer so that data could be sampled every 10ms during movement. At the beginning of each trial, participants centred the stylus inside a start circle that was displayed on the monitor. One to two seconds later, a target circle appeared 25, 64, or 100mm away from the start circle after which the participant moved the stylus as quickly and accurately as possible to the target circle. The target circle and the position of the stylus on the digitising tablet remained visible to the participant throughout the trial.

To test the functional status of open- and closed-loop processing mechanisms with LHD, simple aiming movements were separated into an initial component which consisted of the acceleration and deceleration phases and a secondary component which consisted of single or multiple lower velocity movements to move the arm to the target (see Fig. 1). Recall that the initial component has been associated more with open-loop processing and the secondary component with closed-loop processing. In addition, movement amplitude was varied to provide a converging test between these two hypotheses. Specifically, Wallace and Newell (1983) showed that the performance of smaller amplitude movements was not compromised by the removal of visual feedback whereas the performance of larger amplitude movements was altered when visual feedback was eliminated. Hence, it was hypothesised that if the left hemisphere regulates open-loop processing, the kinematics of the initial movement component should be disrupted with LHD, especially for shorter amplitude movements. Alternatively, if the left hemisphere regulates closed-loop processing the kinematics of the secondary movement component should be disrupted with LHD, especially for longer amplitude movements.

The reaction time to initiate the movement, and the velocity and the error of the initial and secondary components were analysed for each movement amplitude. The results showed that simple aiming movements were carried out normally by individuals with RHD. By contrast, individuals with LHD were slower to initiate the movement and showed deficits in the production of the initial but not the secondary movement component. Specifically, velocity was lower and error was greater for the initial component of the movement in individuals with LHD relative to the control group. This was true regardless of the movement amplitude. Although the duration of the secondary component was also longer in the LHD group, this was attributed to the greater distance the LHD group had to traverse during the secondary component because of their greater error at the end of the initial component.

These findings were consistent with the hypothesis that the left hemisphere regulates some aspect of open-loop processing. One possible mechanism of apraxia may be a breakdown in the ability to maintain or modify a motor program in the absence of visual cues about limb position. Although a greater disruption in shorter amplitude movements was also predicted, it is possible that the range of movement amplitudes in this study may have involved primarily open-loop processing. However, alternative explanations of the results were possible due to some limitations with the study design. Importantly, the availability of visual feedback was not manipulated so that disturbances in the kinematics of the initial movement component in the LHD group could not be directly attributed to the reliance upon visual information to update the motor program. Recall that more recent work in neurologically intact individuals suggests that the motor program can be quickly altered during the early phases of simple aiming movements when visual information is removed (Haaland et al., 1993). This implies that closed-loop processing may be common even during early movement phases which could explain why the movement kinematics in the LHD group were not disrupted more for shorter than longer amplitude movements.

We are in the process of conducting a study in our laboratory to clarify the mechanisms underlying simple aiming deficits with LHD and, in the future when sample sizes permit, to understand the extent to which these deficits are specific to ideomotor limb apraxia or more general to LHD. The task and procedures are similar to the simple aiming study described above except for two modifications in the experimental design. First, four visual feedback conditions were randomly presented to participants in order to examine directly how the removal of specific types of visual cues alters the movement trajectory with LHD. In one condition, visual information was available throughout the trial about both the target location and the arm position. In the remaining three conditions, visual information was removed as soon as the participant initiated the movement (i.e. at the end of the reaction time interval; see Fig. 1). In one of these conditions only target location information was removed, in another only arm position information was eliminated, and in the last both types of visual information were removed. In addition, a wider range of movement amplitudes was studied (i.e. 40, 100, 200, and 300mm) to provide a stronger converging, but indirect test of the open- and closed-loop processing hypotheses.

Nineteen individuals with LHD due to stroke and 20 age- and education-matched controls have been studied to date. A reconstruction of the specific lesion sites has not been conducted to date, but the sample included individuals with lesions in the anterior and the posterior regions of the left hemisphere. The preliminary results suggest that LHD disrupts only certain levels of the motor representation. There was no difference between the groups in the percentage of the total movement devoted to the acceleration phase. This is consistent with previous work (Fisk & Goodale, 1988) showing that the scaling or the higher level organisation of the motor representation appears intact with LHD. There was a trend for longer reaction times in the LHD group ($F(1,37) = 3.81, p = .058$), but the mechanism for this is not clear as the effect was similar for all movement amplitudes. Although the velocity and spatial error in both phases of the initial movement were altered by the type of visual feedback, the removal of arm position and/or target location information had the same impact on the acceleration and the deceleration velocity and spatial error in both groups. These findings suggest that LHD does not disrupt the use of extraretinal target

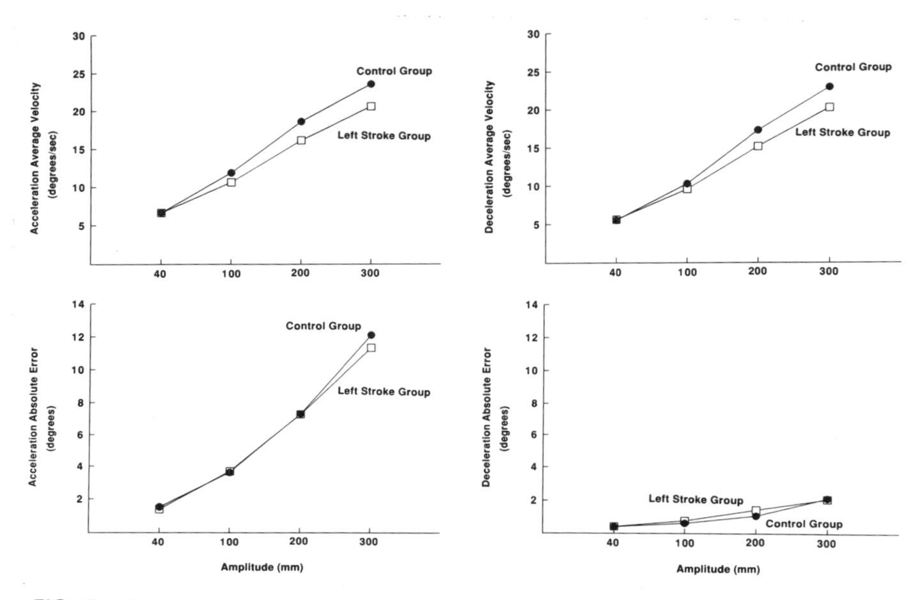

FIG. 2. Average velocity and absolute error for the initial movement component in the left stroke group and the control group. The top graphs show average velocity as a function of movement amplitude for the acceleration and deceleration phases of the initial movement component. The bottom graphs show absolute error as a function of movement amplitude for the acceleration and deceleration phases.

location information during early phases of the movement and/or the maintenance of a motor program in the absence of visual cues about arm position or target location.

Figure 2 displays the preliminary findings for the temporal patterning and spatial error of the initial component of the movement trajectory. Because the effect of visual feedback condition was the same between the groups, the means on these graphs were averaged across this condition. The top graphs of Fig. 2 suggest that, in comparison to the control group, the LHD group showed less of an increase in velocity as movement amplitude increased during both the acceleration and deceleration phases. This observation was supported by the interaction of group $\times$ amplitude for deceleration velocity ($F(3,35) = 4.62$, $p < .01$) and a trend for this effect for acceleration velocity ($F(3,35) = 2.62$, $p < .07$). The bottom graphs of Fig. 2 show that spatial accuracy was similar between the two groups for both phases of the movement. This observation was generally confirmed by the statistical analysis except for an isolated finding in the deceleration phase where there was a group $\times$ amplitude interaction ($F(3,35) = 3.36, p < .05$) showing that spatial error was greater in the LHD group, but only for 200mm movements. Thus, it appears that the group differences in velocity for larger amplitude movements are not correlated with spatial accuracy. Most importantly, these interactions were independent of visual feedback so that they cannot be attributed to group differences in open- or closed-loop processing during the initial phases of the movement. Rather, it appears that LHD disrupts the ability to select optimal movement constraints when the movement context changes, perhaps to adjust for the greater force requirements of longer amplitude movements.

Finally, although there were no group differences in the velocity profiles of the secondary movement component, there was a significant group × visual feedback condition × amplitude interaction for spatial error ($F(9,29) = 2.55$, $p < .05$). Figure 3 shows that there was no group difference in spatial error when

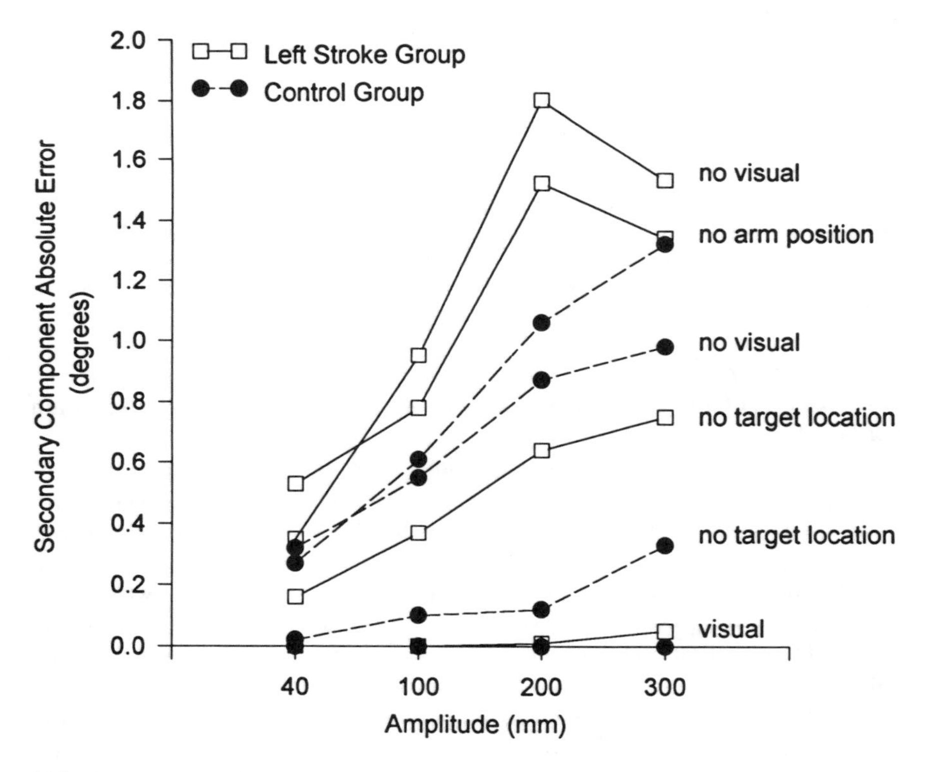

FIG. 3. Absolute error for the secondary movement component in the left stroke and control groups. The graph shows the absolute error as a function of movement amplitude for each visual feedback condition. In the "visual" condition, target location and arm position feedback were available throughout the duration of the movement. In the "no arm position" condition, arm position feedback was eliminated at movement onset. In the "no target location" condition, the location of the target circle was removed at movement onset. In the "no visual" condition, both kinds of feedback were eliminated at movement onset.

all sources of visual information were displayed for the duration of the trial. This finding indicated that LHD does not disrupt the ability to utilise visual information to update the central motor program during movement. No reliable group differences were found in spatial error when arm position information was removed. By contrast, the LHD group showed more spatial error when target location information was eliminated ($F(1,37) = 8.10$, $p < .01$), regardless of movement amplitude. Interestingly, spatial error also increased as movement amplitude increased more in the LHD group than in controls, but only when both sources of information were eliminated ($F(2,35) = 3.58$, $p < .05$).

Our preliminary findings point to two mechanisms that may be disrupted with LHD and possibly also in apraxia. First, there appears to be a disruption in lower levels of the motor representation that are responsible for selecting and implementing movement constraints with changes in movement amplitude. Why does this appear to affect primarily larger amplitude movements? One hypothesis is that more constraints must be selected and implemented as movement amplitude increases because greater forces are required and, at the same time, spatial accuracy must be maintained. This hypothesis is consistent with an early theory (Fitts, 1954) which predicts that the time to execute a movement will depend on the amount of emphasis placed on spatial and temporal precision. Using reciprocal aiming movements and manipulating target size and movement amplitude, Fitts (1954) discovered that movement time could be estimated by a logarithmic trade-off between the distance (D) and target width (W) of rapid aiming movements, that is, MT = A + Blog2(2D/W), where A and B are constants. The log2(2D/W) is the index of difficulty (ID). In our experiment, higher ID levels are associated with longer amplitude movements and, hence, are more complex to program in terms of their temporal and spatial properties. LHD appears to disrupt this process which involves selecting the optimal movement constraints, perhaps the parameterisation of force (Roy & Elliott, 1989), for a particular context. This less than optimal motor program then may lead to a disproportionate dependence on both target and arm position information to execute secondary movements to reach a target, which was our second main finding. Finally, individuals with LHD clearly were more dependent upon target location information, but only to execute lower velocity subcomponents of the movement to hit a target. This may suggest that the ability to maintain a body-centered representation of spatial location is diminished with LHD. It was surprising to us that the removal of arm position information alone did not have a more detrimental effect on the secondary movement because it is also body-centered information. One possibility is that all subjects may utilise proprioceptive cues about arm position, which should not be impaired in the limb ipsilateral to LHD, resulting in less dependence on visual feedback of arm position. However, with LHD the representation of spatial location information may decay more rapidly, although the reason for this is not clear, so that individuals with LHD cannot compensate as effectively as neurologically intact individuals for the removal of target information during the secondary movement component.

In what way might the findings from these studies in individuals with LHD contribute to our understanding of cognitive–motor disturbances in apraxia? Unfortunately, there are no studies to date that have directly compared movement kinematics of apraxic and nonapraxic individuals with LHD. A recent study in our laboratory provides a partial response to this question (Haaland & Harrington, 1994). In a reciprocal aiming task, participants repeatedly made alternating movements to hit targets that varied in size. For all participants, tapping rate increased as the target size increased. However, both the apraxic and nonapraxic individuals showed less of an increase in tapping rate as targets became larger. This study did not have the benefit of a kinematic analysis of movement trajectories, but the findings are consistent with our preliminary results from the simple aiming study that showed diminished velocity for longer amplitude movements with LHD. Specifically, tapping rate deficits with LHD to large but not small targets could be due to problems in adjusting the motor program for changes in the movement context. Larger targets require more rapid movements for optimal performance and, thus, the force of the movement must be adapted accordingly. Interestingly, an impairment unique to the apraxic group was also uncovered in this task. Only the apraxic group showed impaired accuracy for movements to small but not large targets, which may point to degraded program for movements that require a high degree of spatial accuracy. These results suggest that aiming tasks are sensitive to programming deficits manifested in LHD. Although some of these deficits may not be unique to apraxia, this needs to be tested directly by analysing the movement trajectories of apraxic and nonapraxic individuals. It appears, however, that aiming tasks are also useful for revealing deficits that are unique to apraxia, such as impairments in processes that may mediate the control of movement subcomponents requiring spatial precision (Haaland & Harrington, 1994).

These studies portray how disruptions in specific aspects of the motor representation can be investigated in apraxia by decomposing a movement trajectory using kinematic analyses and manipulating variables that ostensibly engage different cognitive operations. An advantage of the simple aiming paradigm is that the degrees of freedom for a movement are constrained so that one can more clearly isolate the cognitive operations involved in planning and implementing actions and, hence, assess the functional status of the motor representation. Nonetheless, it is important to recognise that the study of single joint movements, which are executed in a two-dimensional plane, is also limited in that one cannot entirely capture the intricacies of the action system. It is also important to study the kinematics of multiple joint movements as many have done in neurologically intact individuals. For example, studies of grasping movement kinematics have been enormously informative in terms of identifying subcomponents of multiple joint movements and their organisation (Marteniuk et al., 1987). These types of studies can cast light on the question of whether the cognitive operations supporting representations of arm and hand position are impaired in limb apraxia, particularly when performing multiple joint movements in a three-dimensional space. In addition, kinematic analyses of multiple joint movements will be indispensable in understanding how perceptions of objects are integrated into our pragmatic knowledge of actions, which is another key issue in apraxia research.

MOTOR SEQUENCING IN IDEOMOTOR LIMB APRAXIA

The Organisation and Control of Serial Events

We will now turn our attention to the problem of understanding mechanisms of sequencing. Serial event behaviours introduce more degrees of freedom into the action system and, hence, levels of the motor representation emerge that we have not yet contemplated. But first, let us briefly comment on the relevance of this topic to apraxia. Clinical descriptions of ideomotor limb apraxia do not specifically implicate serial ordering problems, so one might argue that theories of how serial events are centrally represented are not particularly relevant to understanding the disorder. However, many "single" gestures tested in limb apraxia assessments require the sequencing of several component movements. Furthermore, patients with LHD are known to be particularly impaired on sequencing tasks (Jason, 1986; Kimura, 1982; Kimura & Archibald, 1974). Both of these observations provide a strong justification for investigating sequencing in apraxia.

Some of the earliest investigations into sequencing were interested in how serial events comprising complex skills, such as language, writing or playing the piano, are put in the proper order. Lashley (1951) maintained that sequences could be controlled by motor programs rather than by sensory associations between successive events where feedback from one event simply activates the next event. Although many experiments support Lashley's view (see Keele, Cohen, & Ivry, 1990), one phenomenon that substantiates the motor program concept is the effect that anticipating a series of events has on the time to plan and execute a sequence. Specifically, recall that preparatory RT increases with the number of serial events (e.g. number of key press responses, hand postures, or syllables in a word) which shows that the motor program contains higher level information about all responses in the sequence (Harrington & Haaland, 1987; Sternberg et al., 1978). Similarly, the time to execute a movement is influenced by upcoming movements in the sequence (Povel & Collard, 1982; Restle & Brown, 1970). This suggests that movement production is dependent upon the way in which serial events are organised and/or involves the online referencing of a motor program of the sequence.

We mentioned earlier that the concept of a central motor plan assumes that the motor representation has a hierarchical structure. There can be many levels in a hierarchy with higher order or superordinate elements controlling lower level elements. Serial ordering is thought to be accomplished by superordinate elements which activate the temporal unfolding of subordinate elements. In addition to ordering serial events, a hierarchy also controls many other aspects of skills such as organising events into subgroups, which minimises the necessity of relearning new groupings of lower level elements (Gordon & Meyer, 1987). How is this accomplished? A hierarchy is constructed by parsing serial events into subgroups of responses using rules or schemata that organise movements on the basis of their natural rhythmic, spatial or motoric properties (Essens, 1986; Martin, 1972; Povel & Collard, 1982; Restle & Brown, 1970; Summers, Sargent, & Hawkins, 1984). For example, Povel and Collard (1982) studied finger sequences using a task in which keys were labelled 1 through 3 and participants repeatedly executed a key press pattern. The intertap latencies of the sequence 123321 suggested that it was organised into two subordinate patterns, 123 and 321, because latencies were longer between and shorter within these groupings. Thus, hierarchical representations are built by imposing organising principles that optimise movement fluency by controlling the temporal, spatial or motoric patterning of serial events. Organising principles then can be applied to different responses within a similar domain without having to relearn a new schema (Gordon & Meyer, 1987). This reduces the degrees of freedom in the motor system and facilitates the acquisition of new skills.

While other characteristics of the action system emerge when performing serial events, an overview of this literature is beyond the scope of this chapter. However, one topic that merits brief attention is timing. A basic feature of all skilled serial behaviours is their fluency, which is achieved in part by timing serial events to minimise undue delays between responses. Many ideas have been put forth to explain timing mechanisms, but we will limit our discussion to theories about the functional independence of timing mechanisms. Some contend that the same mechanism that controls serial ordering also controls timing (Rumelhart & Norman, 1982), whereas others propose that timing mechanisms control serial ordering (Rosenbaum, 1985). Neither view, however, adequately explains many observations, such as spoonerisms in speech, transposition errors in typewriting, and the fact that the serial order of events such as typewriting remains constant despite changes in the timing of keystrokes. These examples illustrate that serial order errors can occur without a disruption in timing and vice versa (see Summers & Burns, 1990). Thus, it appears that independent mechanisms support timing and serial ordering, although timing clearly becomes an integral part of skilled sequences when rhythmic properties of movements are used to construct hierarchical organisations of serial events (MacKay, 1985; Summers & Burns, 1990).

Mechanisms Underlying Sequencing Impairments in Ideomotor Limb Apraxia

We have seen that the explanation of how serial ordering is accomplished is just one problem that arises when trying to understand the organisation of serial events. Investigations into breakdowns in sequencing, therefore, must contemplate more than a single mechanism. Although sequencing deficits in individuals with LHD could suggest serial ordering problems, they might also reflect a disruption in other layers of the motor representation that are more apt to be revealed when performing multiple movements. For example, sequencing impairments could reflect problems in using hierarchical structures to organise movements (i.e. parsing), dysfunctional timing mechanisms, or problems in transitioning between lower level elements or responses.

Abnormalities in sequencing with LHD have been investigated most thoroughly in studies of nonsymbolic, unfamiliar movements. These studies have generated a number of hypotheses about how actions are represented and controlled by the left hemisphere. There are several reports of greater sequencing deficits with LHD than RHD (Kimura, 1982; Kimura & Archibald, 1974), but LHD also produces deficits in the production of isolated movements (De Renzi, Faglioni, Lodesani, & Vecchi, 1983; Kimura, 1982; Kolb & Milner, 1981). Although this evidence might suggest that sequencing deficits are due to the cumulative effects of problems in planning single movements, even when this factor was controlled impaired sequencing was still evident with LHD (Harrington & Haaland, 1991a). This finding suggests that the representation of movement in the left hemisphere is comprised of more than just hand position or spatial information.

What other levels of the central representation might be disrupted? Jason (1986) contended that a serial ordering deficit hypothesis could not account for motor sequencing impairments in individuals with left or right frontal damage when they were not required to perform movements from memory. By contrast, Roy (1981) has found evidence for serial ordering deficits with LHD. Other investigations into the serial ordering of non-motor tasks have reported deficits in patients with right or left frontal cortical damage (Petrides & Milner, 1982), even when the memory demands of the task were minimised (Shimamura, Janowsky, & Squire, 1990). In the latter studies participants ordered a large number of items so that the serial ordering demands were far greater than in the Jason study, which may partially explain the discrepant findings. This hypothesis clearly needs to be tested more definitively in the motor modality.

Although other hypotheses have been put forth to account for motor sequencing impairments with LHD, interpretations are controversial as to the levels of the motor representation that are disrupted (for a review see Haaland & Harrington, 1990). In addition, few studies have directly examined whether sequencing deficits are specific to the disorder of apraxia or more general to the

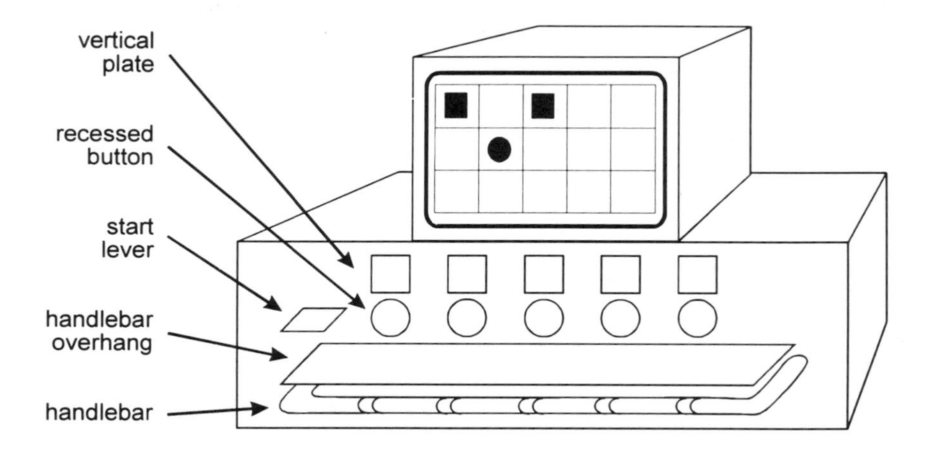

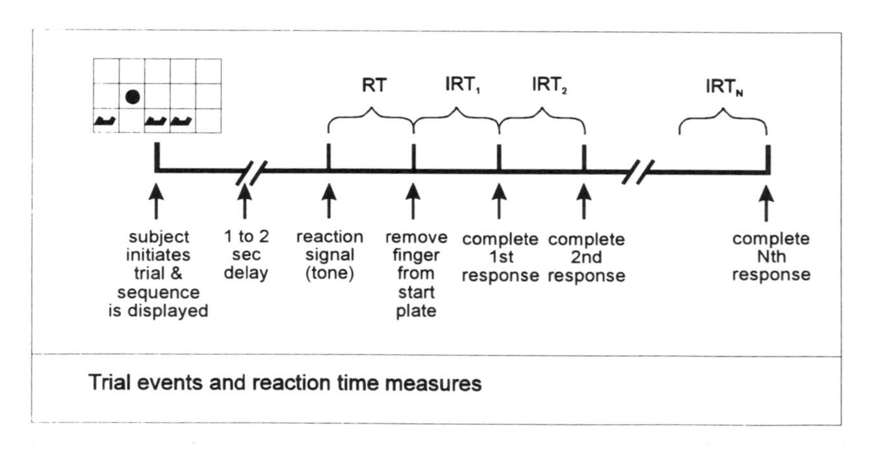

FIG. 4. Diagram of the hand posture sequencing apparatus, the trial events and the reaction time measures. The apparatus contains 15 manipulanda. At the top is a row of five plates, in the middle a row of five recessed buttons, and at the bottom a row of five handlebars. To the left of the manipulanda is a start plate. A monitor displays the stimulus sequence which depicts the type of manipulanda associated with a response and its spatial location on the apparatus. The reaction time measures include: (1) preparatory reaction time (RT), (2) interresponse time (IRT) for each hand posture, and total movement time (MT) which is the sum of all IRTs.

role of the left hemisphere in controlling a wide variety of movements. We conducted a study in our laboratory to address some of these issues (Harrington & Haaland, 1992), using a task in which participants performed sequences of hand postures on the apparatus depicted in Fig. 4. The apparatus was interfaced with a computer and contained 15 manipulanda (i.e. five plates, five recessed buttons, five handlebars). Hand postures consisted of plate (P) responses which required contact with the side of the hand, button (B) responses which required contact with the index finger, and handlebar (H) responses which required a grasp around the bar from underneath. Participants wore gloves equipped with metal contacts to computerise the recording of responses. Participants always moved from the left to the right across the face of the apparatus, and to change hand postures they moved to the right diagonally (up or down) to the next manipulandum. Seventeen normal controls and 16 LHD stroke patients performed the task with their left hand. Using a 15-item limb apraxia assessment (Haaland & Flaherty, 1984), seven of the stroke patients were diagnosed as limb apraxic and nine were nonapraxic.

Figure 4 shows that a hand posture sequence was displayed on a 3 × 5 grid using a pictorial representation of the manipulanda. The bottom of this figure illustrates the trial events which began with the participant resting his or her index finger on the start plate whereupon the sequence immediately appeared on the monitor. After a random delay a tone signaled the participant to begin responding, and upon completion of the last response the visual display terminated. Figure 4 also shows the reaction time measures obtained on each trial which included preparatory RT, interresponse time (IRT) for each posture in the sequence, and movement time (MT) which was the total time to execute the entire sequence. In addition, we measured errors due to a wrong or a prolonged response (i.e. longer than 2000ms). Two types of sequences were presented that varied in length from one to five hand postures. Repetitive sequences contained repetitions of the same posture (e.g. P, PP, PPP, PPPP, PPPPP). Heterogeneous sequences were constructed so that the first two responses required a change in hand posture and for longer sequences, the remaining postures consisted of a repetition of the first posture (e.g. PB, PBP, PBPP, PBPPP).

One important objective of the study was to test the hypothesis that ideomotor limb apraxia produces a deficit in organising sequences of movement. Recall that the duration of preparatory RT generally increases with the number of movements in a sequence. There are some exceptions to this finding, however, that reveal how neurologically intact individuals organise serial events into higher order groupings of responses. For repetitive sequences, variations in the number of hand postures has little or no effect on preparatory RT because participants appear to organise these types of sequences into a single grouping whereby adjacent repetitions of a response add little or no time to planning (Harrington & Haaland, 1987; 1991a). By contrast, for heterogeneous sequences the duration of preparatory RT increases up to four hand postures, after which an asymptote is reached. This suggests that individuals parse serial events into subgroups to organise the sequence. For example, each response may be treated as a single unit for sequences of two and three responses (e.g. HB, HBH) whereas longer sequences of four and five postures are parsed (e.g. HBHH, HBHHH) so that the first three responses form one subgrouping (e.g. HBH) and

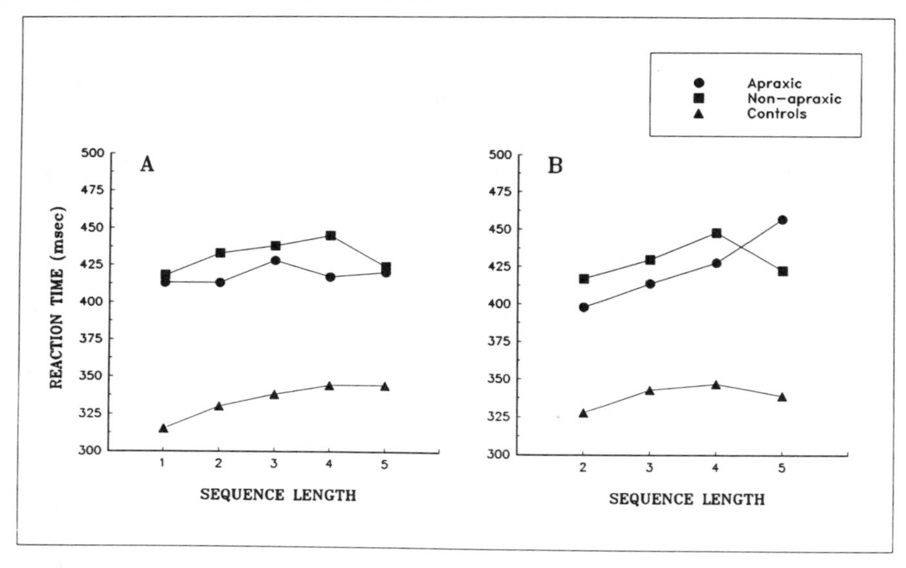

FIG. 5. Mean preparatory reaction times for repetitive (A) and heterogeneous (B) hand posture sequences in apraxic, nonapraxic, and control groups.

the last response(s) form another subgrouping (e.g. H or HH). As for repetitive sequences, a repetition of an adjacent response within a subgrouping does not increase the duration of preparatory RT and hence, an asymptote is reached after four responses. Although other organisations are possible, the point is that parsing is particularly beneficial for longer serial events because a single superordinate unit can control multiple lower order units and thereby reduce processing demands.

If individuals with apraxia are impaired in using higher level organisations to preprogram serial events, variations in sequence length should not have the same impact on preparatory RT as for nonapraxic individuals with LHD and normal controls. The results were consistent with this hypothesis for sequences that were more complex. Figure 5A shows that for repetitive sequences, sequence length had a similar effect in all three groups such that preparatory RT increased in duration between one and two postures, after which an asymptote was reached. This suggested that LHD did not disrupt the ability to form higher level organisations of repetitive serial events. Figure 5B shows that for heterogeneous sequences the preparatory RT continued to increase linearly with sequence length in the apraxic group, whereas an asymptote was reached after four posture sequences in the nonapraxic and the control group. These findings contrast with those for repetitive sequences and suggest that when serial events consist of different movements, limb apraxic patients plan each posture in the sequence as a single element, neglecting to parse longer sequences of responses into subgroupings.

This was the most parsimonious interpretation of the preparatory RT results when we considered the pattern of sequence length effects on MT and errors, as displayed in Figs. 6 and 7. Specifically, impairments in preprogramming should also be apparent during movement when the motor program is referenced. Figures 6 and 7 show that the percentage of change in MT (i.e. relative to the time to execute a single posture) and the percentage of errors was greater in the apraxic group than the other two groups, but only for heterogeneous sequences containing four or five postures. There were no differences between nonapraxic and control groups for either sequence type. This pattern of MT and errors on the longer sequences cannot be due to a greater number of hand posture transitions (e.g. HBHH, HBHHH) because these sequences were structured the same as those with three postures (e.g. HBH), except for an addition of a repeated posture(s) at the end of the sequence. It is unlikely that serial ordering deficits could explain these findings since the visual display of events and their order was available throughout the trial in order to minimise serial ordering problems and thus, better assess the organisation of the motor representation.

Several other hypotheses were tested in this study concerning the functional status of lower levels of the motor representation in limb apraxia, but no other dissociations were found between apraxic individuals and the other two groups

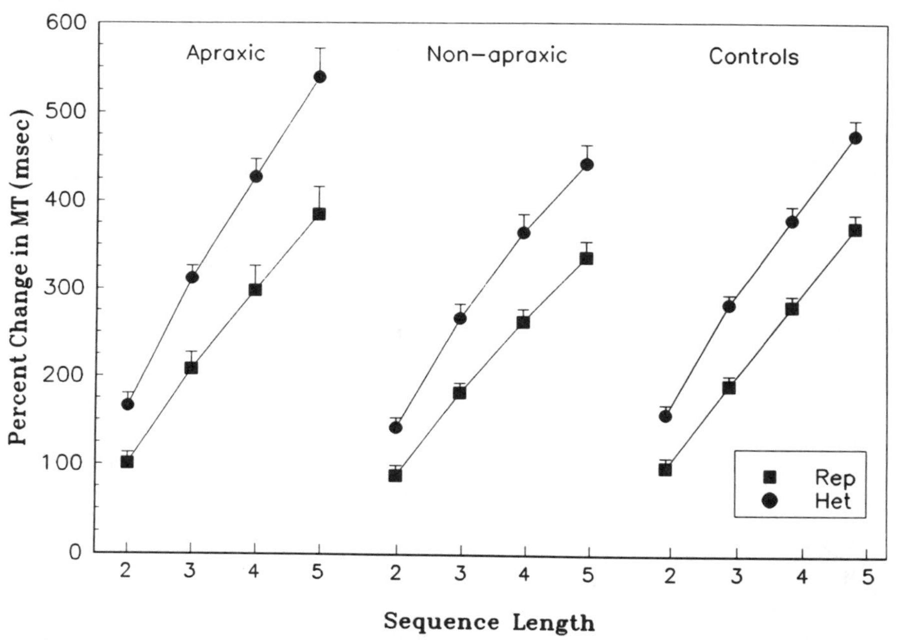

FIG. 6. The mean percent change in movement time (MT) for repetitive (rep) and heterogeneous (het) sequences. Percent change was calculated using the following formula: (MTn–MT1)/MT1 where n represents a particular sequence length and MT1 is the time to executive a single hand posture.

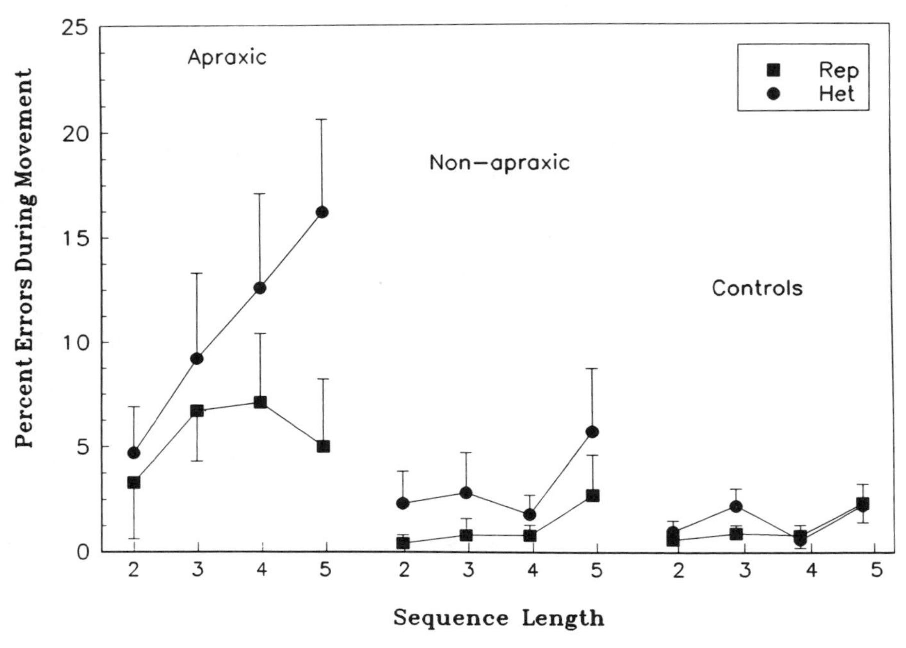

FIG. 7. The mean percentage of errors during movement for repetitive (rep) and heterogeneous (het) sequences.

(for details see Harrington & Haaland, 1992). In particular, the hypotheses that apraxia would be associated with deficits in changing between different postures (Kimura & Archibald, 1974) and in regulating the timing of individual responses (Harrington & Haaland, 1991a) were not supported. LHD apraxic and nonapraxic patients showed similar impairments (i.e. slower IRTs but not greater error rates) in transitioning from one posture to the next regardless of whether the response was a repetition of a previous hand posture or a change was made to a different hand posture. A deficient timing mechanism may be one potential explanation for this finding. This possibility was suggested by another finding involving the effect of the sequence length on the duration of IRTs. Specifically, in the control group the time to execute a single posture (i.e. IRT) was not affected by the number of other postures contained within the sequence. By contrast, in the LHD apraxic and nonapraxic groups the time to execute a single posture was faster when it was contained within a longer than a shorter sequence, but only for repetitive sequences and adjacent repetitive postures in heterogeneous sequences (e.g. PBPP, PBPPP). This negative relationship between IRTs and sequence length may suggest that, during movement, reference is made to a level of the motor representation that is concerned with the scheduling or timing of when to execute a response. For instance, each posture in a sequence may be associated with an absolute time at which to execute the movement. Successive postures that are different would be expected to have different timing requirements and hence, would be response specific. When successive events are repetitions of the same posture, the same timing rule may be applied to all postures so that they are timed simultaneously and the length of a sequence has no effect on IRTs. When timing is dysfunctional, it proceeds more serially and, thus, movement onset is postponed to ensure that the computation for the initiation of a response coincides with the activation of other response parameters. With LHD, the initiation of a posture contained in a longer sequence is faster because there is more time to complete the scheduling of the remaining postures. Response execution is delayed for a posture that is contained within shorter sequences because cognitive operations occur closer in time and must be at least partially completed before a particular hand posture can be activated. The proposed timing deficit explanation is admittedly ad hoc, but consistent with mounting evidence from studies of brain-damaged adults, language-impaired children, and neurologically intact individuals that shows that the left hemisphere is superior to the right hemisphere in efficiently processing rapid, temporal information (Hammond, 1982; Tzeng & Wang, 1984).

In summary, this study furnishes some preliminary clues about several basic and seemingly separate functional levels of the motor representation. Some of these levels appear to be disrupted exclusively in ideomotor limb apraxia and others are impaired with LHD regardless of whether patients are apraxic. Importantly, the dissociation between the apraxic and nonapraxic groups was found only for organising sequential hand postures. This finding suggests that ideomotor apraxia may indeed be a disorder of sequencing, although it is not necessarily due to a primary disruption in serial ordering mechanisms. In our study we could not identify a specific cortical area that was associated with limb apraxia, possibly due to limitations in our lesion quantification methods. However, there is preliminary evidence that some of the levels of the motor representation disclosed in this study may be supported by different neural subsystems. For instance, right-hemisphere damage was not associated with deficits in any aspect of sequencing hand postures (Harrington & Haaland, 1991a) which is consistent with a large body of research showing that the left hemisphere is specialised for representing and controlling movement. In addition, the organisation of hand posture sequences is not disrupted in Parkinson's disease (PD) as it appears to be in ideomotor limb apraxia. Rather, PD patients have a fundamental difficulty in generating a plan for serial events, as evidenced by the finding that sequence length has less of an effect on preparatory RT in PD patients relative to controls (Harrington & Haaland, 1991b). By contrast, individuals with PD also show problems in transitioning between different but not repeated hand postures (i.e. longer IRTs) whereas individuals with LHD are equally slow regardless of whether a hand posture is repeated or changed. This pattern of results suggests that the basal ganglia and its thalamocortical connections may regulate mechanisms that are responsible for switching between different motor programs. However, similar to patients with LHD, PD appears to disrupt the scheduling or timing of movements. These findings may be due to the fact that many stroke patients also incur subcortical damage or a disconnection of cortical and subcortical systems. The cerebellum, however, is also involved in explicit timing operations, particularly those that concern the reproduction and perception of time intervals (Ivry, 1993; Keele & Ivry, 1990). Further research into the temporal aspects of movement is needed in order to determine whether the disruptions with LHD that were attributed to the scheduling or timing of serial events engage levels of the motor representation that are distinct from those that sustain the cognitive operations that are involved in the reproduction and perception of time intervals. Nonetheless, these patterns of findings across different neurological patient groups suggest that various levels of the motor representation are distributed in different areas of the brain.

LIEPMANN'S THEORY OF APRAXIA

Although experimental approaches derived from cognitive psychology are promising for identifying the mechanisms of apraxia, a model that specifically addresses this disorder has yet to emerge from this framework.[2] This is because research in this tradition with apraxic individuals is relatively recent and limited. By contrast, several models of apraxia have been put forth by scientists whose conceptual framework is grounded in neurology and neuropsychology (e.g. Heilman, 1979; Rothi, Ochipa, & Heilman, 1991; Roy & Square, 1985). These models are discussed in other chapters of this book. However, because Liepmann's (1913) conceptualisation of apraxia continues to influence current thinking, it is valuable to consider the utility of his theory in the context of the findings that are beginning to emerge from cognitive psychology.

First, let us consider the components of Liepmann's (1913) motor representation. Liepmann contended that voluntary movements were the result of a mental idea of an action and the motor consequences that were elicited by the idea. The ideational sketch of movement or movement formulae (i.e. space–time plan) was thought to consist of information about the purpose of the movement and the ways of performing an action (e.g. participating body parts, speed and timing, space directives, temporal ordering), and these ideas were represented by visuo-kinaesthetic engrams that were stored largely in left occipito-parietal cortex.

---

[2]Several models of motor control and motor learning in neurologically intact individuals have been generated from motor control theory and cognitive psychology. One model that deserves mention here is MacKay's (1985) because it encompasses a realm of cognitive operations that contemporary theorists consider key for supporting motor control and learning. In addition, MacKay has contemplated some of the potential implications his model may have for explaining the mechanisms of apraxia.

Liepmann also contended that in order to carry out skilled movements it was necessary to have established kinaesthetic memories which consist of traces of innervatory patterns that control the kinaesthetic schemes of limb movements. Innervatory patterns were thought to be physiological, not available to consciousness, and stored in the left precentral and postcentral cortex. Liepmann believed that to perform a highly skilled movement, a space–time plan must be retrieved and associated, via cortical connections traversing the left infero-parietal cortex, with the innervatory pattern. This model was used to account for three distinct types of apraxia. Liepmann viewed ideational apraxia as a disruption in the space-time plan that made the movement impossible because the patient could not construct an idea of the movement. In the case of ideomotor apraxia, the space–time plan was thought to be preserved, but movements were disrupted due to a disconnection between the space–time memories and the innervatory patterns for implementing the movement. On the other hand, a disruption in the innervatory patterns produced limb kinetic apraxia which interfered with the selection of the required muscle synergies to carry out a movement.

Liepmann's ideas about the psychological mechanisms that support different forms of apraxia were largely born out of a functional analysis of behaviour, which continues to dominate the study of apraxia. Clearly Liepmann had a keen appreciation for the different layers of the motor representation and, in particular, the important distinction between knowledge about the purpose of a movement and knowledge about how to implement a movement. Although this continues to be a key distinction in the diagnosis of ideomotor and ideational apraxia, other developments in neurology and physiology (e.g. see Luria, 1980 for a review) suggest that Liepmann's theory does not provide a satisfactory account of the neural substrates of apraxia. From a psychological perspective, the theory also does not furnish a compelling account of the processing mechanisms of apraxia, especially ideomotor apraxia which was defined mostly in terms of its neural basis (i.e. a disconnection syndrome) and symptoms (i.e. a disturbance of purposeful movements). Due to this lack of specificity, the theory does not convincingly disambiguate the cognitive mechanisms of the different types of apraxia.

For example, Liepmann did not consider the specific processes (e.g. preprogramming, temporal ordering) or memory stores (e.g. semantic, motoric) that today are thought to be crucial in specifying the "idea of movement" and also appear to be supported partially by frontal areas of the brain (Luria, 1980). Similarly, work in our laboratory introduces the possibility that ideomotor apraxia may be associated with a deficit in parsing serial motor events to construct higher order motor programs (Harrington & Haaland, 1992). While this parsing deficit appears to be independent of problems in formulating an idea of the action, it points to the involvement of a high level of cognitive processing in ideomotor apraxia that was not explicit in Liepmann's theory. This finding does not easily fit into Liepmann's innervatory pattern component which views such memories as physiological processes that do not require cognitive processing[3]. If higher level cognitive processing was not disrupted in ideomotor apraxia, there would be no explanation for why the structure of a movement sequence alters the duration of the preparatory RT period. As another example, our work suggests that the ability to select optimal response parameters (i.e. velocity, force) for simple aiming movements may be impaired with LHD and perhaps apraxia, but only for certain movement goals (i.e. long amplitude movements). Liepmann's theory does not explicitly address this issue, which has been fundamental in motor control theory for explaining how the motor system reduces the many potential options available for achieving the same movement goal. Moreover, even if the ability to select optimal movement constraints was supported by Liepmann's notion of kinetic memories, it is not clear how the theory would explain why a disruption was not found for all movement goals (i.e. short and long amplitude movements). A more parsimonious explanation of the findings is provided by appealing to the cognitive processing requirements of the movement context.

Finally, recent work indicates that some of the processes that Liepmann captured in the space–time representation (e.g. timing and rhythm, temporal ordering, space directives) have been dissociated functionally and appear to be supported by different neural subsystems, some of which have not been commonly associated with ideational or ideomotor apraxia (e.g. cerebellum, basal ganglia, prefrontal cortex). As we alluded to earlier, the cerebellum appears to be responsible for regulating internal timing mechanisms in motor, perceptual, and auditory modalities (Ivry, 1993). The basal ganglia, supplementary motor area (SMA), and prefrontal areas of the cortex play key roles in the construction of programs for sequential motor events (Harrington & Haaland, 1991; Roland, Larsen, Lassen, & Skinhoj, 1980; Seitz, Roland, Bohm, Greitz, & Stone-Elander, 1990). These findings reveal cognitive and neuroanatomical distinctions that could not have been predicted by Liepmann's model, which lumps these types of operations under the umbrella of movement formulae and confines their representation to more posterior areas of the cortex.

In summary, the space–time plan is a broad concept that does not satisfactorily distinguish between functionally and neuroanatomically distinct cognitive operations. The cognitive mechanisms of ideomotor and ideational apraxia were defined by Liepmann more in terms of symptoms (e.g. disruption in

[3]The work of Kelso and Tuller (1984) seems to reflect some of the general ideas of Liepmann's kinetic memories and innervatory patterns. They invoke the concept of coordinative structures which consist of muscle synergies that when activated behave as single functional units. Like kinetic engrams, coordinative structures do not require preprogramming of various aspects of the movement. Coordinative structures are spatially represented, temporarily constrained by the movement goal, and are not limited to the control of "simple" movements.

the idea of a movement or in knowledge of how to carry out a movement) than processes (e.g. organisation of semantic and motoric memory, parsing processes, selection of movement constraints) so that it is difficult to test the theory in this regard. This also has contributed to confusion over the definitions of ideational and ideomotor apraxia, especially in terms of the role that temporal ordering plays in these forms of apraxia (e.g. see De Renzi, 1985) and whether they represent distinct forms of apraxia or simply a continuum of motor impairment (Kleist, 1934; Sittig, 1931). Although Liepmann's distinction between ideatory and movement production disorders will continue to be useful, delineating the cognitive mechanisms that support ideas of the purpose of movement and knowledge of how movements are carried out will be important for future advances in theories of the apraxias. Motor control theory and cognitive psychology offer potentially promising avenues for building upon and extending Liepmann's initial ideas.

SUMMARY AND CONCLUSION

As we have discussed throughout this chapter, the conception of the motor representation as involving multiple layers leaves open the possibility that there may be many reasons for cognitive-motor disturbances in ideomotor apraxia. Although studies of apraxia that have been guided by this theoretical framework are in their infancy, several findings have emerged that suggest some intriguing directions for future research into the disorder. First, kinematic analyses of movement suggested that individuals with LHD may have difficulties applying optimal movement constraints to accommodate changes in the movement context, especially for longer amplitude movements that require greater forces to be generated while still maintaining an acceptable degree of spatial accuracy. This hypothesis needs to be directly examined in a limb apraxic group, which we are in the process of doing in our laboratory. Moreover, it is important to examine how increasing the degrees of freedom of a task (e.g. varying target size or movement direction) affects the ability of apraxic and nonapraxic individuals to impose optimal constraints for programming movements. Second, the ability to maintain a representation of the spatial location of the target appeared diminished with LHD in later phases of the movement where spatial precision was especially important. This finding needs to be examined directly in limb apraxia especially since other results suggest that the accuracy of movements that require greater precision is compromised in the disorder (Haaland & Harrington, 1994). These findings may also be relevant to the observation that imitation of transitive gestures is particularly impaired in limb apraxia. Specifically, if apraxic individuals have difficulty in generating or sustaining a motor program that engineers the control of the precise spatial characteristics of movements, then this may be especially detrimental for imitating transitive movements in which the hand or fingers must be configured expressly in relationship to an object. Kinematic analyses of grasping movements in apraxia could be very illuminating in this regard. In addition, explorations into the kinematics of grasping in limb apraxia may be key in understanding how functional knowledge of actions (e.g. grasping, pinching, manipulating) is integrated with spatial representations of objects and their location (Mishkin, Ungerleider, & Macko, 1983), the latter two of which have been portrayed as the "what" and "where" visuomotor systems (Goodale & Milner, 1992).

Our sequencing experiment suggested that limb apraxia may produce a deficit in parsing serial events into hierarchical structures to organise movements. This higher level disruption in the motor representation could underlie observations of greater deficits with LHD in sequencing gestures than when performing them in isolation (Kimura, 1982). It may also explain reports of diminished learning of hand sequences in patients with ideomotor apraxia (Motomura, Seo, Asaba, & Sakai, 1989). Future research is undoubtedly needed that directly probes the ability of individuals with limb apraxia to benefit from natural spatial, temporal, or motoric properties of movements in building higher level organisations of the motor representation.

The main theme of this chapter has been that a decomposition of movement into subcomponents is helpful in disclosing the cognitive mechanisms that support the motor representation. We believe this approach will contribute to our understanding of ideomotor apraxia because there may be many reasons for the disorder. This may explain why it has been difficult to identify highly specific neuroanatomical correlates of limb apraxia (Alexander, Baker, Naeser, Kaplan, & Palumbo, 1992) and why limb apraxia is sometimes found with RHD in right-handed individuals (Haaland & Flaherty, 1984). If this conjecture is supported by future research, decomposing movements into basic components will likely lead to the identification of the specific neural mechanisms of the disorder. The decomposition approach is also a promising method for disclosing whether the cognitive operations that support movements draw upon mechanisms that are specific or more general to a modality or skill domain. For instance, uncovering specific mechanisms of limb, buccofacial, and verbal apraxia will allow us to directly assess whether there are common mechanisms for the expression of all apraxias, as some contend (Roy & Square-Storer, 1990). Similarly, decomposing movements into their subcomponents should help ascertain whether the representation of some cognitive operations, such as serial ordering, is specific to a modality or more general to all skill domains. This knowledge would contribute to theoretical perspectives on the organisation of memory and, perhaps, also refine clinical descriptions of apraxia in terms of its specificity to a disruption in the motor representation. Disclosing the specific mechanisms of the apraxias may also help disentangle the distinctions and similarities between ideomotor and ideational apraxia. Although clinical descriptions indicate that the reasons for the two types of apraxia differ, theoretical predictions and empirical findings are sometimes contradictory, especially with regard to the differential role of serial ordering mechanisms (De Renzi &

Lucchelli, 1988; Lehmkuhl & Poeck, 1981; Liepmann, 1908) and the mechanisms that support the performance of pantomime and actual object use gestures (De Renzi & Lucchelli, 1988; Haaland & Flaherty, 1984; Roy *et al.*, 1991).

In closing, the study of apraxia must embrace all levels of inquiry, including the approach we have advocated in this chapter. The complexity of the disorder is a compelling reason to draw upon a multidisciplinary approach that encompasses theoretical perspectives from cognitive psychology, neuropsychology, and neurology. In addition, the cornerstone for empirical investigations into apraxia should include case study and group methods as well as human and animal research. Extending the traditional theoretical and empirical views of apraxia to include these domains will likely advance our knowledge of the disorder and yield insight into the neural subsystems that give rise to specific levels of the motor representation.

ACKNOWLEDGEMENTS

This chapter was supported by grants to both authors from the Department of Veterans Affairs. We gratefully acknowledge Rex M. Swanda for his comments on a draft of this chapter.

REFERENCES

Alexander, M.P., Baker, E., Naeser, M.A., Kaplan, E., & Palumbo, C. (1992). Neuropsychological and neuroanatomical dimensions of ideomotor apraxia. *Brain, 115*, 87–107.

Bernstein, N. (1967). *The coordination and regulation of movements*. London: Pergamon.

Cohen, A., Ivry, R.I., & Keele, S.W. (1990). Attention and structure in sequence learning. *Journal of Experimental Psychology: Learning, Memory, and Cognition, 16*, 17–30.

Dee, H.L., Benton, A.L., & Van Allen, M.W. (1970). Apraxia in relation to hemispheric locus of lesion and aphasia. *Transactions of the American Academy of Neurology, 95*, 147–150.

De Renzi, E. (1985). Methods of limb apraxia examination and their bearing on the interpretation of the disorder. In E.A. Roy (Ed.), *Neuropsychological studies of apraxia and related disorders* (pp. 45–64). Amsterdam: North Holland.

De Renzi, E., Faglioni, P., Lodesani, M., & Vecchi, A. (1983). Performance of left brain-damaged patients on imitation of single movements and motor sequences. Frontal and parietal-injured patients compared. *Cortex, 19*, 333–343.

De Renzi, E. & Lucchelli, F. (1988). Ideational apraxia. *Brain, 111*, 1173–1185.

De Renzi, E., Motti, F., & Nichelli, P. (1980). Imitating gestures: A quantitative approach to ideomotor apraxia. *Archives of Neurology, 37*, 6–10.

Duffy, R.J. & Duffy, J.R. (1989). An investigation of body part as object (BPO) responses in normal and brain-damaged adults. *Brain and Cognition, 10*, 220–236.

Essens, P.J. (1986). Hierarchical organization of temporal patterns. *Perception & Psychophysics, 40*, 69–73.

Fisk, J.D., & Goodale, M.A. (1988). The effects of unilateral brain damage on visually guided reaching: Hemispheric differences in the nature of the deficit. *Experimental Brain Research, 72*, 425–435.

Fitts, P.M. (1954). The information capacity of the human motor system in controlling the amplitude of movement. *Journal of Experimental Psychology, 47*, 381–391.

Franks, I.M. & Van Donkelaar, P. (1990). The effects of demanding temporal accuracy on the programming of simple tapping sequences. *Acta Psychologica, 74*, 1–14.

Georgopoulos, A.P., Ashe, A., Smyrnis, N., & Taira, M. (1992). The motor cortex and the coding of force. *Science, 256*, 1692–1695.

Geschwind, N. (1965). Disconnection syndromes in animals and man. *Brain, 88*, 237–294.

Goodale, M.A. & Milner, A.D. (1992). Separate visual pathways for perception and action. *Trends in Neurosciences, 15*, 20–25.

Goodale, M.A., Pelisson, D., & Prablanc, D. (1986). Large adjustments in visually guided reaching do not depend on vision of the hand or perception of target displacement. *Nature, 320*, 748–750.

Goodglass, H. & Kaplan, E. (1963). Disturbance of gesture and pantomime in aphasia. *Brain, 26*, 703–720.

Goodglass, H. & Kaplan, E. (1972). *The assessment of aphasia and related disorders*. Philadelphia: Lea and Febiger.

Gordon, P.C. & Meyer, D.E. (1987). Hierarchical representation of spoken syllable order. In A. Allport, D.G. MacKay, W. Prinz, & E. Scheerer (Eds.), *Language perception and production* (pp. 445–462). New York: Academic Press.

Haaland, K.Y. & Flaherty, D. (1984). The different types of limb apraxia errors made by patients with left vs. right hemisphere damage. *Brain and Cognition, 3*, 370–384.

Haaland, K.Y. & Harrington, D.L. (1989). Hemispheric control of the initial and corrective components of aiming movements. *Neuropsychologia, 27*, 961–969.

Haaland, K.Y. & Harrington, D.L. (1990). Complex movement behavior: Toward understanding cortical and subcortical interactions in regulating control processes. In G.R. Hammond (Ed.), *Cerebral control of speech and limb movements* (pp. 169–200). Amsterdam: North-Holland.

Haaland, K.Y. & Harrington, D.L. (1994). Limb-sequencing deficits after left but not right hemisphere damage. *Brain and Cognition, 24*, 104–122.

Haaland, K.Y., Harrington, D.L., & Grice, J.W. (1993). Effects of aging on planning and implementing arm movements. *Psychology and Aging, 8*, 617–632.

Hammond, G. (1982). Hemispheric differences in temporal resolution. *Brain and Cognition, 1*, 95–118.

Harrington, D.L. & Haaland, K.Y. (1987). Programming sequences of hand postures. *Journal of Motor Behavior, 19*, 77–95.

Harrington, D.L. & Haaland, K.Y. (1991a). Hemispheric specialization for motor sequencing: Abnormalities in levels of programming. *Neuropsychologia, 29*, 147–163.

Harrington, D.L. & Haaland, K.Y. (1991b). Sequencing in Parkinson's disease: Abnormalities in programming and controlling movement. *Brain, 114*, 99–115.

Harrington, D.L. & Haaland, K.Y. (1992). Motor sequencing with left hemisphere damage. Are some cognitive deficits specific to limb apraxia? *Brain, 115*, 857–874.

Heilman, K.M. (1979). Apraxia. In K.M. Heilman & E. Valenstein (Eds.), *Clinical neuropsychology* (pp.159–185). New York: Oxford University Press

Heilman, K.M., Coyle, J.M., Gonyea, E.F., & Geschwind, N. (1973). Apraxia and agraphia in a left-hander. *Brain, 96*, 21–28.

Heilman, K.M., Rothi, L.J., & Valenstein, E. (1982). Two forms of ideomotor apraxia. *Neurology, 32*, 342–346.

Ivry, R. (1993). Cerebellar involvement in the explicit representation of temporal information. In P. Tallal, A.M. Galaburda, R.R. Llinas, & C. von Euler (Eds.), *Temporal information processing in the nervous system* (pp. 214–230). New York: Annals of the New York Academy of Sciences.

Jason, G.W. (1986). Performance of manual copying tasks after focal cortical lesions. *Neuropsychologia, 24*, 181–191.

Jeannerod, M. (1986a). Mechanisms of visuomotor coordination: A study in normal and brain-damaged subjects. *Neuropsychologia, 24*, 41–78.

Jeannerod, M. (1986b). The formation of finger grip during prehension. A cortically mediated visuomotor pattern. *Behavioral Brain Research, 19*, 99–116.

Jeannerod, M. (1988). *The neural and behavioral organization of goal-directed movements.* Oxford: Clarendon Press.

Jeannerod, M., Michel, F., & Prablanc, C. (1984). The control of hand movements in a case of hemianesthesia following a parietal lesion. *Brain, 107*, 899–920.

Keele, S.W. (1968). Movement control in skilled performance. *Psychological Bulletin, 70*, 387–403.

Keele, S.W. (1987). Sequencing and timing in skilled perception and action: An overview. In A. Allport, D.G. MacKay, W. Prinz, & E. Scheerer (Eds.), *Language perception and production: Relationships between listening, speaking, reading and writing* (pp. 463–487). New York: Academic Press.

Keele, S.W., Cohen, A., & Ivry, R. (1990). Motor programs: Concepts and issues. In M. Jeannerod (Ed.), *Attention and performance XIII: Motor representation and control* (pp.77–110). Hillsdale NJ: Lawrence Erlbaum Associates Inc.

Keele, S.W. & Ivry, R.I. (1987). Modular analysis of timing in motor skill. In G.H. Bower (Ed.), *The psychology of learning and motivation*, (Vol. 21, pp. 183–228). New York: Academic Press.

Keele, S.W. & Ivry, R.I. (1990). Does the cerebellum provide a common computation for diverse tasks? A timing hypothesis. *Annals of the New York Academy of Sciences, 608*, 179–207.

Kelso, J.A.S. & Tuller, B. (1984). A dynamical basis for action systems. In M.S. Gazzaniga (Ed.), *Handbook of cognitive neuroscience* (pp. 321–356). New York: Plenum Press.

Kelso, J.A.S. (1995). *Dynamic Patterns: The Self Organization of Brain and Behavior.* Cambridge, Massachusetts: MIT Press.

Kerr, B. (1978). Task factors that influence selection and preparation for voluntary movements. In G.E. Stelmach (Ed.), *Information processing in motor control and learning* (pp.55–69). New York: Academic Press.

Kimura, D. (1977). Acquisition of a motor skill after left-hemisphere damage. *Brain, 100*, 527–542.

Kimura, D. (1982). Left-hemisphere control of oral and brachial movements and their relation to communication. *Philosophical Transactions of the Royal Society of London, B298*, 135–149.

Kimura, D. & Archibald, Y. (1974). Motor functions of the left hemisphere. *Brain, 97*, 337–350.

Kleist, K. (1934). *Gehirnpathologie.* Leipzig: Barth.

Kolb, B. & Milner, B. (1981). Performance of complex arm and facial movements after focal brain lesions. *Neuropsychologia, 19*, 491–503.

Kolb, B. & Whishaw, I.Q. (1985). Can the study of praxis in animals aid in the study of apraxia of humans? In E.A. Roy (Ed.), *Neuropsychological studies of apraxia and related disorders* (pp.203–224). Amsterdam: North-Holland.

Lashley, K.S. (1951). The problem of serial order in behavior. In L.A. Jeffress (Ed.), *Cerebral mechanisms in behavior* (pp. 112–136). New York: Wiley.

Lehmkuhl, G. & Poeck, K. (1981). A disturbance in the conceptual organization of actions in patients with ideational apraxia. *Cortex, 17*, 153–158.

Lehmkuhl, G., Poeck, K., & Willmes, K. (1983). Ideomotor apraxia and aphasia: An examination of types and manifestations of apraxic symptoms. *Neuropsychologia, 21*, 199–212.

Liepmann, H. (1908). *Drei Aufsatze aus dem Apraxiegebiet.* Berlin: Karger.

Liepmann, H. (1913). Motor aphasia, anarthria, and apraxia. *Transactions of the 17th International Congress of Medicine*, Section XI, Part 2 (pp. 97–106). London.

Luria, A.R. (1980). Disturbances of higher cortical functions with lesions of the sensorimotor regions. *Higher cortical functions in man* (pp. 189–245). New York: Basic Books.

MacKay, D.G. (1985). A theory of the representation, organization and timing of action with implications for sequencing disorders. In E.A. Roy (Ed.), *Neuropsychological studies of apraxia and related disorders* (pp.267–308). Amsterdam: North-Holland.

Marteniuk, R.G., Leavitt, J.L., MacKenzi, C.L., & Athenes, S. (1990). Functional relationships between grasp and transport components in a prehension task. *Human Movement Science, 9*, 149–176.

Marteniuk, R.G., MacKenzi, C.L., Jeannerod, M., Athenes, S., & Dugas, C. (1987). Constraints on human arm movement trajectories. *Canadian Journal of Psychology, 41*, 365–378.

Martin, J.G. (1972). Rhythmic (hierarchical) versus serial structure in speech and other behavior. *Psychological Review, 79*, 487–509.

Meyer, D.E., Abrams, R.A., Kornblum, S., Wright, C.E., & Smith, J.E.K. (1988). Optimality in human motor performance: Ideal control of rapid aimed movements. *Psychological Review, 95*, 340–370.

Mishkin, M., Ungerleider, L.G., & Macko, K.A. (1983). Object vision and spatial vision: Two cortical pathways? *Trends in Neurosciences, 6*, 414–417.

Motomura, N., Seo, T., Asaba, H., & Sakai, T. (1989). Motor learning in ideomotor apraxia. *International Journal of Neuroscience, 47*, 125–130.

Mozaz, M.J., Pena, J., Barraquer, L.L., Marti, J., & Goldstein, L.H. (1993). Use of body part as object in brain-damaged subjects. *The Clinical Neuropsychologist*, 7, 39–47.

Petrides, M. & Milner, B. (1982). Deficits on subject-ordered tasks after frontal- and temporal-lobe lesions in man. *Neuropsychologia, 20*, 249–262.

Poeck, K. (1985). Clues to the nature of disruptions to limb praxis. In E.A. Roy (Ed.), *Neuropsychological studies of apraxia and related disorders* (pp.99–110). Amsterdam: North-Holland.

Poeck, K. & Lehmkuhl, G. (1980). Ideatory apraxia in a left-handed patient with right sided brain lesion. *Cortex, 16*, 273–284.

Povel, D.J. & Collard, R. (1982). Structural factors in patterned finger tapping. *Acta Psychologica, 52*, 107–123.

Prablanc, C., Pelisson, D., & Goodale, M.A. (1986). Visual control of reaching movements without vision of the limb: I. Role of retinal feedback of target position in guiding the hand. *Experimental Brain Research, 62*, 293–302.

Restle, F. & Brown, E.R. (1970). Serial pattern learning. *Journal of Experimental Psychology, 83*, 120–125.

Roland, P.E., Larsen, B., Lassen, N.A., & Skinhoj, E. (1980). Supplementary motor area and other cortical areas in organization of voluntary movements in man. *Journal of Neurophysiology, 43*, 118–136.

Rosenbaum, D.A. (1985). Motor programming: A review and scheduling theory. In H. Heuer, U. Kleinbeck, & K.H. Schmidt (Eds.), *Motor behavior: Programming, control, and acquisition* (pp. 1–33). Berlin: Springer-Verlag.

Rosenbaum, D.A. (1991). *Human motor control*. San Diego: Academic Press.

Rosenbaum, D.A., Inhoff, A.W., & Gordon, A.M. (1984). Choosing between movement sequences: A hierarchical editor model. *Journal of Experimental Psychology: General, 113*, 372–393.

Rothi, L.J.G., & Heilman, K.M. (1984). Acquisition and retention of gestures by apraxic patients. *Brain and Cognition, 3*, 426–437.

Rothi, L.J.G., Mack, L., Verfaellie, M., Brown, P., & Heilman, K.M. (1988). Ideomotor apraxia: Error pattern analysis. *Aphasiology, 2*, 381–388.

Rothi, L.J.G., Ochipa, C., & Heilman, K.M. (1991). A cognitive neuropsychological model of limb praxis. *Cognitive Neuropsychology, 8*, 443–458.

Roy, E.A. (1981). Action sequencing and lateralized cerebral damage: Evidence for asymmetries in control. In J. Long & A. Baddeley (Eds.), *Attention and performance IX* (pp. 487–498). Hillsdale, NJ: Lawrence Erlbaum Associates Inc.

Roy, E.A. & Elliott, D. (1989). Manual asymmetries in aimed movements. *Quarterly Journal of Experimental Psychology, 41A*, 501–516.

Roy, E.A. & Square, P.A. (1985). Common considerations in the study of limb, verbal and oral apraxia. In E.A. Roy (Ed.), *Neuropsychological studies of apraxia and related disorders* (pp. 111–159). Amsterdam: North-Holland.

Roy, E.A., & Square-Storer, P.A. (1990). Evidence for common expressions of apraxia. In G. R. Hammond (Ed.), *Cerebral control of speech and limb movements* (pp. 477–502). Amsterdam: North Holland.

Roy, E.A., Square, P.A., Adams, S., & Friesen, H. (1985). Error/movement notation systems in apraxia. *Recherches Semiotiques/Semiotics Inquiry, 5*, 402–412.

Roy, E.A., Square-Storer, P., Hogg, S., & Adams, S. (1991). Analysis of task demands in apraxia. *International Journal of Neuroscience, 56*, 177–186.

Rumelhart, D.E. & Norman, D.A. (1982). Simulating a skilled typist: A study of skilled cognitive-motor performance. *Cognitive Science, 6*, 1–36.

Seitz, R.J., Roland, E., Bohm, C., Greitz, T., & Stone-Elander, S. (1990). Motor learning in man: A positron emission tomographic study. *Neuroreport, 1*, 57–60.

Shimamura, A.P., Janowsky, J.S., & Squire, L.R. (1990). Memory for the temporal order of events in patients with frontal lobe lesions and amnesic patients. *Neuropsychologia, 28*, 803–813.

Sittig, O. (1931). *Über Apraxie*. Berlin: Karger.

Sternberg, S., Monsell, S., Knoll, R.L., & Wright, C.E. (1978). The latency and duration of rapid movement sequences: Comparisons of speech and typewriting. In G.E. Stelmach (Ed.), *Information processing in motor control and learning* (pp. 118–152). New York: Academic Press.

Summers, J.J. & Burns, B.D. (1990). Timing in human movement sequences. In R.A. Block (Ed.), *Cognitive models of psychological time* (pp. 181–206). Hillsdale NJ: Lawrence Erlbaum Associates Inc.

Summers, J.J., Sargent, G.I., & Hawkins, S.R. (1984). Rhythm and the timing of movement sequences. *Psychological Research, 46*, 107–119.

Tzeng, O.J.L. & Wang, W.S.Y. (1984). Search for a common neurocognitive mechanism for language and movements. *American Journal of Physiology, 15*, R904–R911.

Wallace, S.A. & Newell, K.M. (1983). Visual control of discrete aiming movements. *Quarterly Journal of Experimental Psychology A, 35*, 311–321.

Wickens, J., Hyland, B., & Anson, G. (1994). Cortical cell assemblies: A possible mechanism for motor programs. *Journal of Motor Behavior, 26*, 66–82.

10 Disorders of Writing

Steven Z. Rapcsak

INTRODUCTION

Writing occupies a special place within the repertoire of learned motor skills. The purpose of writing is to communicate ideas by producing conventional graphic signs that represent various sized elements of spoken language (i.e. words, syllables, or phonemes). Defined in this way, writing is probably about 5 000 years old (Gelb, 1963) whereas picture drawing, the antecedent of writing, can be traced back almost 35 000 years (Putnam, 1989). Consistent with its relatively recent appearance in evolutionary history, writing is also a late accomplishment on the ontogenetic time scale. In most literate cultures children start to write after the age of six, but the normal adult speed of writing is not achieved until about 15 years of age (Sassoon, Nimmo-Smith, & Wing, 1986). This is long after children master the motor skills necessary for drawing, using various tools, playing musical instruments, and performing sports activities (Van Galen, 1991).

To produce writing, one must transcode abstract linguistic representations into concrete motor instructions to specific effector muscles. This chapter reviews the cognitive and motor processes involved in externalising linguistic information as visible strokes of the pen and examines patterns of breakdown caused by neurological damage.

A COGNITIVE MODEL OF WRITING

The cognitive information-processing model of writing that provides the conceptual framework for our discussion is represented diagrammatically in

Fig. 1. In its functional architecture, the model is similar to the ones proposed by Ellis (1982, 1988) and Margolin (1984). Following the suggestion of Ellis (1988), we distinguish between "central", or linguistic, processes that generate spellings for words or nonwords and "peripheral" processes that convert abstract graphemic information into codes appropriate for various output modalities (e.g. oral spelling, writing, typing, spelling with anagram letters). Peripheral conversion mechanisms specific to writing and their impairments following neurological damage will serve as our main focus.

Central Processes

In the case of familiar words, spelling is accomplished by retrieving the desired entry from the graphemic output lexicon–the memory store of learned spellings. By contrast, plausible spellings for unfamiliar words or nonwords are assembled by a nonlexical phoneme–grapheme conversion procedure. Spellings generated by the lexical or the nonlexical route, represented as strings of graphemes, are next transmitted to the graphemic buffer. The graphemic buffer is a working memory system that temporarily stores abstract graphemic representations while they are being converted into specific letter names or letter shapes by peripheral mechanisms (Caramazza, Miceli, Villa, & Romani, 1987). Thus, the information held in this buffer specifying the identity and serial ordering of graphemes supports spelling by all possible output modalities (but see Lesser, 1990 for an alternative view).

Peripheral Processes

Allographic Conversion. The first step in translating graphemes into writing movements of the pen involves the selection of the appropriate letter shapes. Graphemes can be realised in different case (upper versus lower) or style (print versus script). The various physical forms each grapheme can take are referred to as allographs. According to Ellis (1982), allographs are stored in long-term memory as abstract spatial descriptions that specify letter shape but not the absolute size of the letter. Allographic representations also do not contain information about the sequence of strokes necessary to create the letter form or the specific muscles that are to be recruited for movement execution. Margolin (1984) suggested that abstract letter shape representations or "physical letter codes" are used not only in writing but also in other forms of "visually-based" spelling (i.e. spelling with anagram letters and typing)[1]. We make the additional

---

[1]Other investigators (Black, Behrmann, Bass, & Hacker, 1989; Ellis, 1988), however, have proposed that peripheral conversion mechanisms for typing and spelling with anagram letters diverge from writing at a processing stage prior to the allographic store (indicated by the dotted line in Fig. 1). In our view, it is not possible to adjudicate between the different proposals with complete certainty based on the limited empirical evidence currently available.

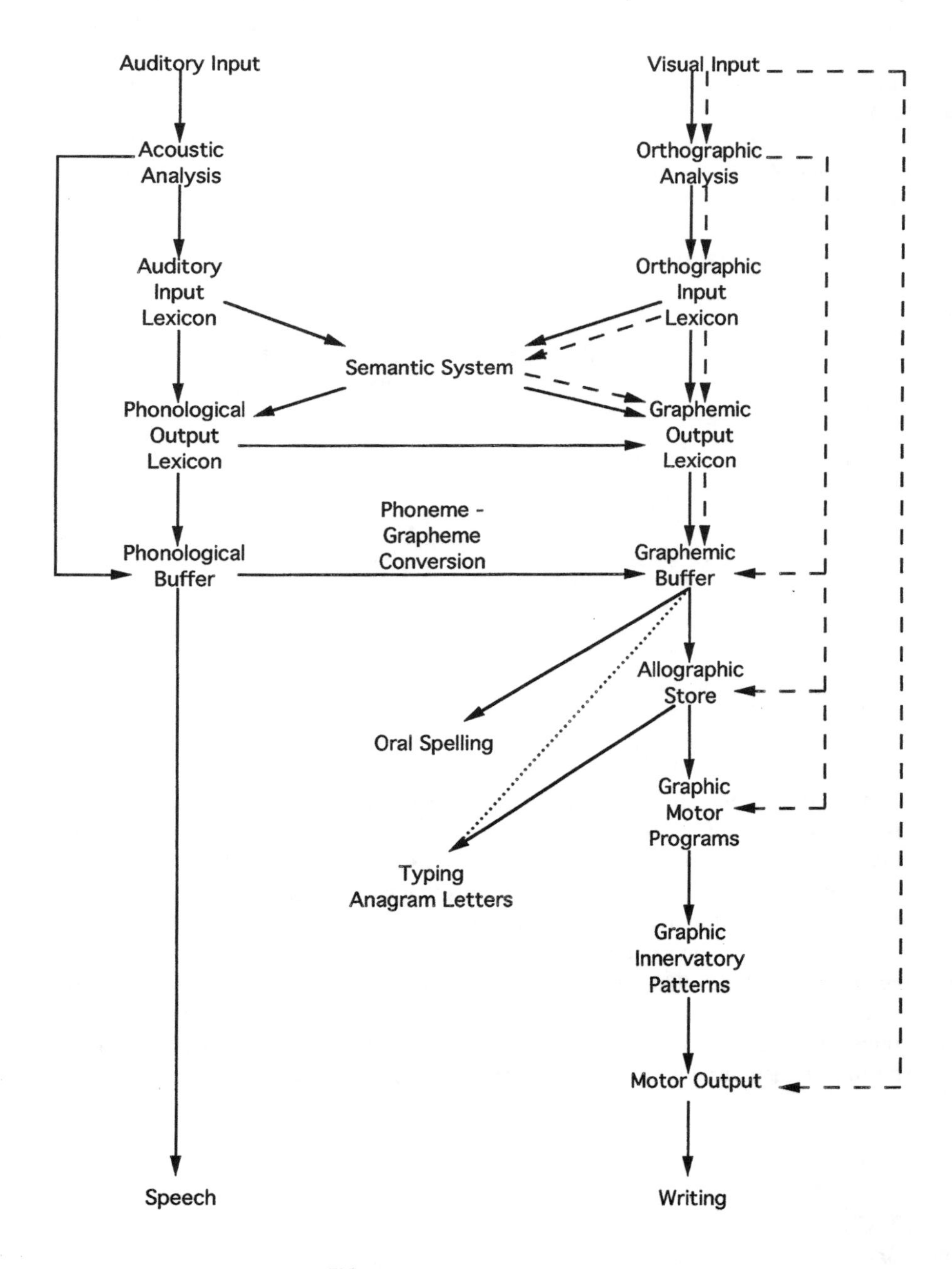

FIG. 1. Cognitive model of writing.

assumption that the spatial representations of letter shapes used in writing can also be activated by visual imagery. However, according to our model, allographic representations do not mediate letter recognition. This is accomplished by the activation of abstract letter recognition units located within the orthographic analysis component of the reading system (Fig. 1).

Graphic Motor Programs. Spatial descriptions of letter shapes generated by the allographic conversion process are subsequently used to activate stored graphic motor programs that guide the execution of the skilled movements of writing. Graphic motor programs are analogous to, but distinct from, visuo-kinaesthetic engrams (Heilman, 1979; Heilman & Rothi, 1985) that contain abstract spatio-temporal codes for other types of learned skilled limb movements. Ellis (1982) proposed that graphic motor programs specify the sequence, direction, and relative size of the strokes necessary to create a given allograph, without specifying absolute stroke size or duration. Although the information about spatio-temporal movement attributes stored in graphic motor programs is letter specific, it is still effector independent in that it does not determine the actual muscles that are to be used to produce the letter. An interesting aspect of writing is that it can be performed by using different muscles of the same limb (e.g. the distal muscles of the hand and wrist when writing with a pen and the proximal muscles of the shoulder and elbow when writing on a blackboard), or by using different limbs altogether (e.g. dominant versus nondominant hand or foot). The observation that overall letter shape remains remarkably constant when produced by different muscle effector systems is consistent with the existence of effector independent abstract motor programs for writing (Merton, 1972; Wright, 1990).

Graphic Innervatory Patterns. The information encoded in graphic motor programs is next translated into graphic innervatory patterns containing sequences of motor commands to specific muscle effector systems. It is at this stage that the appropriate synergies of agonist and antagonist muscles are recruited and concrete movement parameters specifying absolute stroke size, duration and force are inserted into the program (Ellis, 1982, 1988; Margolin, 1984; Van Galen, 1980, 1991). In order to adapt to ever changing contextual requirements, the motor system must have the capacity to compute actual parameters for writing movements "online." Thus, parameter estimation is viewed as a more variable and dynamic process than the retrieval of stored graphic motor programs from long-term memory specifying relatively invariant relationships between abstract spatio-temporal movement attributes (Van Galen, 1980). The preservation of space–time invariance in handwriting despite changing task demands (Viviani & Terzuolo, 1980) bears testimony to the flexibility of the system responsible for generating graphic innervatory patterns. Once the kinematic parameters appropriate to the given biophysical context have been selected, the motor system executes the strokes necessary to create the desired letter as a sequence of rapid ballistic movements.

Afferent Control Systems. Handwriting is a delicate skill that requires continuous visual and kinaesthetic feedback for maximum speed and accuracy. When this feedback is interrupted experimentally, normal individuals make characteristic errors that include a tendency to duplicate or omit letters or strokes in sequences of similar items (Ellis, Young, & Flude, 1987; Kalmus, Fry, & Denes, 1960; Lebrun, 1976; Smith, McCrary, & Smith, 1960; Smyth & Silvers, 1987; Van Bergeijk & David, 1959). Margolin (1984) suggested that sensory feedback played an important role in updating graphic motor programs as to which strokes have already been executed. Sensory feedback is also needed to maintain the correct spacing between letters and words, and to keep the line of writing straight and properly oriented on the page (Lebrun, 1976, 1985; Smyth & Silvers, 1987).

Copying

The relationship between writing and copying is a complex one. In neuropsychological practice, copying is frequently tested in patients with impaired writing. However, in interpreting performance one must keep in mind that copying is not a unitary function. For instance, Ellis (1982) proposed that copying letters or words might be accomplished by at least three separate procedures (represented as broken lines in Fig. 1). In nonlinguistic or "pictorial" copying one reproduces the physical appearance of letters the same way one would reproduce *any* visuospatial pattern. In our model this route is shown as a direct connection between visual input and motor output, bypassing both the central and the peripheral components of the writing system. Since movements of the pen need to be assembled *de novo*, without the processing advantage conferred by stored allographic representations and graphic motor programs, we expect pictorial copying of letters to be relatively slow and fragmented, relying heavily on "closed-loop" visual control. These aspects of performance often give rise to the clinical impression that copying is accomplished in a "slavish" manner typical of unskilled behaviour. Pictorial copying is the procedure normally employed in learning to write, but it may also be resorted to following damage to the peripheral components of the writing system. Patients relying on pictorial copying *exclusively* may be able to reproduce letter shapes adequately, but they should not be able to transcode between various allographic forms of the letter (e.g. transcribe upper case to lower case or vice versa) since this requires access to allographic representations and graphic motor programs. Impairment of the pictorial route should be accompanied by defective copying of all kinds of visual designs and by poor performance on drawing and spatial construction tasks in general. A patient demonstrating selective impairment of pictorial copying with intact writing was described by Cipolotti and Denes (1989).

In contrast to nonlinguistic or pictorial copying, in "graphemic" and "lexical" copying input letter strings are first analysed by the reading system and the appropriate graphic output is subsequently generated by the peripheral components of the writing system. In nonlexical or graphemic copying, a link is postulated between letter recognition units and the graphemic buffer, bypassing both the orthographic input lexicon and the graphemic output lexicon (Ellis, 1982; Morton, 1980). It is conceivable, however, that there are also other connections (Fig. 1) through which letter recognition units can interface directly with the peripheral components of the writing system (e.g. the allographic store or graphic motor programs). Graphemic copying is the procedure used primarily to reproduce single letters, unfamiliar words or nonwords. In copying by the lexical route, the visually presented word first activates its appropriate entry in the orthographic input lexicon. This is followed by the retrieval of the corresponding spelling from the graphemic output lexicon (either directly or via the semantic system), and this information is next transferred to the graphemic buffer for peripheral conversion. Unlike copying by the pictorial route, copying by the graphemic and lexical routes should be facile under normal circumstances since it can take advantage of allographic representations and graphic motor programs stored in long-term memory. Due to their shared output mechanisms, graphemic or lexical copying and writing may show similar patterns of impairment following brain damage.

NEUROPSYCHOLOGICAL DISORDERS OF WRITING

Having presented a cognitive model of writing, we now turn our attention to patterns of breakdown caused by neurological damage. Since our primary focus is writing rather than spelling, only damage to functional components located downstream from the graphemic buffer will be considered. Central to the cognitive neuropsychological approach adopted here is the assumption that individual components of the writing system are also neurologically distinct and thus may be selectively impaired by focal brain pathology. In "pure" cases, processing components above and below the impaired module are assumed to be capable of functioning normally, though this may be difficult to demonstrate for the modules that rely on the damaged component for their input. Although such pure cases are of great theoretical interest, it must be pointed out that mixed patterns of impairment, consistent with simultaneous damage to several processing components, are by no means uncommon in clinical practice (Levine, Mani & Calvanio, 1988; Margolin & Binder, 1984; Zangwill, 1954). The various writing disorders discussed below have been collectively referred to as "peripheral dysgraphias" (Ellis, 1988, 1987; Ellis et al., 1987).

Allographic Disorders

A number of cases have been reported in recent years in which the writing impairment was most likely attributable to dysfunction of the allographic conversion process. Generally speaking, allographic disorders are characterised by an inability to activate or select the letter shapes appropriate for the set of abstract graphemic identities specified by the lexical or nonlexical central spelling routes. Since the location of the damage is downstream from the graphemic buffer, oral spelling should be spared (Fig. 1). Consistent with the hypothesis that abstract letter shape representations may also subserve spelling with anagram letters and typing (Margolin, 1984), these modalities of output have on occasion been noted to be impaired along with writing (Friedman & Alexander, 1989; Kinsbourne & Rosenfield, 1974; Levine et al., 1988; Patterson & Wing, 1989). This observed association could of course also reflect simultaneous damage to functionally independent peripheral conversion mechanisms.

In allographic disorders transcoding between different allographic forms of a letter may be defective (Friedman & Alexander, 1989; Kartsounis, 1992; Patterson & Wing, 1989). Copying, however, can be relatively spared since external visual presentation of letter shapes in some cases may help activate the corresponding allographic representations. As shown in Fig. 1, copying could perhaps also be accomplished through direct connections between letter recognition units and graphic motor programs that bypass the impaired allographic processing stage, or via the pictorial copying route.

The breakdown of the allographic conversion process can take several different clinical forms. For instance, in some cases it may result in complete or partial inability to produce the appropriate letter shapes, even though the graphic motor programs necessary for creating these allographs are presumed to be intact. We refer to letter formation deficits arising from the unavailability of letter shape information as allographic production disorders. Theoretically, it should be possible to distinguish production disorders that reflect a failure to access allographic forms from those that result from damage to letter shape representations within the allographic store itself. In access disorders, allographic forms should be available on some occasion but not on others, leading to considerable variability in performance. When the correct allographs can be retrieved, they should be executed fluently. By contrast, damage to or loss of letter shape representations should lead to consistently poor written production of the corresponding allographs. Damage to or destruction of allographic representations should also be accompanied by poor letter imagery. Such imagery deficits have now in fact been documented in a number of agraphic patients (Crary & Heilman, 1988; Friedman & Alexander, 1989; Levine, et al., 1988).

When asked to write, patients with allographic production disorders may "block" completely and claim that they are unable to produce the desired letters, apparently because they cannot remember or visualise the appropriate letter shapes (Crary & Heilman, 1988; Ellis, 1988; Kartsounis, 1992; Patterson & Wing, 1989). Omission errors are frequent, but partially correct responses may also be observed. Support for the hypothesis that in allographic production disorders the main difficulty lies in remembering rather than in creating letter shapes comes from the temporal analysis of writing in patient D.K. reported by Patterson & Wing (1989). Consistent with a deficit in access to or retrieval of information from the allographic store, D.K.'s writing performance was characterised by abnormally prolonged preparation times for individual letters followed by reasonably normal motor execution times.

Certain patients with allographic production disorders could write upper case letters significantly better than lower case letters (Kartsounis, 1992; Patterson & Wing, 1989), whereas in others the opposite dissociation was documented (De Bastiani & Barry, 1986, cited in Patterson & Wing, 1989). These observations suggest that representations for upper and lower case forms are organised separately within the allographic long-term memory store and may be selectively disrupted by brain damage.

In other cases with presumed damage to the allographic conversion process the main difficulty seemed to have involved selection of the correct allographic forms. Unlike patients with allographic production disorders, patients with allographic selection disorders produce well-formed but incorrect letters. Different types of selection problems have been documented. In the patient reported by De Bastiani & Barry (1989), the disorder manifested itself as an apparently uncontrollable mixing of upper and lower case letters in writing. Other patients with allographic selection disorders produce numerous letter substitution errors but are usually able to keep case constant (Black, Behrmann, Bass, & Hacker, 1989; Friedman & Alexander, 1989; Goodman & Caramazza, 1986; Kapur & Lawton, 1983; Kinsbourne & Rosenfield, 1974; Lambert, Viader, Eustache, & Morin, 1994; Rothi & Heilman, 1981). Black et al. (1989) provided evidence that in patients with substitution errors letter frequency may be an important determinant of performance. It has also been suggested that letter substitution errors in general, and those bearing an obvious physical similarity to the intended letter in particular, may actually be postallographic in origin and may result from a failure to select the correct graphic motor program specified by the allographic code (Black et al., 1989; Ellis, 1982, 1988; Margolin, 1984; Friedman & Alexander, 1989; Lambert et al., 1994). Thus, the selection problem in these cases may be related to impaired transmission of information between the allographic store and graphic motor programs. The physical similarity effect may be attributable to the fact that graphic motor programs containing similar stroke sequences are more likely to be misselected. Since in postallographic selection disorders representations within the allographic store are assumed to be intact and can also be accessed from the graphemic buffer in a normal fashion, we would expect spelling with anagram letters and typing to be spared and letter imagery should also be preserved.

Neuroanatomical Correlates. In virtually all patients with clinical evidence of dysfunction of the allographic conversion process the responsible lesion involved the left parieto-occipital region (Black et al., 1989; Crary & Heilman, 1988; De Bastiani & Barry, 1989; patient 2 in Friedman & Alexander, 1989; Goodman & Caramazza, 1986; Kapur & Lawton, 1983; Kinsbourne & Rosenfield, 1974; Lambert et al., 1994; Levine et al., 1988; Patterson & Wing, 1989). These observations suggest that this cerebral area plays a critical role in activating and selecting the correct letter shape representations for written production and is therefore the most likely cortical location for the putative allographic long-term memory store.

Disorders of Motor Programming: Apraxic Agraphia

In 1867, Ogle described patients with "atactic agraphia" whose writing difficulty consisted of an inability to execute the skilled movements required to produce well-formed letters. According to Ogle, in these individuals ". . . the power of writing even separate letters is lost, sometimes entirely. Here all attempts to write result in a mere succession of up and down strokes, bearing no kind of resemblance to letters." Following the suggestion of Heilbronner (1906), in the more recent neuropsychological literature the term "apraxic agraphia" has been applied to motor programming disorders of writing.

Apraxic agraphia can be defined operationally as a writing disorder characterised by poor letter formation that cannot be attributed to impaired letter shape knowledge or to sensorimotor, basal ganglia or cerebellar dysfunction affecting the writing limb. In apraxic agraphia oral spelling is normal. Since letter shape knowledge is preserved, spelling with anagram letters and typing should be spared and letter imagery should also be intact. Cases of "pure" apraxic agraphia fulfilling most of these criteria have been described by Baxter and Warrington (1986) and by Alexander, Fischer, and Friedman (1992).

Within the framework of our model, the clinical picture of apraxic agraphia arises as a consequence of damage to processing components involved in programming the skilled movements of writing. Possible mechanisms include damage or disconnection of stored graphic motor programs or damage to systems responsible for translating the information contained in graphic motor programs into graphic innervatory patterns to specific muscles. Consistent with the hypothesis that motor programs for writing are distinct from programs for other types of skilled limb movements, apraxic agraphia has been found to be dissociable from limb apraxia (Anderson, Damasio, & Damasio, 1990; Baxter & Warrington, 1986; Coslett, Rothi, Valenstein, & Heilman, 1986; Croisile, Laurent, Michel, & Trillet, 1990; Hodges, 1991; Margolin & Binder, 1984; Papagno, 1992; Roeltgen & Heilman, 1983; Zangwill, 1954).

Although patients with apraxic agraphia can usually hold the pen correctly, they are unable to execute the sequence of strokes necessary to create the letter form specified by the allographic code. Due to the motor programming deficit, the spatio-temporal attributes of writing are severely disturbed and the smooth, facile strokes normally used to produce the required spatial trajectory are replaced by hesistant, awkward, and imprecise movements. False starts and perseverations are common. Patients are usually aware of their writing difficulty and easily become frustrated by repeated errors which they are powerless to correct. In severe cases all attempts at writing may result in an illegible scrawl. Even though individual letters may be unrecognisable, some distinction between various writing styles (e.g. cursive versus print) and between upper and lower case forms may be preserved (Margolin & Binder, 1984; Roeltgen & Heilman, 1983). Frequently observed errors of letter morphology include spatial distortions, stroke omissions resulting in incomplete forms, and insertions of anomalous strokes resulting in nonletters (Margolin & Binder, 1984; Margolin & Wing, 1983) (see Fig. 2). Stroke repetitions, substitutions, and mislocations are also common. In milder cases letters may be recognisable but writing is slow, effortful, and untidy (Hodges, 1991). When letters are legible, it may be possible to demonstrate that written spelling per se is intact. The writing difficulty in apraxic agraphia may be highly specific to letter formation since in some cases numbers could be written correctly (Anderson, Damasio, & Damasio 1990).

Patients with apraxic agraphia can sometimes copy letters significantly better than they can write them spontaneously or to dictation (Fig. 2). According to our model, preserved copying of letters in apraxic agraphia can only be accomplished by the pictorial route and is therefore likely to be carried out in a stroke-by-stroke or "slavish" fashion.

Neuroanatomical Correlates. As a general rule, lesions responsible for apraxic agraphia are located in the hemisphere contralateral to the hand preferred for writing. Thus, in most right-handers the lesion involves the left hemisphere, which is usually also the hemisphere dominant for limb praxis and language. However, neuropsychological evidence suggests that in left-handers and in individuals with mixed or atypical cerebral dominance the hemisphere which controls the programming of the skilled movements of writing may not always be the hemisphere that is also dominant for language and/or limb praxis (Heilman, Gonyea, & Geschwind, 1974; Heilman, Coyle, Gonyea, & Geschwind, 1973; Margolin, 1980; Margolin & Binder, 1984; Rapcsak, Rothi, & Heilman, 1987; Roeltgen & Heilman, 1983; Rosa, Demiati, Cartz, & Mizon, 1991; Tanaka, Iwasa, & Obayashi, 1990; Valenstein & Heilman, 1979; Zangwill, 1954).

Regarding the intrahemispheric location of lesions producing apraxic agraphia, several distinct anatomical patterns can be identified. In a number of cases the lesion involved the parietal lobe (Alexander et al., 1992; Baxter & Warrington, 1986; Coslett et al., 1986, Papagno, 1992; Roeltgen & Heilman, 1983; Valenstein & Heilman, 1979; Zangwill, 1954). Alexander et al. (1992) suggested that within the parietal area the critical lesion site may be in or around the junction of the angular gyrus and the superior parietal lobule. In other cases

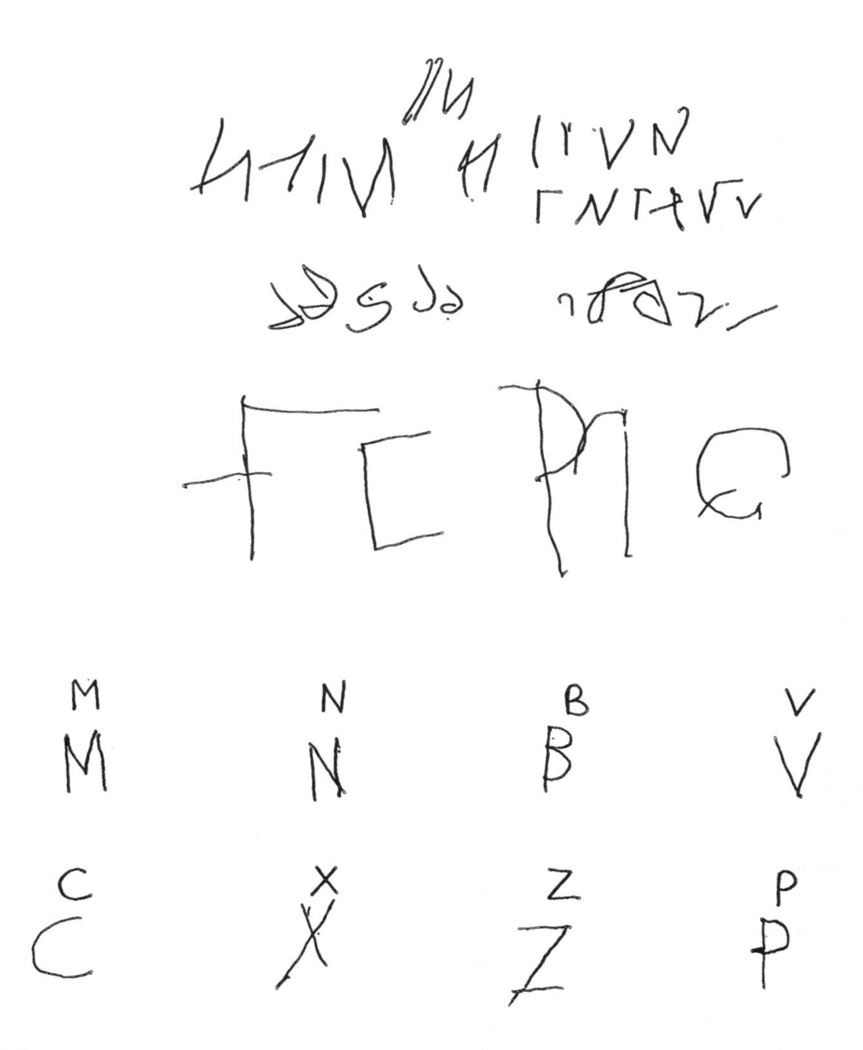

FIG. 2. Writing produced by a patient with apraxic agraphia following left parietal lobe lesion demonstrating severe impairment of letter morphology. In contrast to writing to dictation, copying is relatively spared (bottom rows).

of apraxic agraphia the lesion was anteriorly placed, centring on the premotor cortical area near the foot of the second frontal convolution (Anderson et al., 1990; Hodges, 1991; Gordinier, 1899). This dorsolateral frontal region situated directly above Broca's area is sometimes referred to as Exner's writing centre and was considered by many early investigators to be the cortical location for the motor memories of writing (Exner, 1881; Goldstein, 1948; Henschen, 1922;

Nielsen, 1946; Pitres, 1884). Writing disorders demonstrating the characteristic production features of apraxic agraphia have also been documented following frontal lesions that involved the mesial premotor cortex of the supplementary motor area (SMA) (Rubens, 1975; Watson, Fleet, Rothi, & Heilman, 1986). Finally, there is at least one case report of apraxic agraphia following a hematoma in the white matter of the centrum semiovale, sparing the relevant parietal and frontal cortical regions listed above (Croisile et al., 1990).

Based on the anatomical evidence, we propose that the motor programming of writing is mediated by a distributed neuronal network which includes both posterior and anterior cortical components with distinct functional roles. With analogy to recent neuroanatomical models of limb apraxia (Rothi & Heilman, 1993; Rothi, Ochipa, & Heilman, 1991), we suggest that graphic motor programs are stored in the parietal lobe and that the frontal premotor areas (dorsolateral premotor cortex and SMA) are involved in translating these programs into graphic innervatory patterns. Parietal lesions cause apraxic agraphia by damaging or destroying stored spatio-temporal representations for writing movements, whereas frontal premotor lesions interfere with the generation of the appropriate motor commands to specific muscle effector systems. Lesions located in the deep white matter (e.g. Croisile et al. 1990) may cause apraxic agraphia by disconnecting the parietal and frontal cortical components of the proposed network.

As discussed above, in most right-handers the left hemisphere is dominant for language, limb praxis and writing. When these individuals develop agraphia following left-hemisphere lesions, the writing disorder involves both hands, although the right hand may not be testable due to paralysis. Callosal lesions in right-handers should produce unilateral agraphia of the nondominant hand, since writing with the left hand is believed to require transfer of linguistic and motor information from the left to the right hemisphere across the corpus callosum. Consistent with this hypothesis, a number of right-handed patients with callosal lesions were either completely unable to write with the left hand or produced mostly illegible scrawl (Geschwind & Kaplan, 1962; Graff-Radford, Welsh, & Godersky, 1987; Leiguarda, Starkstein, & Berthier, 1989; Liepmann & Maas, 1907; Rubens, Geschwind, Mahowald, & Mastri, 1977; Sweet, 1941; Watson & Heilman, 1983; Yamadori, Osumi, Imamura, & Mitani, 1988). In some cases it was possible to demonstrate that the left hand could spell words correctly using a typewriter, suggesting that these patients may have suffered from "pure" unilateral apraxic agraphia (Leiguarda et al., 1989; Graff-Radford et al., 1987; Watson & Heilman, 1983). Other patients were considered to have a combination of apraxic and linguistic agraphia, since they were also unable to spell with anagram letters or using a typewriter (Geschwind & Kaplan, 1962; Goldenberg, Wimmer, Holzner, & Wessely, 1985; Liepmann & Maas, 1907).

Unilateral apraxic agraphia is usually associated with ideomotor apraxia of the left hand and typically follows lesions that involve the anterior two-thirds to four-fifths of the corpus callosum, sparing the most caudal portions of the body and the splenium. In some cases the genu was also partially spared. These anatomical observations support the hypothesis that information critical to the motor programming of skilled limb movements is transferred through the body of the corpus callosum (Geschwind, 1975; Liepmann, 1920; Watson & Heilman, 1983), although the exact location of the relevant fibres within this structure has not been determined conclusively (Boldrini, Zanella, Cantagallo, & Basaglia, 1992; Risse, Gates, Lund, Maxwell, & Rubens, 1989; Volpe, Sidtis, Holzman, Wilson, & Gazzaniga, 1982). Since the programming of skilled movements is mediated by an anatomically distributed neuronal network that includes both posterior parietal and frontal premotor cortical regions, we believe that interhemispheric transfer of praxic information could potentially take place at several different sites. Considering what is known about the topography of callosal fibres (Pandya & Seltzer, 1986), exchange of information between the left and right parietal regions is likely to be negotiated through fibres that traverse the posterior part of the body, whereas premotor areas may communicate through callosal fibres located in the anterior or rostral parts of the body. Individuals may differ in the extent to which they can utilise posterior versus anterior connections for interhemispheric transfer both under normal circumstances and following partial callosal damage. In any event, the fact that unilateral apraxic agraphia and ideomotor apraxia are occasionally dissociable following callosal damage (Boldrini et al., 1992; Degos, Gray, Louarn, Ansquer, Poirier, & Barbizet, 1987; Kazui & Sawada, 1993) suggests that motor programs for writing and for other types of skilled movements are transferred through anatomically and functionally independent callosal channels.

In some patients with callosal lesions involving the posterior portions of the body and the splenium, unilateral linguistic agraphia of the left hand was observed without significant limb apraxia (Gersh & Damasio, 1981; Kawamura, Hirayama, & Yamamoto, 1989; Sugishita, Toyokura, Yoshioka, & Yamada, 1980; Yamadori, Nagashima & Tamaki, 1983; Yamadori, Osumi, Ikeda, & Kanazawa, 1980). Unlike patients with unilateral apraxic agraphia, these patients were generally able to produce legible letters which, however, were linguistically incorrect. Thus, in these cases the posterior callosal lesion seemed to have interfered with the transfer of linguistic information relevant to writing, but motor programs for writing and for other types of skilled limb movements were presumably able to cross more anteriorly (Gersh & Damasio, 1981; Watson & Heilman, 1983).

Unfortunately, the neuropsychological interpretation of left-hand writing performance in patients with partial callosal damage is not entirely free of complications. One difficulty concerns the fact that, although it is certainly reasonable to attribute impaired left-hand performance to defective callosal transfer, preserved writing abilities cannot be accepted *prima facie* as proof that the relevant motor and/or linguistic commands originated in the left hemisphere and were subsequently transferred to the right hemisphere through spared fibres of the corpus callosum. This is because such spared abilities in right-handed patients may simply reflect individual variations in the intrinsic capacity of the right hemisphere to produce writing with the left hand (Rapcsak, Beeson, & Rubens, 1991). In addition, it has been demonstrated that following callosal disconnection the isolated left hemisphere may still be capable of at least some crude motor control of left-hand writing through the use of ipsilateral pathways (Levy, Nebes, & Sperry, 1971). It is conceivable that unskilled but linguistically correct typing and spelling with anagram letters could also be accomplished via ipsilateral left-hemisphere connections to the left hand. Another problem relates to a certain lack of precision in our ability to define the exact nature of the information that is unavailable to the right hemisphere in cases of unilateral agraphia following partial callosal damage. For instance, patients with unilateral agraphia of the left hand have been described who produced well-formed but incorrect letters in writing to dictation (see previous discussion). The standard theoretical interpretation offered for this behavioural observation is that the callosal lesion interfered with the transfer of linguistic information from the left to the right hemisphere. However, this general formulation does not specify in cognitive terms the level at which the flow of information between the various functional components of the writing system was interrupted. According to our cognitive model of writing, letter substitution errors in left-hand writing may result from at least two different types of processing impairments caused by interhemispheric disconnection. One possibility is that these errors reflect the inability to select the appropriate right-hemisphere letter shape representations for the strings of graphemes specified by the spelling system of the left hemisphere (i.e. an allographic selection disorder). In this case left-hand typing and spelling with anagram letters might also be impaired. Alternatively, as suggested by Zesiger and Mayer (1992), letter substitution errors may result from a disconnection between the left-hemisphere allographic store and right-hemisphere graphic motor programs (i.e. a postallographic selection disorder). In this case left-hand typing and spelling with anagram letters might be spared (Zesiger & Mayer, 1992). Despite the theoretical and methodological difficulties involved, we believe that detailed analysis of writing and spelling in callosal patients using the framework provided by information-processing models of writing is likely to advance significantly our understanding of the types of linguistic and motor codes transferred through the channels of the corpus callosum.

Nonapraxic Disorders of Motor Execution

As we have seen previously, the inability to translate the abstract spatio-temporal information contained in graphic motor programs into motor commands to specific muscles can give rise to the syndrome of apraxic agraphia. However, according to our model, dysfunction of systems responsible for generating graphic innervatory patterns may also result in the insertion of incorrect kinematic parameters into otherwise intact graphic motor programs, leading to nonapraxic errors of movement force, speed and amplitude. A typical example of this kind of motor execution disorder is the micrographia of patients with Parkinson's disease. As its name implies, micrographia is characterised by an overall diminution of letter size (see Fig. 3). During writing, letter size may decrease progressively from the beginning to the end of the line. Patients are usually unable to increase the size of their writing voluntarily or can do so for brief periods of time only (McLennan, Nakano, Tyler, & Schwab, 1972). Writing speed is also significantly reduced. Except in severe cases, letters in micrographic production remain recognisable and there are no stroke-level errors of the kind seen in apraxic agraphia, attesting to the preservation of information at the level of allographic representations and graphic motor programs (Fig. 3). These production features also suggest that the motor system is capable of activating the correct muscles in the appropriate sequence, even though actual movement parameters cannot be calibrated accurately. Based on a quantitative analysis of handwriting in patients with Parkinson's disease, Margolin & Wing (1983) concluded that micrographia was primarily attributable to an inability to generate the forces necessary to maintain proper letter size. Thus, the writing impairment may be one particular manifestation of the general difficulty in initiating and maintaining the forces required to execute purposeful movements in patients with this neurological disorder (Marsden, 1982).

Poor penmanship is also readily apparent in the writing produced by patients with cerebellar dysfunction (Gilman, Bloedel, & Lechtenberg, 1981). In cerebellar disorders the programmed sequences of rapid alternating muscle contractions necessary to produce writing cannot be executed in a smooth and automatic fashion. Consequently, movements of the pen tend to be slow and disjointed, requiring deliberate effort and concentration on the patient's part. Due to the inadequate control of movement force, amplitude and timing, writing trajectory is frequently irregular and erratic.

Neuroanatomical Correlates. The characteristic abnormalities of voluntary movement in Parkinson's disease are caused by basal ganglia dysfunction. A full discussion of the role of the basal ganglia in movement control is beyond the scope of this chapter, but it is clear that these subcortical structures are critically involved in the automatic execution of learned motor sequences (Marsden, 1982). Converging evidence from anatomical and neurophysiological studies suggests that the basal ganglia work in concert with premotor cortical regions involved in higher order motor programming. Functional integration between the basal ganglia and cortical motor areas is accomplished through a major re-entrant basal ganglia-thalamocortical motor loop (Alexander, DeLong, & Strick, 1986). Specifically, the putamen receives projections from dorsolateral premotor cortex, SMA and sensorimotor cortex and projects back through the

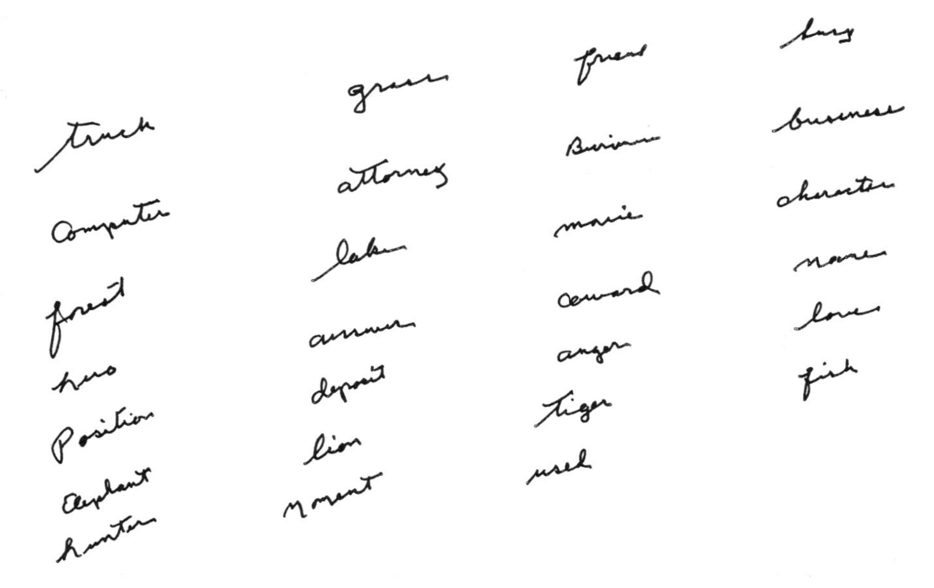

FIG. 3. Micrographia in a patient with Parkinson's disease. Letter morphology is generally well-preserved, even though letter size is diminished.

globus pallidus and the ventrolateral nucleus of the thalamus to the SMA. Operating as a functional unit, the cortical and subcortical components of this distributed neuronal network play a central role in the programming and initiation of voluntary movements and in controlling movement speed and amplitude (Alexander et al., 1986; Brooks, 1986; Goldberg, 1985). Thus, the basal ganglia-thalamocortical motor loop is a likely neural substrate for the system that generates innervatory patterns for learned skilled limb movements, including graphic innervatory patterns for writing movements. This hypothesis is supported by regional cerebral blood flow and positron emission tomography (PET) studies that demonstrated increased metabolic activity in the SMA, basal ganglia and primary sensorimotor cortex during the performance of various writing tasks (Lauritzen, Henriksen, & Lassen, 1981; Mazziotta, Phelps, & Wapenski, 1985).

The smooth and skilful execution of learned movement sequences also requires the functional integrity of the cerebellum. Similar to the basal ganglia, the cerebellum is intimately linked to cortical premotor and motor areas through re-entrant neuronal circuitry (Brooks, 1986). Cerebellar disorders lead to an inability to control movement direction, force, speed and amplitude (Gilman et al., 1981; Holmes, 1939), suggesting that the cortico-cerebellar motor loop is involved in the selection and implementation of kinematic parameters for skilled limb movements. The cerebellum also plays an important role in monitoring motor performance by comparing premotor commands for the intended movement with feedback information about the actual movement taking place (Brooks, 1986). Thus, the cerebellum is involved in error detection and in adjusting the evolving movement to changing contextual requirements.

Impaired Utilisation of Sensory Feedback: Afferent Dysgraphia

Lesions that interfere with the ability to utilise sensory feedback to control the execution of writing movements give rise to a characteristic clinical picture that has been referred to as "afferent dysgraphia" (Lebrun, 1976, 1985; Ellis, 1982, 1988; Ellis et al., 1987). Writing errors in patients with afferent dysgraphia are similar to those produced by normal controls under experimental conditions that prevent the efficient use of visual and/or kinaesthetic feedback (see above). Typical errors include repetitions and omissions that are especially likely to occur when writing sequences of similar or identical letters (e.g. double letters in words like "rubber") or strokes (e.g. letters containing repeated strokes as in "m" or "w"). In addition to the tendency to duplicate or omit letters and strokes, patients with afferent dysgraphia may also have difficulty in keeping the line of writing straight and in maintaining the proper spacing between letters and words. On occasion, letters and words may be written on top of each other (Lebrun, 1985).

Consistent with the hypothesis that patients with afferent dysgraphia are unable to use sensory feedback to monitor and correct their writing errors, Ellis et al. (1987) were able to show that, unlike normal individuals, the afferent dysgraphic patient V.B. was no more error prone in writing with her eyes closed than with her eyes open. Thus, the presence or absence of visual feedback had no appreciable effect on her writing performance. Furthermore, when V.B.'s forefinger was passively moved to form a letter while her eyes remained closed, she was very poor in using kinaesthetic feedback to identify the letter. Under these conditions, she also could not reliably distinguish correctly formed letters from incorrect ones. Taken together, these observations suggest that V.B. was unable to attend to visual and kinaesthetic feedback to control writing.

Neuroanatomical Correlates. Afferent dysgraphia is usually seen following nondominant parietal lobe lesions and is therefore frequently accompanied by prominent left-sided spatial neglect (Ellis, 1982, 1988; Ellis et al., 1987; Lebrun, 1985; Marcie & Hécaen, 1979). However, Ellis et al. (1987) provided evidence that neglect-related writing errors (e.g. leaving an unusually wide margin on the left side of the page) and feedback-related errors (e.g. letter and stroke repetitions or omissions) may be dissociable. Features of afferent dygraphia have also been noted on occasion in the written productions of patients with left parietal lobe damage (see analysis of Zangwill's original case in Margolin, 1984; Papagno, 1992) and, consistent with the suggestion of Ellis et al. (1987), in these cases feedback-related errors were observed without significant contralateral spatial neglect. Based on the available evidence, it would seem that the parietal lobes in general, and the right parietal lobe in particular, play a critical role in monitoring and integrating visual and kinesthetic sensory feedback for the proper control of writing movements.

CONCLUSION

Writing is a complex function that requires the coordinated activity of linguistic, motor, spatial, and perceptual systems for optimum performance. Considering the sheer number and diversity of the cognitive operations involved, it is perhaps not altogether surprising that writing takes a long time to master and that it remains a fragile skill highly susceptible to disruption by brain damage.

In this chapter we started out by sketching the outline of a cognitive information-processing model of normal writing. According to this model, the conversion of abstract graphemic representations into writing movements of the pen is accomplished through a series of hierarchically organised processing stages. An attempt was made to ascribe a distinct functional role to each proposed processing module, although some (e.g. graphic innervatory patterns) were assigned several related operations which may prove amenable to further fractionation. Following the traditional approach of cognitive neuropsychology, we next examined patterns of breakdown caused by focal brain lesions and sought an explanation of abnormal writing performance in terms of damage to individual functional components of the model. We distinguished four major types of "peripheral" writing impairments: disorders of allographic conversion, apraxic agraphia, nonapraxic disorders of motor execution and afferent dysgraphia. Neuroanatomical correlates of these clinical syndromes were reviewed in order to identify the neural systems involved in mediating the various aspects of writing. It was suggested that in most right-handed individuals the left hemisphere played a dominant role not only in generating the appropriate spellings, but also in activating and selecting the correct allographic representations and in programming the skilled movements of writing. By contrast, the right hemisphere is involved primarily in controlling the spatial aspects of writing and in monitoring visual and kinaesthetic feedback.

We proposed that, similar to other types of learned manual motor skills, the programming of writing movements is guided by abstract spatio-temporal movement representations stored in the parietal lobe. The information encoded in parietal graphic motor programs is translated into graphic innervatory patterns to specific muscles by frontal premotor regions that work in concert with the basal ganglia and the cerebellum in selecting the appropriate kinematic parameters and in controlling movement execution. Throughout this chapter we have seen that the motor programming of writing and limb praxis have a lot in common, both in terms of the psychological mechanisms involved and with respect to the neuroanatomical systems implicated. In fact, following left-hemisphere damage apraxic agraphia and ideomotor limb apraxia may coexist in the same patient. However, neuropsychological studies have demonstrated convincingly that the two types of motor programming disorders are dissociable, suggesting that writing and limb praxis are subserved by partially overlapping but functionally independent left-hemisphere neuronal networks.

It should be readily apparent from our discussion that the careful and detailed documentation of impaired and preserved spelling and writing abilities in neurological patients has made an invaluable contribution to our understanding of the cognitive and motor processes involved in writing. Because properly conducted neuropsychological investigations play such a critical role in advancing knowledge and in validating (or invalidating) theories of cognition, we would like to close with a few general comments regarding assessment.

It is important to bear in mind that in examining patients with agraphia the evaluation should not be limited to written production only. Oral spelling, typing, and spelling with anagram letters should also be routinely investigated in a parallel fashion in order to clarify the relationship among the various output modalities. Patients should be asked to write in different case or style and to transcode between various allographic forms (e.g. transcribe from upper to lower case or from print to script). Assessment of copying should include letters, words, nonwords, and nonlinguistic visual patterns as stimuli. The need for a systematic and theoretically motivated examination of spelling and writing errors cannot be overemphasised, since this kind of analysis plays an essential role in determining the exact nature of the functional impairment.

To date, letter imagery has only been examined in a small number of agraphic patients. Invariably, the imagery tasks used required patients to answer questions about the physical appearance of letters (e.g. whether or not the letter contained straight or curved lines). Thus, these imagery tasks investigated letter shape knowledge that is thought to be represented at the level of the allographic long-term memory store. However, we believe that it would be equally important to test mental imagery for the sequence and direction of the strokes necessary to create a given allograph (e.g. knowing that in writing the capital letter "B" one first produces the straight vertical segment and that this is followed by the addition of the curved segments at the top and at the bottom using semicircular movements of the pen executed in a clockwise direction; or that the letter "C" requires only a single semicircular stroke executed in counterclockwise direction). This type of imagery task would presumably probe knowledge represented at the level of graphic motor programs. It is conceivable that, parallel with their writing impairment, patients with apraxic agraphia have defective imagery for the sequence and direction of the strokes needed to form letters, even though general letter shape knowledge may be relatively preserved. Furthermore, it may be the case that patients with apraxic agraphia following parietal lesions that damage graphic motor programs have impaired imagery for stroke-level information, but patients with apraxic agraphia following frontal lesions have spared imagery since in these individuals graphic motor programs are believed to be intact. Clearly, the role of mental imagery in writing deserves further investigation.

As far as the motor execution of writing is concerned, it must be acknowledged that precise information regarding the spatial and temporal characteristics of writing movements is very difficult to obtain from visual inspection alone. Objective computer-assisted analysis of pen-point spatial trajectory recorded from a digitiser writing tablet would prove extremely useful in studying the kinematic attributes of writing in neurological patients. This technique has been used extensively in research with normal individuals and has provided important insights into the motor control of handwriting. Finally, we believe that newer neuroimaging modalities like PET scanning and functional magnetic resonance imaging (fMRI) offer an exciting opportunity for studying the functional neuroanatomy of writing.

REFERENCES

Alexander, G.E., DeLong, M.R., & Strick, P.L. (1986). Parallel organisation of functionally segregated circuits linking basal ganglia and cortex. *Annual Review of Neuroscience, 9,* 357–381.

Alexander, M.P., Fischer, R.S., & Friedman, R. (1992). Lesion localisation in apractic agraphia. *Archives of Neurology, 49,* 246–251.

Anderson, S.W., Damasio, A.R., & Damasio, H. (1990). Troubled letters but not numbers: Domain specific cognitive impairments following focal damage in frontal cortex. *Brain, 113,* 749–766.

Baxter, D.M. & Warrington, E.K. (1986). Ideational agraphia: A single case study. *Journal of Neurology, Neurosurgery and Psychiatry, 49,* 369–374.

Black, S.E., Behrmann, M., Bass, K., & Hacker, P. (1989). Selective writing impairment: Beyond the allographic code. *Aphasiology, 3,* 265–277.

Boldrini, P., Zanella, R., Cantagallo, A., & Basaglia, N. (1992). Partial hemispheric disconnection of traumatic origin. *Cortex, 28,* 135–143.

Brooks, V. B. (1986). *The neural basis of motor control.* New York: Oxford University Press.

Caramazza, A., Miceli, G., Villa, G. & Romani, C. (1987). The role of the graphemic buffer in spelling: Evidence from a case of acquired dysgraphia. *Cognition, 26,* 59–85.

Cipolotti, L. & Denes, G. (1989). When a patient can write but not copy: Report of a single case. *Cortex,* 25, 331–337.

Coslett, H.B., Rothi, L.J.G., Valenstein, E., & Heilman, K.M. (1986). Dissociations of writing and praxis: Two cases in point. *Brain and Language, 28,* 357–369.

Crary, M.A. & Heilman, K.M. (1988). Letter imagery deficits in a case of pure apraxic agraphia. *Brain and Language, 34,* 147–156.

Croisile, B., Laurent, B., Michel, D., & Trillet, M. (1990). Pure agraphia after deep left hemisphere hematoma. *Journal of Neurology, Neurosurgery & Psychiatry, 53,* 263–265.

De Bastiani, P. & Barry, C. (1986). After the graphemic buffer: Disorders of peripheral aspects of writing in Italian patients. Cited in K. Patterson & A.M. Wing (1989). Processes in handwriting: A case for case. *Cognitive Neuropsychology, 6,* 1–23.

De Bastiani, P. & Barry, C. (1989). A cognitive analysis of an acquired dysgraphic patient with an allographic writing disorder. *Cognitive Neuropsychology, 6,* 25–41.

Degos, J.D., Gray, F., Louarn, F., Ansquer, J.C., Poirier, J., & Barbizet, J. (1987). Posterior callosal infarction: Clinicopathological correlations. *Brain, 110,* 1155–1171.

Ellis, A. W. (1982). Spelling and writing (and reading and speaking). In A. W. Ellis (Ed.), *Normality and pathology in cognitive functions.* London: Academic Press.

Ellis, A.W. (1988). Normal writing processes and peripheral acquired dysgraphias. *Language and Cognitive Processes, 3,* 99–127.

Ellis, A.W. Young, A.W., & Flude, B.M. (1987). "Afferent dysgraphia" in a patient and in normal subjects. *Cognitive Neuropsychology,* 4, 465–486.

Exner, S. (1881). *Untersuchungen über die Localisation der Functionen in der Grosshirnrinde des Menschen.* Wien: W. Braunmüller.

Friedman, R.B. & Alexander, M.P. (1989). Written spelling agraphia. *Brain and Language, 36,* 503–517.

Gelb, I.J. (1963). *A study of writing.* Chicago: University of Chicago Press.

Gersh, F. & Damasio, A. R. (1981). Praxis and writing of the left hand may be served by different callosal pathways. *Archives of Neurology, 38,* 634–636.

Geschwind, N. (1975). The apraxias: Neural mechanisms of disorders of learned movement. *American Scientist, 63,* 188–195.

Geschwind, N. & Kaplan, E. (1962). A human cerebral deconnection syndrome: A preliminary report. *Neurology, 12,* 675–685.

Gilman, S., Bloedel, J.R. & Lechtenberg, R. (1981). *Disorders of the cerebellum.* Philadelphia: F.A. Davis.

Goldberg, G. (1985). Supplementary motor area structure and function: Review and hypotheses. *Behavioral and Brain Sciences, 8,* 567–616.

Goldenberg, G., Wimmer, A., Holzner, F., & Wessely, P. (1985). Apraxia of the left limbs in a case of callosal disconnection: the contribution of medial frontal lobe damage. *Cortex, 21,* 135–148.

Goldstein, K. (1948). *Language and language disturbance.* New York: Grune & Stratton.

Goodman, R.A. & Caramazza, A. (1986). Dissociation of spelling errors in written and oral spelling: The role of allographic conversion in writing. *Cognitive Neuropsychology, 3,* 179–206.

Gordinier, H.C. (1899). A case of brain tumor at the base of the second left frontal convolution. *American Journal of Medical Science, 117,* 526–535.

Graff-Radford, N. R., Welsh, K., & Godersky, J. (1987). Callosal apraxia. *Neurology, 37,* 100–105.

Heilbronner, K. (1906). Über isolierte apraktische Agraphie. *Münchener medizinische Wochenschrift, 53,* 1897–1901.

Heilman, K. M. (1979). Apraxia. In K. M. Heilman & E. Valenstein (Eds.), *Clinical neuropsychology.* New York: Oxford University Press.

Heilman, K.M., Coyle, J.M., Gonyea, E.F., & Geschwind, N. (1973). Apraxia and agraphia in a left-hander. *Brain, 96,* 21–28.

Heilman, K.M., Gonyea, E.F., & Geschwind, N. (1974). Apraxia and agraphia in a right-hander. *Cortex, 10,* 284–288.

Heilman, K. M., & Rothi, L. J. G. (1985). Apraxia. In K.M. Heilman & E. Valenstein (Eds.), *Clinical neuropsychology.* New York: Oxford University Press.

Henschen, S.E. (1922). *Klinische und anatomische Beitrage zur Pathologie des Gehirns,* (Vol. 7). Stockholm: Nordiska Bokhandeln.

Hodges, J.R. (1991). Pure apraxic agraphia with recovery after drainage of a left frontal cyst. *Cortex,* 27, 469–473.

Holmes, G. (1939). The cerebellum of man. *Brain, 62,* 1–30.

Kalmus, H., Fry, D.B., & Denes, P. (1960). Effects of delayed visual control on writing, drawing and tracing. *Language and Speech, 3,* 96–108.

Kapur, N. & Lawton, N.F. (1983). Dysgraphia for letters: A form of motor memory deficit. *Journal of Neurology, Neurosurgery and Psychiatry, 46*, 573–575.

Kartsounis, L.D. (1992). Selective lower-case letter ideational dygraphia. *Cortex, 28*, 145–150.

Kawamura, M., Hirayama, K., & Yamamoto, H. (1989). Different interhemispheric transfer of Kanji and Kana writing evidenced by a case with left unilateral agraphia without apraxia. *Brain, 112*, 1011–1018.

Kazui, S. & Sawada, T. (1993). Callosal apraxia without agraphia. *Annals of Neurology, 33*, 401–403.

Kinsbourne, M. & Rosenfield, D.B. (1974). Agraphia selective for written spelling: An experimental case study. *Brain and Language, 1*, 215–225.

Lambert, J., Viader, F., Eustache, F., & Morin, P. (1994). Contribution to peripheral agraphia: A case of post-allographic impairment? *Cognitive Neuropsychology, 11*, 35–55.

Lauritzen, M., Henriksen, L., & Lassen, N. A. (1981). Regional cerebral blood flow during rest and skilled hand movements by Xenon-133 inhalation and emission computerised tomography. *Journal of Cerebral Blood Flow and Metabolism, 1*, 385–389.

Lebrun, Y. (1976). Neurolinguistic models of language and speech. In H. Whitaker & H.A. Whitaker (Eds.), *Studies in neurolinguistics*, (Vol.1). New York: Academic Press.

Lebrun, Y. (1985). Disturbances of written language and associated abilities following damage to the right hemisphere. *Applied Psycholinguistics, 6*, 231–260.

Leiguarda R., Starkstein, S., & Berthier, M. (1989). Anterior callosal hemorrhage: A partial interhemispheric disconnection syndrome. *Brain, 112*, 1019–1037.

Lesser, R. (1990). Superior oral to written spelling: Evidence for separate buffers? *Cognitive Neuropsychology, 7*, 347–366.

Levine, D.N., Mani, R.B., & Calvanio, R. (1988). Pure agraphia and Gerstmann's syndrome as a visuospatial-language dissociation: An experimental case study. *Brain and Language, 35*, 172–196.

Levy, J., Nebes, R. D., & Sperry, R. W. (1971). Expressive language in the surgically separated minor hemisphere. *Cortex, 7*, 49–58.

Liepmann, H. (1920). Apraxie. *Ergebnisse der Gesamten Medizin, 1*, 516–543.

Liepmann, H., & Maas, O. (1907). Fall von Linksseitiger Agraphie und Apraxie bei rechsseitiger Lähmung. *Journal für Psychologie und Neurologie, 10*, 214–227.

Marcie, P. & Hécaen, H. (1979). Agraphia: writing disorders associated with unilateral cortical lesions. In K. M. Heilman & E. Valenstein (Eds.), *Clinical neuropsychology*. New York: Oxford University Press.

Margolin, D.I. (1980). Right hemisphere dominance for praxis and left hemisphere dominance for speech in a left-hander. *Neuropsychologia, 18*, 715–719.

Margolin, D. I. (1984). The neuropsychology of writing and spelling: Semantic, phonological, motor and perceptual processes. *Quarterly Journal of Experimental Psychology, 36A*, 459-489.

Margolin, D.I. & Binder, L. (1984). Multiple component agraphia in a patient with atypical cerebral dominance: An error analysis. *Brain and Language, 22*, 26–40.

Margolin, D.I. & Wing, A.M. (1983). Agraphia and micrographia: Clinical manifestations of motor programming and performance disorders. *Acta Psychologica, 54*, 263–283.

Marsden, C. D. (1982). The mysterious motor function of the basal ganglia: The Robert Wartenberg Lecture. *Neurology, 32*, 514–539.

Mazziotta, J. C., Phelps, M. E., & Wapenski, J. A. (1985). Human cerebral motor system metabolic responses in health and disease. *Journal of Cerebral Blood Flow and Metabolism*, Supplement 1, S213–S214.

McLennan, J.E., Nakano, K., Tyler, H.R., & Schwab, R.S. (1972). Micrographia in Parkinson's disease. *Journal of the Neurological Sciences, 15*, 141–152.

Merton, P.A. (1972). How we control the contraction of our muscles. *Scientific American, 226*, 30–37.

Morton, J. (1980). The logogen model and orthographic structure. In U. Frith (Ed.), *Cognitive processes in spelling,* London: Academic Press.

Nielsen, J. M. (1946). *Agnosia, apraxia, aphasia: Their value in cerebral localisation.* New York: Paul B. Hoeber.

Ogle, J.W. (1867). Aphasia and agraphia. *Report of the Medical Research Council of Saint-George's Hospital (London), 2,* 83–122.

Pandya, D.N. & Seltzer, B. (1986). The topography of commissural fibres. In F. Leporé, Ptito, M., & Jasper, H.H. (Eds.), *Two hemispheres—One brain: Functions of the corpus callosum.* New York: Alan R. Liss.

Papagno, C. (1992). A case of peripheral dysgraphia. *Cognitive Neuropsychology, 9,* 259–270.

Patterson, K. & Wing, A.M. (1989). Processes in handwriting: A case for case. *Cognitive Neuropsychology, 6,* 1–23.

Pitres, A. (1884). Considerations sur l'agraphie. *Revue de Medicine, 4,* 855–873.

Putnam, J.J. (1989). The search for modern humans. *National Geographic, 174,* 438–477.

Rapcsak, S.Z., Beeson, P.M., & Rubens, A.B. (1991). Writing with the right hemisphere. *Brain and Language, 41,* 510–530.

Rapcsak, S.Z., Rothi, L.J.G., & Heilman, K.M. (1987). Apraxia in a patient with atypical cerebral dominance. *Brain and Cognition, 6,* 450–463.

Risse, G.L., Gates, J. Lund, G. Maxwell, R., & Rubens, A.B. (1989). Interhemispheric transfer in patients with incomplete section of the corpus callosum: Anatomical verification with magnetic resonance imaging. *Archives of Neurology, 46,* 437–443.

Roeltgen, D.P. & Heilman, K.M. (1983). Apractic agraphia in a patient with normal praxis. *Brain and Language, 18,* 35–46.

Rosa, A., Demiati, M., Cartz, L., & Mizon, J.P. (1991). Marchiafava-Bignami disease, syndrome of interhemispheric disconnection, and right-handed agraphia in a left-hander. *Archives of Neurology, 48,* 986–988.

Rothi, L.J. & Heilman, K.M. (1981). Alexia and agraphia with spared spelling and letter recognition abilities. *Brain and Language, 12,* 1–13.

Rothi, L.J.G. & Heilman, K.M. (1993). Apraxia. In K.M. Heilman & E. Valenstein (Eds.), *Clinical neuropsychology.* New York: Oxford University Press.

Rothi, L. J. G., Ochipa, C., & Heilman, K. M. (1991). A cognitive neuropsychological model of praxis. *Cognitive Neuropsychology, 8,* 443–458.

Rubens, A.B. (1975). Aphasia with infarction in the territory of the anterior cerebral artery. *Cortex, 11,* 239–250.

Rubens, A.B., Geschwind, N., Mahowald, M.W., & Mastri, A. (1977). Posttraumatic cerebral hemispheric disconnection syndrome. *Archives of Neurology, 34,* 750–755.

Sassoon, R., Nimmo-Smith, I., & Wing, A.M. (1986). An analysis of children's penholds. In H.S.R. Kao, G.P. Van Galen, & R. Hoosain (Eds.), *Graphonomics: Contemporary research in handwriting.* Amsterdam: North Holland.

Smith, W.M., McCrary, J.W., & Smith, K.U. (1960). Delayed visual feedback and behavior. *Science, 132,* 1013–1014.

Smyth, M.M. & Silvers, G. (1987). Functions of vision in the control of handwriting. *Acta Psychologica, 65,* 47–64.

Sugishita, M., Toyokura, Y., Yoshioka, M., & Yamada, R. (1980). Unilateral agraphia after section of the posterior half of the truncus of the corpus callosum. *Brain and Language, 9,* 215–225.

Sweet, W. H. (1941). Seeping intracranial aneurysm simulating neoplasm. *Archives of Neurology and Psychiatry,* 45, 86–104.

Tanaka, Y., Iwasa, H., & Obayashi, T. (1990). Right hand agraphia and left hand apraxia following callosal damage in a right-hander. *Cortex, 26,* 665–671.

Valenstein, E. & Heilman, K.M. (1979). Apraxic agraphia with neglect-induced paragraphia. *Archives of Neurology, 36,* 506–508.

Van Bergeijk, W.A. & David, E.D. (1959). Delayed handwriting. *Perceptual and Motor Skills, 9,* 347–357.

Van Galen, G.P. (1980). Handwriting and drawing: A two stage model of complex motor behavior. In G.E. Stelmach & J. Requin (Eds.), *Tutorials in motor behavior.* Amsterdam: North Holland.

Van Galen, G.P. (1991). Handwriting: Issues for a psychomotor theory. *Human Movement Science, 10,* 165–192.

Viviani, P. & Terzuolo, C. (1980). Space-time invariance in learned motor skills. In G.E. Stelmach & J. Requin (Eds.), *Tutorials in motor behavior.* Amsterdam: North Holland.

Volpe, B. T., Sidtis, J. J., Holtzman, J. D., Wilson, D. H., & Gazzaniga, M. S. (1982). Cortical mechanisms involved in praxis: Observations following partial and complete section of the corpus callosum in man. *Neurology, 32,* 645–650.

Watson, R.T., Fleet, W.S., Rothi, L.J.G., & Heilman, K.M. (1986). Apraxia and the supplementary motor area. *Archives of Neurology, 43,* 787–792.

Watson, R. T., & Heilman, K. M. (1983). Callosal apraxia. *Brain, 106,* 391–403.

Wright, C.E. (1990). Generalised motor programs: Reevaluating claims of effector independence in writing. In M. Jeannerod (Ed.), *Attention and Performance,* (Vol. XXIII). Hillsdale: Lawrence Erlbaum Associates Inc.

Yamadori, A., Nagashima, T., & Tamaki, N. (1983). Ideogram writing in a disconnection syndrome. *Brain and Language, 19,* 346–356.

Yamadori, A., Osumi, Y., Ikeda, H., & Kanazawa, Y. (1980). Left unilateral agraphia and tactile anomia: Disturbances seen after occlusion of the anterior cerebral artery. *Archives of Neurology, 37,* 88–91.

Yamadori, A., Osumi, Y., Imamura, T., & Mitani, Y. (1988). Persistent left unilateral apraxia and a disconnection theory. *Behavioral Neurology, 1,* 11–22.

Zangwill, O.L. (1954). Agraphia due to a left parietal glioma in a left-handed man. *Brain, 77,* 510–520.

Zesiger, P. & Mayer, E. (1992). Left unilateral dysgraphia in a patient with an atypical pattern of handedness: A cognitive analysis. *Aphasiology, 6,* 293–307.

11 Apraxia of Speech: Another Form of Praxis Disruption

Paula A. Square, Eric A. Roy, and Ruth E. Martin

INTRODUCTION

This chapter is concerned with motor speech disturbances arising from left-hemisphere brain damage. Aphasia, the language disorder due to acquired dominant hemisphere brain damage, is not the focus. Speech and language, although interdependent for verbal expression, are different behaviours. Speech is a sensorimotor behaviour consisting of the evocation of movement formulae and the translation and modification of movement formulae into innervatory patterns and finally the innvervation of muscles of the vocal tract from diaphragm to lips for the purposes of communication. Language is the selection and manipulation of representational symbols, such as words, graphemes and phonemes, and the application of rule systems which govern the arrangement and connection of the representational symbols. Internally formulated language is expressed via motoric processes such as speech, writing, or signing. Thus, verbal expression, like limb gesture, is comprised of both conceptual and production processes (Rothi, Ochipa, & Heilman, 1991; Roy & Square, 1985a). The ability to manipulate aspects of language, including sentence structure and word usage among others, and the internal evocation of phonemic and graphemic representations, comprise the conceptual or linguistic level of verbal expression.

This chapter will focus only on the manner in which aspects of speech production processes are altered as a result of damage to the left hemisphere. That is, apraxia of speech, or the disorder arising from the disruption of the processes of evocation, translation and/or possibly modification of motor programs, is the focus. The dysarthrias, or the motor speech disorders arising from muscle weakness, aberrations of tone or peripheral biomechanical incoordination, will not be discussed although these may also arise from central damage.

The discussion which follows provides evidence which indicates that language and sensorimotor speech processes may be independently disrupted as a result of left-hemisphere brain damage. This dissociation of processes demonstrates the separateness of apraxia of speech from aphasia in the same way as dissociation studies of limb apraxia and aphasia have shown the separateness of those disorders (cf. Square-Storer, Roy, & Hogg, 1990). This is followed by a general discussion of the neurophysiology of motor control. Preliminary evidence is then provided which demonstrates that several variations of motor speech implementation difficulties may arise from left-hemisphere damage. Hypotheses regarding the nature and variations of motor speech disorders subsequent to left-hemisphere brain damage are then put forth. The first of these hypotheses is that apraxia of speech may arise from parietal lobe lesions. The second is that the motor speech disruption arising from frontal cortical-subcortical lesions is most likely a combination of apraxia and dysarthria, especially if the lesion extends subcortically. The third is that Broca's area per se is just one of many brain regions which play a role in the regulation of left hemisphere motor speech control. The final sections of this chapter review the similarities between speech, buccofacial, and limb apraxia with regard to lesion sites associated with the disorders and resulting behaviours as measured both behaviourly and physiologically. The chapter concludes with a brief review of motor speech treatments for apraxia of speech. These converging lines of evidence lend credance to the argument that "apraxia of speech" is a motor disorder which indeed is another form of praxis disruption.

Definition

The term "apraxia of speech" (AOS) came into popular usage due to the influence of Frederic Darley (1968) and his colleagues at Mayo Clinic. Analogous terms for the disorder which appear in the contemporary literature are verbal apraxia, speech apraxia, phonetic disintegration and aphemia. The Mayo group also provided the most complete and most frequently accepted contemporary description of AOS. It was described as an articulatory-prosodic speech disorder "... due to an impaired ability to program the positioning of the speech musculature ... and the sequencing of muscle movements ..." (Darley, Aronson, & Brown, 1975a). This definition included two key elements of volitional motor behaviour typically described as being governed by the left hemisphere for limb, oral nonverbal, and speech movements: posture generation and seriation of gestures (cf. Kimura, 1982; Square, 1995).

Disruptions to these two processes, although almost always co-occurring in individuals diagnosed with apraxia, may indeed have their basis in separate and distinct motor cognitive processes. An impaired ability to produce postures has been interpreted by some as resulting from a degraded internalised schemata for the three-dimensional space in which the posture is to be performed (MacNeilage, 1970). In the case of speech and nonverbal oral actions, the three-dimensional spatial coordinate system would be defined as the physical boundaries of the vocal tract. Movements within this intrapersonal space, of course, are not guided by vision as are limb movements. Thus, a proprioceptive memory and/or intact processes for accessing memory of the spatial coordinate system are essential for the production of accurate volitional posturing of oral verbal and nonverbal movements. Selection of the appropriate amplitude and timing of muscle contractions to achieve a volitional posture within this space (e.g. speech sound such as "o" or lip rounding) could be disturbed, then, because of a degraded internalised schemata of the spatial coordinate system or inability to access the spatial coordinate system.

The ability to seriate muscle contractions to achieve accurate continuous action patterns is another key element of praxis. Disability in the realm of sequencing or seriation of muscle contractions leads to inaccurate control of skilled action sequences but not to inaccurate or uncoordinated automatic vegetative actions (e.g. swallowing).[1] Rote speech such as counting from one to five would also be relatively unimpaired. It has been hypothesised that a degraded memory for actions (Rochon, Caplan, & Waters, 1990; Roy, Brown, Winchester, Square, Hall, & Black, 1993) and/or a disruption of internalised oscillatory (rhythmical) mechanisms (Grillner, 1982; Kent, 1984; Martin, 1972; Square, 1994; Thelan, 1981) may underlie disruptions of the seriation of muscle contractions for propositional speech.

That the disorder of AOS is due to brain damage sustained to the dominant, usually left, hemisphere has never been denied. Darley (1977) believed, like his predecessors who also described a disorder of articulated speech due to left-hemisphere damage (Broca, 1861; Wernicke, 1885), that the speech disruption likely was due to damage to Broca's area and its environs. AOS was described as a disorder distinct from aphasia but one which often co-occurred with aphasia.

AOS was described by the Mayo group as also differing from the dysarthrias (Darley et al., 1975a; Darley, Aronson, & Brown, 1975b). Localisation of brain damage differed as well as characteristics of the function of the oromotor musculature when engaged in both speech and nonspeech tasks. Examination of the speech musculature in patients with AOS reveals relatively normal muscle tone, strength and/or range of movement; when coexisting impairments in these parameters exist, they are so mild that they cannot account for the existing

---

[1]The traditional definition of apraxia focuses on dissolution of skilled motor acts with preservation of automatic and vegetative motor functions. Readers should be aware, however, that Tuch and Nielsen (1941) wrote of "apraxia of swallowing."

speech impairment. Impairments of tone, strength and range of muscle movement, on the other hand, are hallmark features of the dysarthrias. Furthermore, abnormal oral reflexes are absent in AOS but in the one dysarthria also ascribed to cortical damage, pseudobulbar or spastic dysarthria, abnormal oral reflexes are a hallmark feature (Darley et al., 1975a).

The most striking difference between AOS and each of the dysarthrias, however, is the resulting speech pattern (Darley et al., 1975a, 1975b). Specifically, speech errors due to AOS are described as being remarkably variable with regard to both loci and type. That is, patients are described as sometimes articulating speech targets correctly and other times incorrectly (Johns & Darley, 1970) and, when errors are committed in the same word, the locus and the nature of the error is likely to change over repeated attempts (see Mlcoch, Darley, & Noll, 1982). The frequency of the inconsistent speech errors is thought to relate in part to the nature of the speech tasks in which the patient is engaged. While disrupted speech is less evident in automatic speech tasks (e.g. counting), repetition of longer and more phonetically complex speech productions as well as self-formulated utterances are disrupted with regard to initiation, kinetic melody, and articulatory accuracy.

Apractic speech is described as being marked not only by inconsistent articulation errors but also by slow rate and prosodic deviations or abnormalities of rhythm and the stress patterns of speech. The nature of the prosodic abnormalities continues to be debated but Darley and colleagues (1975a) felt that they were "... possibly in compensation ..." for anticipated articulatory errors.

The perceptual characteristics of disordered speech embraced under the umbrella of "apraxia of speech" as defined by the Mayo group thus included: articulatory errors perceived to be predominantly substitutions but also including distortions, sound omissions, and additions; trial and error groping both preceding speech initiation and within utterances; dysprosodic speech marked by slow rate and/or abnormal excess and equal stress; islands of error-free speech; more difficulty on longer and phonetically more complex utterances and a remarkable awareness of speech errors accompained by a robust effort to self-correct. (cf. Darley et al., 1975a; Square-Storer & Roy, 1989; Wertz, LaPointe, & Rosenbek, 1984).

Kent and Rosenbek (1983) offered a refined definition of AOS based upon acoustic analysis of the speech of seven patients diagnosed as having AOS and minimal aphasia; each also had either transient or chronic hemiplegia. They described AOS as a central disruption of motor speech control in which "... errors in sequencing, timing, coordination, initiation and vocal tract shaping ..." (p. 245) occur.

Contemporary Controversy

Debates regarding the nature of speech errors which sometime accompany aphasia have raged in the neurological literature since the 1860s (Broca, 1861;

Dejerine, 1901; Marie, 1906; Wernicke, 1874; 1885). The controversy has focused historically on the question of whether the speech errors observed in aphasic patients are an inherent part of impaired language (aphasia) or the result of damage to the speech motor control system. Stated differently, the debate has centred on whether the speech errors in aphasia are due to dysfunction of central formulation, the latter encompassing disrupted linguistic processing, or are the result of defective sensorimotor implemetation of utterances.

Debates regarding the nature of the disorder accelerated in the 1970s with the two diametrically opposed philosophical camps arguing even more forcefully, using data derived from neuropsychological and neurolinguistic investigations. Those who advocated that the speech errors were linguistic in nature presented three lines of evidence. First, the speech errors in aphasia which were perceived to be sound substitutions, could be shown to be systematic and consistent with linguistic theory which, in turn, argued for a breakdown at the level of phonological knowledge (Blumstein, 1973a; 1973b; Fry, 1959; Klich, Ireland, & Weidner, 1979). Second, a series of studies demonstrated an interaction between the occurrence of speech errors and increasingly complex language demands (Dunlop & Marquardt, 1977; Farmer, O'Connell, & Jesudowich, 1979; Hardison, Marquardt, & Peterson, 1977; Martin, 1974; Martin & Rigrodsky, 1974a; 1974b; Sasanuma & Fujimura, 1971, 1972), thereby leading to the explanation that the speech breakdown had its basis in language demands which exceeded the linguistic processing abilities of aphasic patients. Third, aphasic patients who demonstrated articulatory errors also committed graphemic and phonemic errors in writing and reading, respectively; this finding was interpreted as an indication that phonological impairments were the common source of all three types of errors (Hécaen & Consoli, 1973).

The conclusions derived from most of these studies, however, appear to have been fundamentally flawed due to the types of individuals studied: aphasics with speech errors. Rarely were the performances of control groups consisting of aphasic individuals without speech errors considered. More importantly, left-hemisphere lesioned individuals presenting only with a prosodic-articulatory disorder and without aphasia were never studied. Thus, the conclusions drawn from these studies are highly suspect in that the confounding factor of aphasia, per se, could have accounted for the individuals' deviant linguistic parameters.

During the same decade, a significant number of studies were undertaken, the results of which were interpreted as evidence that the articulatory-prosodic speech errors made by aphasic patients were motoric in nature. Shankweiler and Harris (1966), Johns and Darley (1970), and Trost and Canter (1974) were the foremost advocates of a motor explanation. Variability in speech production was highlighted by Johns and Darley (1970) as an indication of impairment to motoric regulation of speech, especially at the level of motor programming, thus substantiating their choice of the label for the disorder, apraxia of speech. Others suggested that at least some of the speech errors were due to motoric simplification processes (Marquardt, Reinhart, & Peterson, 1979; Shankweiler & Harris, 1966), but other errors were motorically more complex with all types of errors being highly variable (Shankweiler & Harris, 1966). Others pointed to a disability for coordinating the spatial and temporal paramenters of speech movements (Freeman, Sands, & Harris, 1978; Itoh, Sasanuma, Hirose, Yoshioka, & Ushijima, 1978; Itoh, Sasanuma, & Ushijima, 1979 Jaffee & Duffy, 1978; Sands, Freeman, & Harris, 1978; Shankweiler, Harris, & Taylor, 1968) with many of these investigators preferring the term "apraxia of speech." Finally, some researchers advocated that the articulatory-prosodic speech disorder could occur independent of aphasia. In such cases of dissociation, the lesion site was generally shown to be subcortical, occurring deep to Broca's area and thus interrupting the final outflow of motor impulses. The term "aphemia" was the preferred one for the speech disorder (Goodglass & Kaplan, 1972; Schiff, Alexander, Naeser, & Galaburda, 1983). When, however, the articulatory-prosodic speech disorder occurred in association with a nonfluent aphasic syndrome (Broca's, global and conduction aphasia), it was thought by this latter group to be unquestionably caused by the "aphasia" itself (e.g. Goodglass & Kaplan, 1972). This concept continues to be held to this day by several eminent aphasiologists and neurolinguists (Bellugi, 1980; Goodglass, 1993).

A number of lines of investigation argued against the disorder being based in linguistic (aphasia) impairment. One line of evidence was the relative preservation of speech perception in patients with "pure" AOS (Aten, Johns, & Darley, 1971; Deal & Darley, 1972; Lozano & Dreyer, 1978; Square-Storer, Darley, & Sommers, 1988) indicating that psycholinguistic processing at this level was intact in patients with "pure" apraxia of speech (Square-Storer et al., 1988) and in patients with predominately apraxia of speech with coexisting mild aphasia (Aten et al., 1971; Deal & Darley, 1972; Lozano & Dreyer, 1978). Such processing, however, was deviant among aphasic patients (Aten et al., 1971) both with and without coexisting motor speech impairment (Square-Storer et al., 1988). Another line of evidence arguing for the separate nature of the speech and language impairments resided in the evidence which indicated that a dissociation between motor control and representational content subsequent to left-hemisphere brain damage existed. Kimura and Archibald (1974) concluded that the speech disorders associated with aphasia designated by some as "apraxic" may, indeed, be solely motoric in nature.

Contemporary controversy and confusion regarding the nature of the articulatory-prosodic speech errors has abated over the last decade. First, in-depth studies of patients with left-hemisphere brain damage resulting in the articulatory-prosodic speech disorder but not aphasia have appeared in the literature (e.g. Square-Storer, et al., 1988; cf. Square-Storer et al., 1990). Second, there have been major advances in our knowledge of the neurophysiological control of motor behaviour which have implicated the probable role of numerous brain regions in oromotor behaviour (cf. Square & Martin, 1994). In addition, there have been a number of studies which have demonstrated that some patients with left-hemisphere brain damage demonstrate significant abnormalities in the motor control of the oral facial musculature when it is engaged in both nonverbal and speech behaviour, especially in aphasia of the nonfluent type (Broca's and global) and sometimes, also, in conduction aphasia (cf. McNeil & Kent, 1990; Square-Storer, Qualizza, & Roy, 1989). Finally, a growing number of studies have demonstrated that an inherent part of the speech disorder which frequently accompanies aphasia is the dissolution of the ability to coordinate temporally the actions of the various speech articulators (cf. McNeil & Kent, 1990; Square-Storer, 1987). The dysregulation of temporal control of multiple muscle contractions throughout the vocal tract is one likely cause of many of the articulatory and prosodic deviancies (Freeman, Sands, & Harris, 1978; Hardcastle, Morgan, & Clarke, 1985; Itoh et al., 1978; 1979; Washino, Kasai, Uchida, & Takeda, 1981).

Apraxia of Speech and Phonemic Paraphasia

The distinction between phonemic paraphasia and apraxia of speech has long been a source of confusion. Awareness of speech errors and the attempt to self-correct are clinical characteristics which differentiate the two disorders. Because of the lack of awareness of speech errors, individuals with phonemic paraphasia do not self-correct. This lack of attempt to self-correct results in fluent speech output which is not marked by reattempts of utterances.

We can only speculate as to the cognitive and perceptual processing differences which may underlie phonemic paraphasia and apraxia of speech. In the traditional view of phonemic paraphasia, a degraded ability to recall lexical items and their phonological (speech sound) structures were seen as the underlying disabilities. In more posterior lesioned aphasic individuals, there is also a reduced ability to self-monitor speech output using auditory feedback and, thus, distorted speech output is emitted fluently. In apraxia of speech with no or minimal aphasia, on the other hand, internalised lexical and phonological templates are evoked but the choice of muscles to be contracted (spatial) and the order of muscle contraction (temporal) is incorrectly programmed, i.e. the space–time formulae for muscle contractions for utterances are incorrect. The apractic speaker recognises his speech errors due to preserved abilities to monitor his output via the auditory modality. This, in turn, results in attempts to correct the motor action but, because of degraded internalised spatial and temporal schemata for movements complicated perhaps by reduced oral sensory feedback, attempts at correction are variably successful and speech is marked by numerous reattempts.

The literature from the last two decades has abounded with controversy regarding the linguistic reality of the differences between phonemic paraphasia and apractic speech errors. Using broad phonetic transcription as the basis of notation for phoneme errors, speech sound errors appeared remarkably similar for patients in the two diagnostic categories (Blumstein, 1973a, 1973b). As pointed out by Buckingham and Yule (1987), however, the use of this system of broad notation has resulted in false evaluation of the speech errors produced by left-hemisphere damaged patients and an overstatement of the differences between phonetic (motor) and phonemic (rules for sound structure of words) errors. Indeed, it has been put forward by these investigators that some phonemic paraphasic errors may be the result of subtle motor control deficits and the crude method of broad phonetic transcription has led to a false interpretation that all speech errors in fluent aphasia are the result of deviant linguistic encoding at the level of the rules of phonology. Buckingham (1992) stresses, however, the cognitive basis of most phonemic paraphasias.

Roy and Square (1985a) and Square-Storer and Roy (1989) believe that apraxia in all modalities may have either a cognitive and/or production basis. Thus, our current position regarding phonemic paraphasia is that these speech errors are likely to have their basis in both cognitive–linguistic and motor processes with the former predominating. In what has been traditionally called apraxia of speech, the programming of muscle contractions for posture and action production is predominately deviant. Nonetheless, the occurrence of AOS in the absence of aphasia is rare; thus, aphasic individuals with prominent AOS likely have a predominant motor control deficit for speech coupled with a less prominent phonological mapping deficit (see Square-Storer, Roy, & Hogg, 1990 for a more complete discussion).

THE PURE FORM OF APRAXIA OF SPEECH

The pure disorder of articulated speech due to left-hemisphere lesion in the absence of aphasia has been studied by several investigators (Adams, McNeil, & Weismer, 1989; McNeil & Adams, 1991; McNeil, Caliguiri, & Rosenbek, 1989; McNeil, Liss, Tseng, & Kent, 1990; McNeil, Weismer, Adams, & Mulligan, 1990, Nebes, 1975; Square, 1981; Square-Storer & Apeldoorn, 1991; Square-Storer et al., 1988). The patients described as having "pure" apraxia of speech did not demonstrate coexisting aphasia as measured using standardized tests for the disorder. The existence of patients with aphemia or apraxia of speech occurring without aphasia provides confirming evidence that the disorders are independent. Nonetheless in patients with AOS, as in patients with limb apraxia, aphasia almost always co-occurs. That is, few patients present with only apraxia and no aphasia. One possible interpretation of the dissociation of left-hemisphere non-dysarthric disorders of speech (apraxia of speech) and language impairment (aphasia) is that different systems may mediate these two functions. What has received little discussion in the literature until recently, however, are the neurophysiological mechanisms of speech motor control overseen by the dominant hemisphere. The purpose of the next section is to review this information.

NEUROPHYSIOLOGICAL BASIS OF LEFT-HEMISPHERE SPEECH MOTOR CONTROL

A comprehensive description of the hypothesised left-hemisphere neural correlates of motor speech production has been recently presented by Square and Martin (1994). Viewing speech production as the product of multiple functional processes (Gracco, 1991), Square and Martin (1994) attempted to describe the component processes regulating oromotor nonspeech and speech control based upon the evidence from recent neurophysiological studies of oral motor control in both humans and nonhuman primates. The left-hemisphere cortical motor areas proposed as being of prime importance to orofacial motor control and possibly motor speech control include: the primary motor cortex or area 4; premotor cortex including area 44 and lateral area 6; the supplementary motor area or mesial area 6 of the premotor cortex; and the parietal lobe. Other areas highlighted as being significant to the control of speech production are the motor tracts deep to primary and premotor areas; the basal ganglia; the cerebellum; internal capsule; cingulate cortex; and the insula. Finally, the substantial influence of the basal ganglia and cerebellum on frontal lobe regulation of motor behaviour and the role of the thalamus as a relaying mechanism for input to and from the parietal lobe and the basal ganglia to the frontal lobe cannot be excluded in any discussion of speech motor control processes subserved by the dominant hemisphere.

Table 1 summarises the information presented by Square and Martin (1994) with minor modifications. Information is included with regard to both left-hemisphere cortical and subcortical functional regions which may be involved in the control of speech production, the speech symptoms which have been reported to result from lesions to each region, and the probable diagnostic speech classification into which lesioned patients might fall. A brief discussion of the functional contribution of each cortical region follows. Although these cortical regions are discussed separately, the concepts of multiple functional linkages and dynamic parallel processing between the cortical structures and subcortical motor centers and cerebellum must be highlighted. That is, there is a dynamic relaying of information between structures; damage to any one component will likely result in a predictable motor control impairment but the resulting speech patterns may also be affected by hypometabolism of remote motor regions (Metter, Hanson, Kempler, Jackson, Mazziota, & Phelps, 1987a; Metter, Kempler, Jackson, Hanson, Riege, Camras, Mazziota, & Phelps, 1987b).

The Primary Motor Area (Area 4)

It has been suggested that a role of the primary motor cortex is to fine-tune movement synergies preprogrammed at subcortical levels through the utilisation of afference. It has also been suggested that sensory input may underlie the "time critical" processes of orofacial motor execution (Gracco & Abbs, 1987). Lesions

TABLE 1. Functional Effects on Motor Speech Resulting from Damage to Various Sites in the Dominant Hemisphere

Lesion Site	*Suggested Function*	*Analogous Existing Speech Classification*	*Speech Symptoms*	*Perceived Quality*
Frontal Cortical Lesion Primary motor cortex (area 4)	Movement execution Activation of distal musculature for fractional movements	Pseudobulbar palsy Upper motoneuron dysarthria	Slow, effortful speech Impaired spatial targeting within vocal tract	Dysarthric Spastic (?) Paretic
Nonprimary motor cortex (Premotor area (areas 6, 44)	Organisation of motor output in relation to sensory information. Programming of complex, multiple movement sequences "Kinetic melody". Preparation for movement	Apraxia of speech, aphemia "Small" Broca's asphasia Anterior operculum syndrome (with bilateral damage)	Slow, effortful speech Articulatory struggle and groping Articulatory distortions Dysprosody	Apractic Dysarthric
Supplementary motor area (area 6)	Movement initiation Scaling of motor output Motor sequencing Organisation and preparation for internally generated motor behaviours	Akinetic mutism Proxysmal speech	Initiation difficulty Dysprosody Akinesia	Dysfluent Dysarthric Dysphonic Initially mute
Frontal Subcortical Lesion Basal ganglia	Facilitation or inhibition of cortically initiated movements Modulation of frontal lobe functions. Role in preparation for movement	Dysarthria, hyperkinetic	Articulatory inaccuracy Prosodic excess Prosodic insufficiency Reduced volume	Dysarthric sometimes "Dystonic"

White matter underlying Broca's area	Region of afferent and efferent projections with Broca's area.	Dysarthria Aphemia	Slow, effortful speech Impaired articulation	Dysarthric Apraxic
Cingulate cortex	Role in engaging neocortex for ideational or propositional speech Role in learned responses to emotional states	Possible mutism		
Insula	Premotor association areas for orofacial behaviours	Apraxia of speech		Apraxia of speech
Internal capsule	Carries ascending and descending projections from motor cortex	Dysarthria	Slow speech Impaired articulation	Dysarthric
Parietal Cortical Lesion				
Midparietal lobe	Complex motor sequencing	"Apraxia of speech"	Repetitions, reattempts Articulation errors No slowness; word, syllables, and vocalic nuclei of normal duration	"Apraxic"

Note: From Square, P.A. & Martin, R.E. (1994). Neuromotor speech impairment in aphasia. In R. Chapey (Ed.), *Language intervention strategies in adult aphasia* (Vol. 3). Baltimore MD: Williams and Wilkins.

to area 4 disrupt the execution of fractionated movements of the distal musculature (Freund, 1987) and the use of sets of muscles for executing certain movements (Evarts, 1986). The primary motor cortex receives diverse input from both cortical and subcortical regions (see Square & Martin, 1994).

Unilateral damage to the orofacial region of primary motor cortex is traditionally thought to result in a mild transient weakness of the contralateral tongue, lips, and lower face and transient imprecision of speech articulation (Abbs & Welt, 1985; Darley et al., 1975a). Recently, cases of persisting unilateral upper motor neuron dysarthria (UUMN) have been reported by Hartmann and Abbs (1992) and Duffy and Folger (1986). It may be that UUMN results from the dysregualtion of the control of fractionated oral movements.

Premotor Area (Brodmann's Area 6)

The premotor lateral convexity area 6 is one region which comprises the association motor cortex. It has been suggested by Wise (1985) that the role of this area is to use sensory inputs to organise and guide motor behaviour. Freund (1987) reported that humans with lesions to this area demonstrated a loss of kinetic melody, dissolution of context for composite movements and disintegration of complex skilled movements. The dissolution of learned skilled purposeful movements is consistent with our traditional definition of apraxia (Liepmann, 1977; Rothi, Ochipa, & Heilman, 1991).

The function of the premotor area has been described somewhat differently by Goldberg (1985). He characterised this area as a "protomotor" area which has derived phylogenetically from insular cortex and which functions to specify movements at a relatively abstract level. Movement specification at an abstract level appears to be a concept consistent with Liepmann's concept of internalised "movement formulae" or the time–space forms of movements. Unlike Goldberg (1985), Liepmann attributed the arousal of time–space forms of movement to the parietal lobe. Nonetheless, Liepmann's movement formulae may be conceptualised as analogous to motor programs. Motor programs may be defined as the internally generated spatio-temporal specifications for skilled movements. While Goldberg (1985) seems to suggest that lateral area 6 may be active in the abstract specification of temporal and spatial aspects of skilled movement, Liepmann (1908) ascribed the function of translation of movement formulae into innervatory patterns to this frontal region.

Lateral area 6 receives input from cortical sensory and association areas, especially posterior parietal association cortex and secondary somatosensory cortex (Godschalk, Lemon, Kuypers, & Ronday, 1984) and from visual and auditory areas (Wise, 1985). Extensive cerebellar input is also received. Schell and Strick (1984) have suggested that cerebellar input is so great that the cerebellum drives lateral area 6.

To our knowledge, there are no studies of the effects on speech from isolated damage to lateral area 6. More extensive damage to the inferior region of area 4 and subadjacent white matter with extension into the Rolandic operculum and insula (area 6) but sparing Broca's area (area 44) and underlying white matter results generally in a speech deficit without accompanying aphasia (Alexander, Benson, & Stuss, 1989; Baum, Blumstein, Naeser, & Palumbo, 1990; Damasio, 1991; Lecours, & Lhermitte, 1976; Levine, & Mohr, 1979; Pellat, Gentil, Lyard, Vila, Tarel, Moreau, & Benabio, 1991; Schiff et al., 1983; Tonkonogy & Goodglass, 1981). A very similar speech deficit has been reported to occur following a subcortical lesion of the anterior limb of the internal capsule between the head of the caudate and the putamen (Schiff et al., 1983). The speech disorder is referred to by some as aphemia (Pellat et al., 1991; Schiff et al., 1983), by others as phonetic disintegration (Lecours & Lhermitte, 1976), and by yet others as apraxia of speech (Darley, 1968; Darley et al., 1975a). It is characterised by a slow rate, disrupted prosody, and distorted and effortful articulation.

Pars Opercularis or Brodmann's Area 44

Brodmann's area 44 which is part of the frontal operculum is also considered as motor association cortex. Relatively little is known about its inputs and outputs. Abbs and Welt (1985) described area 44 as receiving projections from the posterior parietal and temporal cortical regions with output to regions adjacent to the cranial motor nuclei.

The role of area 44 in the control of motor speech is far less understood than traditionally believed. Although area 44 has historically been acknowledged as the centre of articulated speech, there are numerous lines of evidence which contradict this long-held notion. Broca (1861) described the deficit of articulated speech due to left-hemisphere damage as a disorder in which "memory for the motor images of words" was lost. He believed that this disorder of the "faculty to articulate words" was due to damage to the third frontal convolution which later became known as Broca's area. As the field of aphasiology evolved, the syndrome of Broca's aphasia was described, leading to great confusion regarding the motor versus the linguistic role of the third frontal convolution in the control of verbal expression. Broca's aphasia is marked by agrammatism, the predominant use of substantive words, verbal stereotypies and effortful production. Because Broca's aphasia, a linguistic disorder, progressively became attributed to Broca's area but was usually accompanied by aphemia, i.e. apraxia of speech, a motor speech disorder, the distinction between an impairment of articulated motor speech processes and impairments in the manipulation of syntax and control of semantic aspects of language became blurred and confused.

The work of Mohr and colleagues (Mohr, Pessin, Finkelstein, Funkenstein, Duncan, & Davis, 1978) had some effect on reinvoking the distinction between deficits of articulated speech due to left-hemisphere damage and deficits of language. Using early imaging techniques, autopsy, and clinical records, this group determined that the disorder of apraxia of speech resulted from circumscribed lesions to area 44, Broca'a area, while the syndrome of Broca's aphasia resulted from larger lesions encompassing not only Broca's area but also relatively large areas surrounding area 44. In contrast, other evidence that Broca's area may not result in the articulatory-prosodic speech disorder called apraxia of speech, comes from statistical predictive models based on results of PET studies of the roles played by different frontal lobe regions in the regulation of fluency (Metter, Hanson, Jackson, Kemplar, & Van Lancker, 1991). This model predicts that Broca's area has no direct influence on fluency, but fluency is almost always disrupted to some degree in AOS. Finally, reports of lesioned patients by both Levine and Mohr (1979) and Tanabe and Ohigashi (1982) have called into question the role of the third frontal convolution in control of motor speech production. From these studies, it was concluded that a much larger lesioned area affecting the inferior two-thirds of the Rolandic cortex as well as the Rolandic operculum was necessary to induce speech apraxia of frontal lobe origin.

Studies using neuroimaging techniques such as MRI and CT imaging have also resulted in persuasive evidence that area 44 is not damaged in all cases of Broca's aphasia (Dronkers, Shapiro, Redfern, & Knight, 1992; Kertez, Harlock, & Coates, 1979; Mazzocchi & Vignolo, 1979; Poeck, de Bleser, & von Keyerlingk, 1984). Furthermore, the recent work by Dronkers and colleagues (Dronker et al., 1992) has demonstrated that damage to Broca's area may result in chronic aphasia syndromes other than Broca's aphasia, especially anomic aphasia, without accompanying motor speech impairment.

Of greater importance to the discussion at hand, however, is that the work of Dronkers and colleagues which has indicated that area 44 is not necessarily damaged in patients with the specific diagnosis of "apraxia of speech." That is, area 44 may or may not be damaged in cases of left-hemisphere motor speech impairment.

The Supplementary Motor Cortex (Area 6)

Area 6 on the mesial surface of the hemispheres is discussed separately from lateral area 6 because it appears to have some distinct functions. It, too, is considered a protomotor area (Goldberg, 1985), but unlike lateral area 6 which derives from insula, mesial area 6 derives from the cingulate cortex. It has been suggested that SMA is active in setting a preparatory state or the motor set which precedes the execution of complex motor sequences (see Kurata, 1992; Roland, Larsen, Lassen, & Skinhoj, 1980; Wise & Strick, 1984), especially for movements which are internally generated or self-initiated (Mushiake, Inase, & Tanji, 1991). In humans, electrical stimulation of some regions of SMA results in complex movements of the hand, arm, and fingers (Penfield & Welch, 1951). SMA receives extensive cortical input, especially from the parietal lobe, as well as major subcortical input from the basal ganglia via the thalamus (see Square & Martin, 1994). SMA projections represent a cascading system to a number of different levels including primary motor area bilaterally, premotor, prefrontal, parietal and cingulate cortex (Jurgens, 1985).

Lesions to the left SMA and/or cingulate cortex may result in a transient mutism and akinesia (Alexander et al., 1989; Damasio & Geschwind, 1984). SMA lesions have also resulted in palalalia or sudden involuntary rhythmic repetitions of syllables, words, and phrases (Alajouanine, Castaigne, Sabouraud, & Contamin, 1959; Jonas, 1981; Nagafuchi, Aoki, Niizuma, & Okita, 1991; Wallesch, 1990). Other abnormal speech symptoms may include echolalia (Alexander et al., 1989; Jonas, 1981), "stuttering" (Jonas, 1981), hesitations, perseverations, reduced volume, and aphonia (Jonas, 1981). Petrovici (1983) described the speech patterns of several patients with tumours of the left SMA. He labelled the disorder "aphemia" and highlighted the characteristics of blocks, reiterations, and reduced speech drive. It was concluded from the results of this study that SMA, as one of its functions, serves as a motor control centre, especially with regard to the sequencing of speech behaviours.

The Insula

The insula is a small island of cortex within the cerebral hemispheres that lies mesial to the frontal, parietal, and temporal lobes. The insula has been implicated as a region which is significant in oromotor behaviour and functions as a premotor association area. Evidence suggesting the involvement of the insula in oromotor behaviour has come from both animal and human studies. For example, in the awake monkey, electrical stimulation of four cortical regions has been shown to evoke chewing-like movements. These four areas include the: (1) face region of the primary motor cortex, i.e. face MI; (2) face region of the primary somatosensory cortex; (3) cortical masticory area 6, located lateral to the face region of the primary motor area; and (4) a deep cortical masticatory area located on the inferior face of the frontal operculum, i.e. the insula (Huang et al., 1989).

That the insula is a significant structure in speech praxis has been highlighted by Dronkers and colleagues. They have shown that the one damaged structure common to all patients studied with AOS was not located on the lateral surface of the cortex at all but instead was found to reside in the insula (Dronkers, in press; Dronkers, Shapiro, Redfern, & Knight 1992). Of interest here is the reminder that the insula is an area from which lateral area 6 phylogenetically derives (Goldberg, 1985). Area 6 is thought to be responsible for specifying movements at a relatively abstract level (Goldberg, 1985).

Parietal Lobe

It has been suggested that a general pattern of sensory convergence characterises the organisation of the parietal lobe. That is, information from primary sensory areas is conveyed to immediately adjacent, first unimodal parasensory association cortex, which in turn sends inputs to more distal, second unimodal parasensory areas. Further, several polymodal association areas located at the junctions between unimodal association areas have been proposed. For example, polysensory association areas include the intraparietal sulcus and the temporo-parietal area including the superior temporal sulcus (Freund, 1987). Extensive links between the parietal and frontal lobes exist and most likely serve to mediate the integration of sensory input for the preparation of motor behaviour.

Historically, there has been speculation that damage to the parietal lobe is also responsible for a motor speech disorder referred to variably as "incomplete anarthria" (Marie, Foix, & Bertrand, 1917), "phonetic disintegration" (Alajouanine, Ombredane, & Durand, 1939), "afferent motor aphasia" (Luria, 1966), "verbal apraxia" (Canter, 1969), "apraxia of speech" (Deutsch, 1979; Disimoni & Darley, 1977; Square, Darley, & Sommers, 1982; 1988) and "apraxia of vocal expression" (Denny-Brown, 1965). Buckingham (1979) explained that those who advocated centre lesion theories of apraxia of speech postulate that memories for movements are stored both in Broca's area and the parietal lobe, the latter possibly including the inferior post-Rolandic sensory area for speech (Luria, 1973) or the supramarginal gyrus (Denny-Brown, 1958; Kleist, 1916). Destruction of these areas, thus, may result in a disturbance of praxis for vocal tract musculature in the production of speech.

Deutsch (1979) reported results of a preliminary study which strongly inferred that two types of "apraxia of speech" may exist—one due to frontal lobe lesions and the other due to temporo-parietal lesions. The speech characteristics of 18 patients with left-hemisphere lesions, all of whom demonstrated symptoms of apraxia of speech, were studied. The sites of lesion, frontal or temporo-parietal, of 88% of the patients could be predicted statistically based upon three dimensions: percentage of monosyllabic articulation errors; the total number of polysyllabic errors; and sequencing errors. It was concluded that two types of speech apraxia exist, a frontal and a parietal syndrome.

Recently, Square-Storer and Apeldoorn (1991) have also put forward the notion that there may be different presentations of the motor speech disorder which results from left-hemisphere frontal, parietal and subcortical brain damage. All have commonly come to be known as "apraxia of speech." Based upon neuroimaging information and acoustic analyses, an apraxia of speech due to left parietal lobe damage was identified which differed from the apraxia of speech syndromes associated with lesions to the frontal quadrilateral space and more frontal cortical lesions. This study will be described more fully below. What is becoming increasingly evident, however, is that left parietal regions, for which futher investigations are needed for specification of exact localisation, may play a significant role in the control of speech praxis.

Lenticular Zone

Wernicke (1885) and Lichtheim (1884–1885) appear to have been the first to describe subcortical motor aphasia. Dejerine's (1901) discussion of pure subcortical motor aphasia and Marie's (1906) discussion of "anarthria" brought the quadrilateral space or lenticular zone to the forefront as a typical site of lesion for "pure" apraxia of speech. It was Marie's (1906) belief that "anarthria," a synonym for what Broca (1861) called "aphemia", was caused by subcortical lesions extending deep into the white matter of the left hemisphere which affected principally the lenticular nucleus. Goodglass and Kaplan (1972) have discussed aphemia as a subcortical motor aphasia (sic) in which disordered articulation marked by awkwardness and a slow rate of production exists; no language impairment accompanies the speech impairment. Indeed, with the advent of CT scanning, such subcortical lesions resulting in a pure motor speech disorder have been verified (Naeser, Alexander, Helm-Estabrooks, Levine et al., 1982; Square, 1981; Square, Darley, & Sommers, 1982; Square & Mlcoch, 1983; Square-Storer & Apeldoorn, 1991).

Lesions to the basal ganglia typically result in dysarthrias of the hypokinetic and hyperkinetic types (Darley et al., 1975a). The speech disorders resulting from these dysarthrias, however, include many symptoms of apraxia of speech such as initiation difficulties, reattempts, excess and equal stress, sound substitutions, and distortions and irregular articulatory breakdown. These characteristics may co-occur with symptoms which are decidedly dysarthric including a slow rate of articulation which is severely laboured and effortful and sometimes explosive. Reports by Alajouanine, Ombredane, and Durand (1939) and Square and Mlcoch (1983) have described the speech as having some characteristics of apraxia in combination with dysarthric qualities similar to those associated with dystonic (slow hyperkinetic) speech characteristics.

It may be unlikely that the apraxia of speech symptoms result directly from the damage to the basal ganglia. Positron emission tomography (PET) studies have served to increase our understanding of the functional anatomy of speech. Metter and colleagues undertook a series of resting PET studies of cerebral metabolism in aphasic patients in the 1980s. One of the most striking findings from these studies was that structural lesions to the basal ganglia, particularly the caudate nucleus, have a remote effect on the metabolism of the frontal lobe (Metter, Riege, Hanson, Jackson, Kempler, & Van Lancker, 1988a; Metter, Riege, Hanson, Phelps, & Kuhl, 1988b; Metter, Riege, Hanson, Phelps, & Kuhl, 1982). These results indicate that deep subcortical left-hemisphere lesions most likely result in speech impairments with characteristics consistent with slow hyperkinetic dysarthria and others consistent with those of "apraxia of speech," the latter possibly being due to hypometabolism of the frontal lobe, especially in the premotor and prefrontal regions. The dysarthric qualities which result could be due to structural damage to the lenticular zone itself or may be due to remote effects to area 4.

In the section which follows, a small sample of patients identified as having pure apraxia of speech will be discussed. Each also had lesion sites verified by CT scan.

PURE APRAXIA OF SPEECH LESION SITES AND SPEECH SYMPTOMS

The acoustic and perceptual speech patterns of left-hemisphere damaged (LHD) speakers with minimal aphasia were first described acoustically by Alajouanine, Ombredane & Durand (1939). They concluded that the motor speech disorder that results from LHD is a variable combination of apraxia, paresis and dystonia. This is logical given that sites of lesion most frequently reported are frontal cortex, parietal cortex and lenticular zone. We hypothesise that frontal cortical damage may give rise to both apraxic and paretic components; parietal cortical damage most likely gives rise to apraxic components; and the lenticular zone and its environs most likely give rise to dystonic components via damage to the basal ganglia, paretic components via damage to motor tracts coursing through the white matter and internal capsule, and apraxic characteristics due to the remote effects on the frontal lobe.

Perceptual and acoustic studies of LHD patients without aphasia and with varying sites of lesions has resulted in the identification of dramatically different speech patterns (Square, et al., 1982; Square & Mlcoch, 1983; Square-Storer & Apeldoorn, 1991) and, thus, the advancement of several hypotheses. First, AOS as defined by Darley (1968) and Darley and colleagues (1975a) is most likely to result following parietal lobe lesions as identified by CT scans (Square-Storer and Apeldoorn, 1991). The role of the dominant parietal lobe in motor speech control, specifically speech movement sequencing, has been noted historically using varying paradigms of study (Canter, 1969; Deutsch, 1984; Itoh et al., 1979; Kimura, 1982; Luria, 1966). Indeed, the "pure" apraxic speaker used as an acoustic exemplar of the disorder as described by the Mayo group and who later came to be known as the "Tornado man" (Darley, Aronson, & Brown, 1975b) had a parietal lobe lesion.

The speech of the two nonaphasic parietal-lesioned AOS speakers, one with a lesion of the temporo-parietal region and one with a lesion in the region of the postcentral sulcus, was studied by Square and colleagues. Speech patterns were marked by initiation difficulties and by severe groping, expressed both audibly and visually (Square, 1981; Square et al., 1982; Square-Storer & Apeldoorn, 1991). Groping generally preceded the initiation of an utterance but was also notable within utterances, especially in polysyllabic utterances (Square, 1981). In addition, speech was marked by numerous off-target approximations of phonemes with repeated attempts to self-correct (Square, 1981; Square et al., 1982). Absolute and relative durations of phonemes as well as fundamental frequency as measured acoustically were well within normal limits except in some cases of self-correcting reattempts (Square-Storer & Apeldoorn, 1991). Both perceptually and acoustically, there were few instances of syllable dissociation but some occurrences of syllable segregation, especially on reattempted initiating syllables (see Kent & Rosenbek, 1983, for definitions). This is the type of speech pattern which Square-Storer and Apeldoorn (1991) labelled pure apraxia of speech.

If indeed a disorder of speech due to damage to sensory association areas of the parietal lobe results in a disruption of the evocation of movement formulae (Liepmann, 1977) and, if indeed, regions of the left frontal lobe, particularly lateral area 6 and insula, play a role in the translation of movement formulae into innervating patterns for speech (Liepmann, 1908), then we would speculate that lesions to both areas of the left hemisphere may result in perceptual and acoustic speech symptomotology consistent with the characteristics associated with apraxia of speech (Darley et al., 1975a). In our lab, however, a "pure apractic speaker" with strictly a Broca's lesion or a lesion confined to lateral area 6 or insula has never been studied.

In another of our patients, a hemiplegic patient with a lesion to the left lenticular zone affecting especially the head of the caudate, speech was also marked by initiation difficulties with trial and error groping both preceding utterances and within utterances. There, however, was a predominance of distortion errors which marked both consonant and vowel productions and a markedly slow rate due to perceived prolongation of speech sounds and syllable dissociation (Square, 1981). In addition, there was a salient and overriding effortful quality to this patient's speech. The perceived slowness was acoustically verified by long absolute durations of phonemes and syllables, relatively equal durations of syllables, significant syllable dissociation, abnormal f^o (fundamental frequency) contours of multisyllabic words, phrases, and sentences, and numerous vowel distortions (Mlcoch & Square, 1984; Square, 1981; Square et al., 1982; Square-Storer & Apeldoorn, 1991). All of the latter deviancies were pervasive, i.e. unrelenting. Speech was also marked by some explosiveness due to bursts of loudness (Square et al., 1982; Square & Mlcoch, 1983). This patient also had mild phonatory harshness and extremely mild yet pervasive hypernasality; these latter characteristics also occur in pseudobulbar dysarthria (Darley et al., 1975a). The presentation of this patient's speech led to a second hypothesis: LHD, especially damage to the left lenticular zone, may result in a motor speech syndrome which is a combination of the initiating and sequencing deviancies coupled with "paretic–dystonic" features (Mlcoch & Square, 1984; Square-Storer & Apeldoorn, 1991).

Paretic and dystonic speech features occur in the dysarthrias associated with frontal cortical and subcortical lesions (Darley et al., 1975a). Primary motor cortex damage which may disrupt the fine control of voluntary movements, likely disrupts the execution of fractionated distal movements or the use of sets of muscles for executing certain movements (Evarts, 1986). Since primary motor cortex damage frequently coexists with premotor damage, we may speculate that both dysarthric and apraxic speech symptoms are likely to arise. Evidence which suggests that a left-hemisphere "dysarthria' may exist has been reported by Duffy and Folger (1986) and Hartman and Abbs (1992). Hartman and

Abbs (1992) refer to the "dysarthria" as unilateral upper motor neuron dysarthria (UUMN). Associated with UUMN dysarthria are phonatory, articulatory and rate deviancies similar to those associated with pseudobulbar dysarthria, the latter of which is due to bilateral motor strip damage (Darley et al., 1975a). Subcortical extension of frontal lesions is usual and involves the white matter through which sensorimotor tracks course, the left basal ganglia (BG), the internal capsule, and sometimes even the thalamus. Each of these subcortical areas have either basic motor functions or are conduits through which pyramidal and extrapyramidal motor tracts travel (see Square & Martin, 1994, for a review). Finally, there may also be subtle influences on speech output due to cerebellar dysfunction. Recent PET studies have demonstrated that a secondary result of frontal damage is contralateral cerebellar hypometabolism (see, for example, Metter, 1987b). Therefore, ataxic-like symptoms could complicate further the clinical presentation of frontal lesions which result in some symptoms of "apraxia of speech." Indeed, a case of ataxico-apraxic speech has been reported (Rosenbek, McNeil, Teetson, Odell, & Collins, 1981).

A fourth patient, the data for whom have not been published, had a fronto-parietal left hemisphere lesion which was predominantly parietal with its greatest volume being subcortical. She also had a right hemiparesis. Her speech was marked by all the characteristics present in the parietal lesioned subject discussed above. She, like the patient with a lesion to the lenticular zone, had "paretic" features but her speech was not nearly as effortful and explosive as that of the basal ganglia lesioned patient. This presentation led to another hypothesis: "dysarthric" qualities including slow, effortful speech may result from damage to the motor tracts coursing through the white matter underlying motor and/or sensory facial regions in the left hemisphere. In conclusion, we hypothesise that symptoms typically thought to be "apractic," including initiation difficulties marked by groping, off-target productions and attempts to self-correct, and a disability for producing longer sequences of speech, are most likely consistent with cortical damage to the dominant parietal lobe. Other "dysarthric-like" qualities such as slow effortful and even explosive speech may or may not compound this disorder.

Lesion Sites Common to Limb, Buccofacial, and Verbal Apraxia

Historically, arguments have been put forward supporting either the parietal lobe (Heilman, 1979a, 1979b; Kleist, 1934) or frontal lobe (Geschwind, 1965; Goldstein, 1911; Haaxman & Kuypers, 1975; Wilson, 1908) as the principal praxis centres in the control of limb movements. Indeed, several individuals also advanced the notion that both the parietal and frontal lobes resulted in limb apraxic syndromes which were qualitatively different (Faglioni, 1979; Kimura, 1982; Luria, 1966) and much in keeping with the hypotheses of variations in motor speech control due to left-heimsphere damage as put forth in the previous section. Rothi et al. (1991) point out, however, that "Although Geschwind (1965) thought the convexity of the premotor cortex was important for (limb) praxis, apraxia has not been reported from a lesion limited to the frontal cortical area, and the function of this area remains unknown" (p.448). Faglioni and Basso (1985) concluded in their review of neural mechanisms underlying limb apraxia that "The bulk of the data currently available support the opinion that apraxia is critically associated with damage to the region extending from the parietal lobe to the premotor area with special emphasis on the supramarginal gyrus and to a lesser extent on the premotor cortex" (p. 12). They further concluded that the supramarginal gyrus most likely is a praxis centre rather than being merely a relay station of integrated sensory information to the frontal lobe. Finally, based upon the results of a study by Basso, Luzzatti, and Spinnler (1980) it was concluded that specific lesion site does not dictate the type of limb apraxia (Faglioni & Basso, 1985).

With regard to buccofacial apraxia, more anterior lesions have been implicated, especially those affecting frontal and central opercula, insula, first temporal convolution, centrum semiovale and putamen (Mintz, Raade, & Kertesz, 1989; Raade, Rothi, & Heilman, 1991; Tognola & Vignolo, 1980). Nonetheless, posterior regions have been implicated as of prime importance in cases in which production of buccofacial sequences and not postures is affected (Kimura, 1982; Mateer, 1978; Mateer & Kimura, 1977).

The preliminary results of speech studies indicate that fine-grained acoustic analyses reveal different characteristics and clusters of speech deviancies associated with "apraxia of speech" with differing sites of lesion including left cortical parietal, left subcortical-cortical parietal and left subcortical frontal involving the lenticular zone (Square, 1981; Square et al., 1982; Square-Storer & Apeldoorn, 1991).

The role of the SMA in the control of limb praxis, especially for the use of tools, has been inferred from the case studies reported by Watson, Fleet, Rothi, and Heilman (1986). A notable difference between the performances of patients with parietal lobe lesions and those with SMA lesions was that the patients with lesions to SMA could both discriminate and comprehend pantomines unlike the parietal lesioned patients. Thus, the mechanisms underlying the apraxias due to these varying sites of lesion could be different. It will be recalled from the discussion of SMA above that speech apraxia has also been identified subsequent to lesions to this region (Petrovici, 1983).

COMMONALITIES IN DISRUPTIONS TO MOVEMENTS IN THE APRAXIAS

The commonalities in error types committed during limb, oral nonverbal, and speech motor output by patients clinically diagnosed as having ideomotor limb, buccofacial apraxia, and apraxia of speech have been presented by Roy and

Square (1985a) and Square-Storer and Roy (1989). The spatial and temporal errors committed across the three praxis modalities are remarkably similar when studied using error notation systems applied to the analysis of limb movements (e.g. Haaland & Flaherty, 1984; Rothi, Heilman, Mack, Verfaellie, & Brown, 1987; Roy & Square, 1985b) and oral nonverbal movements (e.g. Poeck & Kerschensteiner, 1975; Roy & Square, 1985b). Observational studies of types of speech errors have relied largely on the use of broad and narrow phonetic transcription (e.g. Johns & Darley, 1970; Square, 1981; Trost & Canter, 1974) and have allowed comparisons with errors committed in the limb and oral nonverbal modalities as described by error notation analysis. Square-Storer and Roy (1989) summarised the observed commonalities of movement errors in the three apraxias as follow: difficulties initiating movements; errors in spatial targeting; difficulties coordinating substructures, e.g. hand and wrist, lips and jaw; abnormal rates of movement; motor augmentation; omissions of subcomponents of movements; disorders of sequencing; and perseverative behaviours. No attempt has been made to relate the above types of errors to lesion sites. Nonetheless, persuasive evidence has been presented by Kimura and colleagues that the production of individual movements is disrupted subsequent to frontal damage whereas the sequencing of elements of more complex movements is affected by both frontal and parietal lobe lesions (Kimura & Archibald, 1974; Kimura, 1982, Mateer & Kimura, 1977; Mateer, 1978) over the three modalities.

Finer-grained analyses of movement disruption have been undertaken for patients diagnosed as having apraxia of speech using acoustic and physiologic measures and for patients with ideomotor limb apraxia using kinematic measures. Again, striking similarities exist. In particular, both types of apraxia are marked by both spatial and temporal disruptions of movement. In apraxia of speech, acoustic (Kent & Rosenbek, 1983; Square-Storer & Apeldoorn, 1991) and electropalatographic studies (Hardcastle et al. 1985; Washino et al., 1981) have highlighted the spatio-temporal disruptions. The prominence of temporal dyscoordination of laryngeal and supralaryngeal events (Freeman et al., 1978; Fromm, Abbs, McNeil, & Rosenbek, 1982; Itoh, Sasanuma, Tatsumi, Murakami, Fukusato, & Suzuki, 1982; Kent & Rosenbek, 1983) has been clearly demonstrated acoustically. Finally, temporal dyscoordination of movements of the supralaryngeal articulators, such as velum and tongue, lips and jaw, etc. (Fromm et al., 1982; Itoh et al., 1979; Itoh, Sasunuma, Hirose, Yosioka, & Yushigima, 1980; Tseng, McNeil, Adams, & Weismer, 1990) has been highlighted in physiological investigations.

While far fewer in number, studies indicating that spatio-temporal dyscoordination is characteristic of limb apraxia also exist (Charlton, Roy, Marteniuk, MacKenzie, & Square-Storer, 1988; Poizner, Mack, Verfaellie, Rothi, & Heilman, 1990; Poizner & Kegl, 1992). Charlton et al. (1988) reported that in a prehension grasping task, temporal dyscoordination between the transport component of the arm and grasp component of the hand was evident. Poizner and colleagues (1990, 1992) have pointed to three types of disruption in movement coordination in limb apraxia: impaired spatial orientation; spatio-temporal decoupling; and disturbed joint use. Roy, Brown, and Hardie (1993) have highlighted intersubject variability in performance, especially for joint use, and have emphasised the importance of examining individual differences among limb apraxic patients just as Square-Storer and Apeldoorn (1991) have emphasised the importance of describing individual differences among patients with apraxia of speech.

In conclusion, the use of both error notation systems and more fine-grained analyses, including acoustic, physiologic, and kinematic have highlighted the similarities among limb, buccofacial, and speech apraxia. These movement disruptions associated with apraxia of speech and their similarities to those observed in limb apraxia provide further support for the concept that the disorder, apraxia of speech, is indeed another form of praxis disruption. Furthermore, speech, limb, and buccofacial apraxia frequently coexist (Square-Storer, Roy, & Hogg, 1990). The association among the apraxias may reflect some common disruptions to motor control in that similar types of errors and performance deficits are observed over the three modalities (Roy & Square-Storer, 1990).

MOTOR TREATMENTS FOR APRAXIA OF SPEECH

Treatment approaches for apraxia of speech can be conceptualised as being directed towards three different levels of disability: postural shaping of the vocal cavity to produce speech sounds; enhancement of kinaesthetic awareness in order to gain control for making subtle shaping adjustments in similar yet contrasting movements; and establishing rate and rhythmic schemata for the production of longer movement sequences such as phrases and sentences (Square & Martin, 1994). Postural shaping techniques are directed toward enhancing a patient's awareness of the location and timing of muscle contractions in order to produce postures of the supraglottal articulators requisite for target speech sounds. Traditional techniques have relied upon focusing the patient's attention on the spatial requirements of shaping by providing visual models coupled with enhanced auditory stimulation, a technique known as integral stimulation (Rosenbek, Lemme, Ahern, Harris, & Wertz, 1973). Tactile stimulation may also be used to posture the oral cavity. Such techniques include phonetic placement (Van Riper, 1963; Rosenbek, 1978), motokinaesthetic therapy (Stichfield & Younge, 1938) and PROMPT, a dynamic tactile-kinaesthetic cueing technique developed by Chumpelik (1984). It has been demonstrated that PROMPT has robust effects for reestablishing control of both volitional postural shaping of the oral tract for the production of syllabic speech units and for enhancing the sequencing of muscle contractions for longer speech units (Square-Storer & Chumpelik (Hayden), 1989; Square-Storer, Hayden (Chumpelik), & Adams, 1985; Square, Chumpelik (Hayden), Morningstar, & Adams, 1986).

Therapy directed towards enhancement of kinaesthetic awareness for making subtle distinctions in vocal tract configuration is a second level at which treatment may be directed. Subtle configuration changes are taught within the context of contrasting words which have similar phonological structures but which vary minimally with regard to speech sound structure (e.g. me, pea, knee, key . . .; pin-bin; no-bow . . .). Advocates of such therapeutic intervention stress the importance of heightening the patient's conscious awareness of the feel of the contrasting movements, i.e. enhancement of kinaesthetic awareness, in order to establish enhanced volitional control. Wertz, LaPointe, and Rosenbek (1984) provide a detailed discussion of the use of imitation of contrastive speech drills.

For many patients with apraxia of speech, it is not necessary to rebuild speech from the bottom up, that is from the reestablishment of volitional control of speech postures to the volitional control of minimal contrasts. Instead, a more top-down approach to treatment may be extremely effective for establishing volitional motor control of functional phrases and sentences. Such top-down methods emphasise slowing rate and/or enhancing the melodic line of functional phrases and sentences. Examples of effective methods which focus on slowing rate of speech control include prolonged speech (Southwood, 1987), finger counting for slow placing of speech (Simmons, 1978), and application of vibrotactile stimulation to the hand to indicate to patients the relative duration of syllables in polysyllabic words (Rubow, Rosenbek, & Collins, 1982). Square (1994), among others, has hypothesised that a slow rate of speech production probably enhances sensory feedback for jaw (mandible) positioning for syllabic nuclei or the vowels, thereby allowing the system more time to formulate the readjustments of the vocal tract required for upcoming postures and/or to organise upcoming programs for muscle contractions.

Methods which enhance speech melody have also been reported to be highly facilitative of speech accuracy for some patients with apraxia of speech. We speculate that the melodic methods are especially useful for individuals whose greatest deficit resides in an inability to sequence muscle contractions in longer utterances rather than for those whose primary deficit resides in volitional postural shaping for speech sounds. Examples of melodic methods include contrastive stress drills (Wertz et al, 1984), singing (Keith & Aronson, 1975); Melodic Intonation Therapy (Albert, Sparks, & Helm, 1973; Naesser & Helm-Estabrooks, 1985; Sparks & Deck, 1986; Sparks, Helm, & Albert, 1974) and PROMPT (Chumpelik, 1984; Square, Hayden, & Adams, 1985; Square, Hayden, Morningstar, & Adams, 1986; Square-Storer & Hayden, 1989).

The differential effect of these motor treatment approaches on individual patients with apractic speech output seems to substantiate that, indeed, at least two primary disabilities occur in apractic speakers—deficits of postural shaping and deficits for sequencing muscle contractions in speech movement sequences. Although some have speculated that apraxia is a unitary disorder that varies on a continuum of severity (Wertz et al., 1984), there appears to be growing support for the notion that two areas of disability may exist with postural shaping deficits complicated by sequencing deficits being more reflective of frontal lobe dysfunction and deficits which are primarily sequencing ones being more reflective of parietal lobe lesions. Both disruptions respond well to motor speech treatments but are relatively unaffected by aphasia therapy.

CONCLUSION

The existence of the disorder, apraxia of speech, has been a controversial issue in the neurological and neuropsychological literature. This chapter has attempted to provide a number of converging lines of evidence which support the hypothesis that the left hemisphere is specialised for the control of skilled motor behaviour over all modalities of output—limb, oral nonverbal, and oral verbal. Similarities between apraxia of speech and limb apraxia abound with regard to site of lesion and the characteristics of motor disruption observed and measured. Despite the overriding similarities, intersubject variability has been emphasised. Whether individual differences in movement performances among apraxic patients exist because there are different types of apraxia is a question which remains. Alternatively, might it be that apraxia in any modality is likely to coexist with other subtle motor disruptions such as paresis, dystonia, or ataxia thereby resulting in a variety of manifestations? Future research, particularly that which emphasises structural–functional relationships, may hold the answers to these questions.

REFERENCES

Abbs, J.H. & Welt, C. (1985). Neurophysiologic processes of speech movement control. In N. Lass (Ed.), *Handbook of speech–language pathology and audiology*, (154–170). Toronto: B. Decker.

Adams, S.G., McNeil, M.R., & Weismer, G. (1989). Speech movement velocity profiles in neurogenic speech disorders. *Journal of the American Speech and Hearing Association, 31*, 113.

Alajouanine, T., Castaigne, P., Sabouraud, O., & Contamin, F. (1959). Palilalie paroxystique et vocalizations itératives au cours de crises épileptiques par lésion intéressant l'aire motrice supplémentaire. *Revue Neurologique, 101*, 186–202.

Alajouanine, T., Ombredane, A., & Durand, M. (1939). *Le syndrome de désintegration phonétique dans l'aphasie*. Paris: Masson.

Albert, M.L., Sparks, R.W., & Helm, N.A. (1973). Melodic Intonation Therapy for aphasia. *Archives of Neurology, 29*, 130–131.

Alexander, M.P., Benson, D.F., & Stuss, D.T. (1989). Frontal lobes and language. *Brain and Language, 37*, 656–691.

Aten, J.L., Johns, D.L., & Darley, F.L. (1971). Auditory perception of sequenced words in apraxia of speech. *Journal of Speech and Hearing Research, 14*, 131–143.

Basso, A., Luzzatti, C., & Spinnler, H. (1980). Is ideomotor apraxia the outcome of damage to well-defined regions of the left hemisphere? Neuropsychological study of CAT correlation. *Journal of Neurology, Neurosurgery, and Psychiatry, 43*, 118–126.

Baum, S.R., Blumstein, S.E., Naeser, M.A., & Palumbo, C.L. (1990). Temporal dimensions of consonant and vowel production: An acoustic and CT scan analysis of aphasic speech. *Brain and Language, 39*, 33–56.

Bellugi, U. (1980). The structuring of language: Clues from similarities between signed and spoken language. In U. Bellugi & M. Studdert-Kennedy (Eds.), *Signed and spoken language: Biological constraints on linguistic form*, (pp. 115–140). Dahlem Konferenzen, Weinheim/Deerfield Beach, FL: Verlag Chemie.

Blumstein, S. (1973a). *A phonological investigation of aphasic speech*. The Hague: Moulton.

Blumstein, S. (1973b). Some phonological implications of aphasic speech. In H. Goodglass & S. Blumstein (Eds.), *Psycholinguistics and aphasia*. Baltimore: John Hopkins.

Broca, P. (1861). Remarques sur le siège de la faculté de langage suivies d'une observation d'aphémie. *Bulletin de la Société d'Anatomie, 6* (2e sèrie), 330–357.

Buckingham, H.W. (1979). Explanation of apraxia with consequences for the concept of apraxia of speech. *Brain and Language, 8*, 202–226.

Buckingham, H.W. (1992). The mechanisms of phonemic paraphasia. *Clinical Linguistics and Phonetics, 6*(1 and 2), 41–63.

Buckingham, H.W. & Yule, G. (1987). Phonemic false evaluation: Theoretical and clinical aspects. *Clinical Linguistics and Phonetics, 1* (2), 113–125.

Canter, G.J. (1969). The influence of primary and secondary verbal apraxia on output disturbances in aphasic syndromes. Paper presented to the American Speech and Hearing Association, Chicago, Illinois.

Charlton, J., Roy, E.A., Marteniuk, R.G., MacKenzie, C.L., & Square-Storer, P.A. (1988). Disruptions to reaching in apraxia. *Society for Neuroscience Abstracts, 14*, 1234.

Chumpelik, D.A. (1984). The PROMPT system of therapy. *Seminars in Speech and Language. 5*, 139–156.

Damasio, A.R. (1991). Aphasia. *The New England Journal of Medicine, 326*, 531–539.

Damasio, A.R. & Geschwind, N. (1984). The neural basis of language. *Annual Review of Neuroscience, 7*, 127–147.

Darley, F.L. (1968). Apraxia of speech: 107 years of terminological confusion. Paper presented to the American Speech and Hearing Association, Denver, Colorado.

Darley, F.L. (1977). A retrospective review of aphasia. *Journal of Speech and Hearing Disorders, 42*, 161–169.

Darley, F.L., Aronson, A.E., & Brown, J.R. (1975a). *Motor Speech Disorders—Audio Tapes*. Philadelphia: W.B. Saunders.

Darley, F.L., Aronson, A.E., & Brown, J.R. (1975b). *Motor Speech Disorders—Audio Tapes*. Philadelphia: W.B. Saunders.

Deal, J. & Darley, F.L. (1972). The influence of linguistic and situational variables on phonemic accuracy in apraxia of speech. *Journal of Speech and Hearing Research, 15*, 639–653.

Dejerine, J. (1901). Sémiologie du système nerveux. In C.J. Bouchard (Ed.), *Traité de pathologie générale*, (Vol. 5, pp. 391–471). Paris: Masson.

Denny-Brown, D. (1958). The nature of apraxia. *Journal of Nervous and Mental Disorders, 126*, 9–33.

Denny-Brown, D. (1965). Physiological aspects of disturbances of speech. *Australian Journal of Experimental Biology and Medical Science, 43*, 455–474.

Deutsch, S.E. (1984). Prediction of site of lesion from speech apraxic error patterns. In J.C. Rosenbek et al. (Eds.), *Apraxia of speech: Physiology, acoustics, linguistics and management* (pp. 113–34). San Diego: College Hill Press.

Deutsch, S.E. (1979). *Prediction of site of lesion from speech apraxic error patterns*. Paper presented at the annual convention of the American Speech, Language and Hearing Association, Atlanta, Georgia (unpublished).

Disimoni, F. & Darley, F. (1977). Effects on phoneme duration control of three utterence-length conditions in an apractic patient. *Journal of Speech and Hearing Disorders, 42*, 257–64.

Dronkers, N.F. (in press). A new brain region for coordinating speech coordination. *Nature*.

Dronkers, N.F., Shapiro, J.K., Redfern, B.B., & Knight, R.T. (1992). The role of Broca's area in Broca's aphasia. *Journal of Clinical and Experimental Neuropsychology, 14(1)*, 52–53.

Dronkers, N.F., Shapiro, J.K., Redfern, B.B., & Knight, J.K. (1992). *The third left frontal convolution plays no special role in the function of language: Marie's quadrilateral space revisited.* Paper presented at the 1992 conference of The Academy of Aphasia, Toronto, Ontario (unpublished).

Duffy, J.R. & Folger, W.N. (1986). *Dysarthria in unilateral central nervous system lesion: A retrospective study*. Paper presented at the annual convention of the American Speech, Language and Hearing Association, Detroit (unpublished).

Dunlop, J.M., & Marquardt, T.P. (1977). Linguistic and articulatory aspects of single word production in apraxia of speech. *Cortex, 13*, 17–29.

Evarts, E.V. (1986). Motor cortex outputs in primates. In *Cerebral Cortex, Vol. 5, Sensory Motor Areas and Aspects of Cortical Connectivity*, 217–241.

Faglioni, P. (1979). Specializzazione anatomo-funzionale della corteccia e organizzazione del gesto. Contributo della sperimentazione animale allo studio dell'aprassia. *Cortex, 15* (Suppl. to No. 3), 1–32.

Faglioni, P. and Basso, A. (1985). Historical perspective on neuroanatomical correlates of limb apraxia. In E. A. Roy (Ed.), *Neuropsychological studies of apraxia and related disorders. Advances in psychology, 23*, 3–44. Amsterdam: North-Holland.

Farmer, A., O'Connell, P., & Jesudowich, B. (1979). *Naming and reading errors and response latency in Broca's aphasia*. Paper presented to the annual convention of the American Speech, Language and Hearing Association, Atlanta, Georgia (unpublished).

Freeman, F., Sands, E., & Harris, K. (1978). Temporal coordination of phonation and articulation in a case of verbal apraxia. *Brain and Language, 6*, 106–11.

Freund, H.J. (1987). Abnormalities of motor behavior after cortical lesions in humans. In S. Geiger, F. Plum, & V. Mountcastle (Eds.), *Handbook of physiology, Vol. 5, The nervous system*, (pp. 763–810). Bethesda: American Physiological Society.

Fromm, D., Abbs, J.H., McNeil, M.R., & Rosenbek, J.C. (1982). Simultaneous perceptual-physiological method for studying apraxia of speech. In R.H. Brookshire (Ed.), *Clinical aphasiology*, (Vol. 12, pp. 251–262). Minneappolis, MN: BRK Publishers.

Fry, D.B. (1959). Phonemic substitutions in an aphasic patient. *Language and Speech, 2*, 52–61.

Geschwind, N. (1965). Disconnection syndromes in animals and man. *Brain, 88*, 237–94, 585–644.

Godschalk, M., Lemon, R.N., Kuypers, H.G.J.M., & Ronday, H.K. (1984). Cortical afferents and efferents of monkey postarcuate area: An anatomical and electrophysiological study. *Experimental Brain Research, 56*, 410–424.

Goldberg, G. (1985). Supplementary motor area structure and function: review and hypotheses. *Behavioral and Brain Sciences, 8*, 567–616.

Goldstein, K. (1911). Über Apraxie. *Beihefte zur medizinischen Klinik, 7*, 271.

Goodglass, H. (1993). *Understanding aphasia*. San Diego, CA: Academic Press Inc.

Goodglass, H. & Kaplan, E. (1972). *The assessment of aphasia and related disorders.* Philadelphia, PA: Lea and Febiger.

Gracco, V.L. (1991). Sensorimotor mechanisms in speech motor control. In H. Peters, W. Hulstijn, & W. Starkweather (Eds.), *Speech motor control and stuttering*, (pp. 53–76). New York: Elsevier Science.

Gracco, V.L. & Abbs, J.H. (1987). Programming and execution processes of speech movement control: potential neural correlates. In L.E. Keller & M. Gopnik (Eds.), *Symposium on motor and sensory language processes,* (pp. 165–218). Hillsdale, NJ: Lawrence Erlbaum Associates Inc.

Grillner, S. (1982). Possible analogies in the control of innate motor acts and the production of sound in speech. In S. Grillner, B. Lindblom, J. Labker, & A. Persson (Eds). *Speech motor control.* (pp. 217–230). New York: Pergamon Press.

Haaland, K.Y. & Flaherty, D. (1984). The different types of limb apraxia errors made by patients with left or right hemisphere damage. *Brain and Cognition, 3,* 370–384.

Haaxman, R. & Kuypers, H.G. (1975). Intrahemispheric cortical connextions and visual guidance of hand and finger movements in Rhesus monkey. *Brain, 98,* 239–260.

Hardcastle, W.J., Morgan, R.A., & Clark, C.J. (1985). Articulatory and voicing characteristics of adult dysarthric and verbal apraxic speakers. *British Journal of Disorders of Communication, 20,* 249–70.

Hardison, D., Marquardt, T.P., & Peterson, H.A. (1977). Effects of selected linguistic variables on apraxia of speech. *Journal of Speech and Hearing Research, 22,* 334–343.

Hartman, D.E. & Abbs, J.H. (1992). Dysarthria associated with focal unilateral upper motor neuron lesion. *European Journal of Disorders of Communication, 27,* 187–196.

Hécaen, H. & Consoli, S. (1973). Analyses of language troubles in the course of lesions of Broca's area. *Neuropsychologia, 11,* 377–388.

Heilman, K.M. (1979a). Apraxia. In K.M. Heilman & E. Valenstein (Eds.), *Clinical neuropsychology.* (pp. 159–185). New York: Oxford University Press.

Heilman, K.M. (1979b). The neuropsychological basis of skilled movement in man. In M.S. Gazzaniga (Ed.), *Handbook of behavioural neurobiology, Vol. 2, Neuropsychology,* (pp. 447–461. New York: Oxford University Press.

Huang, C., Hiraba, H., Murray, G.M., & Sessle, B.J. (1989). Topographical distribution and functional properties of cortically induced rhythmical jaw movements in the monkey (Macaca fascicularis). *Journal of Neurophysiology, 61,* 635–650.

Itoh, M., Sasunuma, S., Hirose, H., Yosioka, H., & Yushigima, T. (1980). Abnormal articulatory dynamics in a patient with apraxia of speech. *Brain and Language, 11,* 66–75.

Itoh, M., Sasanuma, S., Hirose, H., Yoshioka, H., & Ushijima, T. (1978). Abnormal articulatory dynamics in a patient with apraxia of speech: X-ray microbeam observation. *Annual Bulletin of the Research Institute of Logopedics and Phonoeatrics, 12,* 87–96. Tokyo: University of Tokyo.

Itoh, M., Sasunuma, S., Tatsumi, I.F., Murakami, S., Fukusako, Y., & Suzuki, T. (1982). Voice onset time characteristics in apraxia of speech. *Brain and Language, 17,* 193–210.

Itoh, M., Sasanuma, S., & Ushijima, T. (1979). Velar movements during speech in a patient with apraxia of speech. *Brain and Language, 7,* 227–239.

Jaffee, D.W. & Duffy, J.R. (1978). *Voice-onset time characteristics of patients with apraxia of speech.* Paper presented at the annual convention of the American Speech and Hearing Association, San Francisco, CA (unpublished).

Johns, D. & Darley, F.L. (1970). Phonemic variability in apraxia of speech. *Journal of Speech and Hearing Research, 13,* 556–583.

Jonas, S. (1981). The supplementary motor region and speech emission. *Journal of Communicative Disorders, 14,* 349–373.

Jurgens, U. (1985). Efferent connections of the supplementary motor area. *Experimental Brain Research, 58,* A1-A2.

Keith, R. & Aronson, A. E. (1975). Singing as therapy for apraxia of speech and aphasia: Report of a Case. *Brain and Language, 2,* 483–488.

Kent, R.D. (1984) Psychobiology of speech development: Coemergence of language and movement system. *American Journal of Physiology, 246,* R888–R894.

Kent, R.D. & Rosenbek, J.C. (1983). Acoustic patterns of apraxia of speech. *Journal of Speech and Hearing Research, 25*, 231–249.

Kertesz, A., Harlock, W., & Coates, R. (1979). Computer tomographic localization lesion size and prognosis in aphasia and nonverbal impairment. *Brain and Language, 8*, 34–50.

Kimura, D. (1982). Left-hemisphere control of oral and brachial movements and their relation to communication. *Philosophical Transactions of the Royal Society of London, B298*, 135–149.

Kimura, D. & Archibald, Y. (1974). Motor functions of the left hemisphere. *Brain, 97*, 337–350.

Kleist, K. (1916). Postoperative psychosen. *Monographien aus dem Gesamtgebiete der Neurologie und Psychiatrie: Heft 11*, 1879–1960.

Kleist, K. (1934). *Gehirnpathologie vornehmlich auf Grund der Kriegserfahrungen*. Leipzig: Barth.

Klich, R.J., Ireland, J.V., & Weidner, W.E. (1979). Articulatory and phonological aspects of consonant substitutions in apraxia of speech. *Cortex, 15*, 451–470.

Kurata, K. (1992). Somatotopy in the human supplementary motor area. *Trends in Neuroscience, 15*, 159–160.

Lecours, A.R. & Lhermitte, F. (1976). The "pure form" of the phonetic disintegration syndrome (pure anarthria): Anatomo-clinical report of a historical case. *Brain and Language, 3*, 88–113.

Levine, D.N. & Mohr, J.P. (1979). Language after bilateral cerebral infarctions: Role of the minor hemisphere in speech. *Neurology, 29*, 927–938.

Lichtheim, L. (1884–1885). On aphasia. *Brain, 7*, 433–484.

Liepmann, H. (1908). *Drei Aufsatze aus dem Apraxiegebiet. Volume 1*. Berlin: Karger.

Liepmann, H. (1977). The syndrome of apraxia (motor asymboly) based on a case of unilateral apraxia. (A translation from *Monatschrift für Psychiatrie und Neurologie 1900, 8*, 15–44). In D.A. Rottenberg & F.H. Hockberg (Eds.), *Neurological classics in modern translation*. New York: Macmillan Publishing Co.

Lozano, R.A. & Dreyer, D.R. (1978). Some effects of delayed auditory feedback on dyspraxia of speech. *Journal of Communication Disorders, 11*, 407–415.

Luria, A.R. (1966). *Higher cortical functions in man*. New York: Basic Books.

Luria, A.R. (1973). Neuropsychological studies in the U.S.S.R. A review. *Proceeding of the National Academy of Sciences of the United States of America, 70(4)*, 1270–03.

MacNeilage, P.F. (1970). The motor control of the serial ordering of speech. *Psychological Review, 77*, 182–196.

Marie, P. (1906). La troisieme circovolution frontale gauche ne joue aucur role special dans la fonction du language. *Semaine médicale, 26*, 241–7.

Marie, P., Foix, C., & Bertrand, I. (1917). Topographie cranio-cérébrale. *Annales de Médecine, 55*.

Marquardt, T.P., Reinhart J.B., & Peterson, H.A. (1979). Markedness analysis of phonemic substitution errors in apraxia of speech. *Journal of Communication Disorders, 12*, 481–494.

Martin, A.D. (1974). Some objections to the term apraxia of speech. *Journal of Speech and Hearing Disorders, 39*, 53–64.

Martin, A.D. & Rigrodsky, S. (1974a). An investigation of phonological impairment in aphasia. Part I. *Cortex, 10*, 317–328.

Martin, A.D. & Rigrodsky, S. (1974b). An investigation of phonological impairment in aphasia. Part II. Distinctive feature analysis of phonemic commutation errors in aphasia. *Cortex, 10*, 329–346.

Martin, J.G. (1972). Rhythmic (hierarchical) versus serial structure in speech and other behavior. *Psychological Review, 79*, 487–509.

Mateer, C. (1978). Impairments of nonverbal oral movements after left hemisphere damage: A follow-up analysis of errors. *Brain and Language, 6*, 334–341.

Mateer, C. & Kimura, D. (1977). Impairments of nonverbal oral movements in aphasia. *Brain and Language, 4*, 262–276.

Mazzocchi, F. & Vignolo, L.A. (1979). Localization of lesions in aphasia: Clinical-CT scan correlations in stroke patients. *Cortex, 15*, 627–654.

McNeil, M.R. & Adams, S.G. (1991). A comparison of speech kinematics among apraxic, conduction aphasic, ataxic dysarthric and normal geriatric speakers. *Clinical Aphasiology, 19*, 279–294.

McNeil, M.R., Caligiuri, M., & Rosenbek, J.C. (1989). A comparison of labio-mandibular kinematic durations, displacements, velocities and dysmetrias in apraxic and normal adults. *Clinical Aphasiology, 18*, 173–194.

McNeil, M.R. & Kent, R.D. (1990). Motoric characteristics of adult aphasic and apraxic speakers. In G.R. Hammond (Ed.), *Advances in Psychology: Cerebral Control of Speech and Limb Movements* (pp. 349–386). New York: Elsevier/North-Holland.

McNeil, M.R., Liss, J., Tseng, C-H, & Kent, R.D. (1990). Effects of speech rate on the absolute and relative timing of apraxic and conduction aphasic sentence production. *Brain and Language, 38*, 135–158.

McNeil, M.R., Weismer, G., Adams, S., & Mulligan, M. (1990). Oral structure nonspeech motor control in normal, dysarthric, aphasic and apraxic speakers: Isometric fine force and static fine position. *Journal of Speech and Hearing Research, 33*, 255–268.

Metter, E.J., Hanson, W.R., Jackson, C.A., Kempler, D. & Van Lancker, D. (1991). Brain glucose metabolism in aphasia: a model of the interrelationship of frontal lobe regions on fluency. In T.E. Prescott (Ed.) *Clinical Aphasiology, 19*, 69–76. Austin, TX: Pro-Ed.

Metter, E.J., Hanson, W.R., Kempler, D., Jackson, C., Mazziota, J., & Phelps, M. (1987a). Left prefrontal glucose metabolism in aphasia. In R. Brookshire (Ed.), *Clinical Aphasiology, 17*, 300–312. Minneapolis, MN: BRK Publishers.

Metter, E.J., Kempler, D, Jackson, C., Hanson, W., Riege, W., Camras, L., Mazziota, J., & Phelps, M.E. (1987b). Cerebellar glucose metabolism in chronic aphasia. *Neurology, 37*, 1599–1606.

Metter, E.J., Riege, W.R., Hanson, W.R., Jackson, C., Kempler, D., & Van Lancker, D. (1988a). Subcortical structures in aphasia: Analysis based on FBG, PET and CT. *Archives of Neurology, 45*, 1229–1234.

Metter, E.J., Riege, W.R., Hanson, W.R., Phelps, M.E., & Kuhl, D.E. (1988b). Evidence for a caudate role in aphasia from FBG positron computed tomography. *Aphasiology, 2*, 33–43.

Metter, E.J., Riege, W.R., Hanson, W. R., Phelps, M. E., & Kuhl, D. E.(1982). Role of the caudate nucleus in aphasic language: Evidence from FDG-PET. *Neurology, 32*, A92.

Mintz, T., Raade, A.S., & Kertesz, A. (1989). *Lesion size and localization in buccofacial apraxia: A retrospective analysis*. Paper presented at the meeting of the Canadian Association of Speech–Language Pathologists and Audiologists, Toronto, Ontario.

Mlcoch, A., Darley, F.L., & Noll, D. (1982). Articulatory consistency and variability in apraxia of speech. In R. Brookshire (Ed.), *Clinical aphasiology conference proceedings*, (pp. 50–53). Minneapolis, MN: BRK Publishers.

Mlcoch, A.G. & Square, P.A. (1984). Apraxia of speech: Articulatory and perceptual factors. In *Speech and Language: Advances in basic research and practice* (Vol. 10). New York: Academic Press. (By invitation, referred).

Mohr, J.P., Pessin, M.S., Finkelstein, S., Funkenstein, H., Duncan, G.W., & Davis, K.R. (1978). Broca aphasia: Pathologic and clinical. *Neurology, 28*, 311–324.

Murray, G.M. (1989). *An analysis of motor cortex neural activities during trained orofacial motor behaviour in the awake primate (Macaca fascicularis)*. Unpublished doctoral dissertation, University of Toronto.

Mushiake, H., Inase, M., & Tanji, J. (1991). Neuronal activity in the primate premotor, supplementary, & precentral motor cortex during visually guided and internally determined sequential movements. *Journal of Neurophysiology, 66*, 705–718.

Naeser, M.A., Alexander, M.P., Helm-Estabrooks, N., Levine, H.L., Laughlin, S.A. and Geochwind, N. (1982). Aphasia with predominantly subcortical lesion sites. *Archives of Neurology, 39*, 2–14.

Naesser, M.A. & Helm-Estabrooks, N. (1985). CT scan lesion localization and response to melodic intonation therapy with nonfluent aphasia cases. *Cortex, 21*, 203–233.

Nagafuchi, M., Aoki, Y., Niizuma, H., & Okita, N. (1991). Paroxysmal speech disorder following left-frontal brain damage. *Brain and Language, 40*, 266–273.

Nebes, R.D. (1975). The nature of internal speech in a patient with aphemia. *Brain and Language, 2*, 489–497.

Pellat, J., Gentil, M., Lyard, G., Vila, A., Tarel, V., Moreau, O., & Benabio A.L. (1991). Aphemia after a penetrating brain wound: A case study. *Brain and Language, 40*, 459–470.

Penfield, W. & Welch, K. (1951). *Speech and brain mechanisms*. Princeton, NJ: Princeton University Press.

Petrovici, J.N. (1983). Speech disorders in tumors of the supplementary motor area. *Zentralbl für Neurochirurgie, 44*, 97–104.

Poeck, K., de Bleser, R,. & von Keyerlingk, D.G. (1984). Computed tomography localization of standard asphasic syndromes. In F.C. Rose (Ed.), *Advances in neurology*: (Vol. 42, pp. 71–89). *Progress in Aphasiology*. New York: Raven.

Poeck, K. & Kerschensteiner, M. (1975). Analysis of sequential motor events in oral apraxia. In K.J. Zulch et al. (Eds.), *Cerebral Localization* (pp. 98–111. Berlin: Springer-Verlag.

Poizner, H. & Kegl, J. (1992). Neural basis of language and motor behaviour: Perspectives from American Sign Language. *Aphasiology, 6*, 219–256.

Poizner, H., Mack, L., Verfaellie, M., Rothi, L.J.G., & Heilman, M. (1990). Three-dimensional computer graphic analysis of apraxia: Neural representations of learned movement. *Brain, 113*, 85–101.

Raade. A.S., Rothi, L J. G., and Heilman, K.M. (1991). The relationship between buccofacial and limb apraxia. *Brain and Cognition, 16*, 130–146.

Rochon, E.A., Caplan, D., & Waters, G. (1990). Short-term memory processes in patients with apraxia of speech. *Journal of Neurolinguistics. 5* (2/3), 237–264.

Roland, P.E., Larsen, B., Lassen, N.A., & Skinhoj, E. (1980). Supplementary motor area and other cortical areas in organization of voluntary movements of man. *Journal of Neurophysiology, 43*, 118–136.

Rosenbek, J.C. (1978). Treating apraxia of speech. In D.F. Johns (Ed.), *Clinical management of neurogenic communicative disorders* (pp. 191–241). Boston: Little Brown.

Rosenbek, J.C., Lemme, M.L., Ahern, M.B., Harris, E. and Wertz, R. (1973). A treatment for apraxia of speech in adults. *Journal of Speech and Hearing Disorders, 38*, 462–472.

Rosenbek, J.C., McNeil, M.R., Teetson, M., Odell, K., & Collins, M.J. (1981). A syndrome of neuromotor speech deficit and dysographia. *Clinical Aphasiology, 11*, 309–315.

Rothi, L.J.G., Heilman, K.M., Mack, L., Verfaellie, M., & Brown, P. (1987). *Ideomotor apraxia: Error pattern analysis*. Paper presented to the annual meeting of the International Neuropsychological Society, Atlanta, Georgia (unpublished).

Rothi, L.J.G., Ochipa, C., & Heilman, K.M. (1991). A cognitive neuropsychological model of limb praxis. *Cognitive neuropsychology, 8*, 443–458.

Roy, E.A., Brown, L., & Hardie, M. (1993). Movement variability in limb gesturing. In K. Newell & D. Corcos (Eds.), *Variability in Motor Performance* (pp. 449–474). Champaign, IL: Human Kinetics Publishers.

Roy, E.A., Brown, L., Winchester, T., Square, P., Hall, C., & Black, S. (1993). Memory processes and gestural performance in apraxia. *Adapted Physical Activity Quarterly, 10*, 293–311.

Roy, E.A. & Square, P.A. (1985a). Common considerations in the study of limb, verbal and oral apraxia. In E.A. Roy (Ed.), *Neuropsychological studies of apraxia and related disorders* (pp. 111–161). Amsterdam: North-Holland.

Roy, E.A. & Square, P.A. (1985b). Error/movement notation systems in apraxia. *Recherches Semiotiques/Semiotics Inquiry, 5*, 402–12.

Roy, E.A. & Square-Storer, P.A. (1990). The apraxis-correlative studies of fine motor control and action sequencing. In G. Hammond (Ed.), *Cerebral control of speech and limb movements: Advances in psychology* (pp. 477–502). Amsterdam: Elsevier Science.

Rubow, R.T., Rosenbek, J.C., & Collins, M.J. (1982). Vibrotactile stimulation for intersystemic reorganization in the treatment of apraxia of speech. *Archives of Physical Medicine and Rehabilitation, 63*, 150–153.

Sands, E.S., Freeman, F.J., & Harris, K.S. (1978). Progressive changes in articulatory patterns in verbal apraxia: A longitudinal case study. *Brain and Language, 6*, 97–105.

Sasanuma, S. & Fujimura, O. (1971, 1972). Selective impairment of processing phonetic and non-phonetic transcriptions of words in aphasic patients: Kana and Kanji in visual recognition and writing. *Cortex, 7*, 196–218.

Schell, G.R. & Strick, P.L. (1984). The origin of thalamic inputs to the arcuate premotor and supplementary motor areas. *Journal of Neuroscience, 4*, 539–560.

Schiff, H.B., Alexander, M.P., Naeser, M.A., & Galaburda, A.M. (1983). Aphemia: Clinical–anatomic correlations. *Archives of Neurology, 40*, 720–727.

Shankweiler, D. & Harris, K.S. (1966). An experimental approach to the problem of articulation in aphasia. *Cortex, 2*, 287–292.

Shankweiler, D., Harris, K.S., & Taylor, M.L. (1968). Electromyographic studies of articulation in aphasia. *Archives of Physical Medicine and Rehabilitation, 1*, 1–8.

Simmons, N.N. (1978). Finger counting as an intersystemic reorganizer in apraxia of speech. In R. Brookshire (Ed.) *Clinical aphasiology conference proceedings* (pp. 74–179). Minneapolis: BRK Publishing.

Southwood, H. (1987). The use of prolonged speech in the treatment of apraxia of speech. In R. Brookshire (Ed.) *Clinical aphasiology conference proceedings* (pp. 227–287). Minneapolis: BRK Publishing.

Sparks, R. & Deck, J. (1986). Melodic Intonation Therapy. In R. Chapey (Ed.) *Language intervention strategies in adult aphasia* (pp. 320–332). Baltimore: Williams and Wilkins.

Sparks, R. , Helm, N., & Albert, M. (1974). Aphasia rehabilitation resulting from Melodic Intonation Therapy. *Cortex, 10*, 203–223.

Stichfield, S. & Younge, E. H. (1938). *Children with delayed and defective Speech: Moto-kinesthetic factors and their training.* Stanford, California: Stanford University Press.

Square, P.A. (1981). *Apraxia of speech in adults: speech perception and production.* Unpublished doctoral dissertation, Kent State University.

Square, P.A. (1994). Neuromotor speech impairment accompanying aphasia. *Newsletter of the Special Interest Division 2 (Neurophysiology and Neurogenic Speech Disorders) of the American Speech Language and Hearing Association, 4*(1), 11–16.

Square, P.A. (1995). Apraxia of speech reconsidered. In F. Bell-Berti & L.J. Raphael (Eds.), *Providing speech: Contemporary issues.* Woodbury, NY: American Institute of Physics Press.

Square, P.A., Darley, F.L., & Sommers, R.K. (1982). An analysis of the productive errors made by pure apractic speakers with differing loci of lesions. In R. Brookshire (Ed.), *Clinical aphasiology: Conference Proceedings* (pp. 245–250). Minneapolis: BRK Publishers.

Square, P., Hayden (Chumpelik), D., & Adams, S. (1985). Efficacy of the PROMPT system for the treatment of acquired apraxia of speech. In R. Brookshire (Ed.), *Clinical aphasiology conference proceedings* (pp. 319–320). Minneapolis: BRK Publishers.

Square, P., Hayden (Chumpelik), D., Morningstar, D., & Adams, S. (1986). Efficacy of the PROMPT system for the treatment of acquired apraxia of speech. In R. Brookshire (Ed.) *Clinical aphasiology conference proceedings* (pp. 221–226). Minneapolis: BRK Publishers.

Square, P.A. & Martin, R.E. (1994). The nature and treatment of neuromotor speech disorders in aphasia. In R. Chapey (Ed.), *Language intervention strategies in adult aphasia*, (3rd ed., pp. 467–499). Baltimore: Williams and Wilkins.

Square, P.A. & Mcloch, A.G. (1983). The syndrome of subcortical apraxia of speech: Acoustic analysis. In R. Brookshire (Ed.), *Clinical aphasiology: conference proceedings*, (pp. 239–243). Minneapolis: BRK Publishers.

Square-Storer, P.A. (1987). Acquired apraxia of speech. In H. Winitz (Ed.). *Human communication and its disorders: A review—1987* (pp. 88–166). Norwood, NJ: Ablex.

Square-Storer, P.A. & Apeldoorn, S. (1991). An acoustic study of apraxia of speech in patients with different lesion loci. In C.A. Moore, K.M. Yorkston, & D.R. Beukelman (Eds.), *Dysarthria and apraxia of speech: Perspectives on management* (pp. 271–286. Baltimore, ML: Paul H. Brookes.

Square-Storer, P.A., Darley, F.L., & Sommers, R.K. (1988). Speech processing abilities in patients with aphasia and apraxia of speech. *Brain and Language, 33*(1), 65–85.

Square-Storer, P. & Hayden, D. (1989). PROMPT Treatment. In P. Square-Storer (Ed.) *Acquired Apraxia of Speech in Aphasic Adults* (pp. 190–219). Hove, UK: Lawrence Erlbaum Associates Ltd.

Square-Storer, P.A., Qualizza, L., & Roy, E.A. (1989). Isolated and sequenced oral motor posture production under difficult input modalities by left-hemisphere damaged subjects. *Cortex, 25*, 371–386.

Square-Storer, P.A. & Roy, E. A. (1989). The apraxias: Commonalities and Distinctions. In P. Square (Ed.), *Acquired apraxia of speech in aphasic adults* (pp. 20–63). Hove, UK: Lawrence Erlbaum Associates Ltd.

Square-Storer, P. A., Roy, E.A., & Hogg, S.C. (1990). The dissociation of aphasia from apraxia of speech, ideomotor limb and buccofacial apraxia. In G.R. Hammond (Ed.), *Cerebral control of speech and limb movements. Advances in psychology, 70*, 451–476. Amsterdam: North–Holland.

Tanabe, M. & Ohigashi, Y. (1982). Broca's area and Broca's aphasia: Based on the observations of two cases with lesions involving the Broca's area. *No to Shinkei, 34*, 797–804.

Thelan, E. (1981). Rhythmical behaviour in infancy: An ethological perspective. *Developmental Psychology, 17*, 237–257.

Tognolo, G. & Vignolo, L.A. (1980) Brain lesions associated with oral apraxia in stroke patients: A Clinico-neuroradiological investigation with CT scan. *Neuropsychologia, 18*, 257–272.

Tonkonogy, J. & Goodglass, H. (1981). Language function, foot of the third frontal gyrus, and rolandic operculum. *Archives of Neurology, 38*, 486–490.

Trost, J.E. & Canter, G.J. (1974). Apraxia of speech in patients with Broca's Aphasia. *Brain and Language, 1*, 63–79.

Tseng, C., McNeil, M.R., Adams, S.G., & Weismer, G. (1990). Effects of speaking rate on bilabial asynchrony in neurogenic populations. *Journal of the American Speech and Hearing Association, 32*, 109.

Tuch, B.E. & Nielsen, J.M. (1941). Apraxia of swallowing. *Bulletin of the Los Angeles Neurological Society, 6*, 52–54.

Van Riper, C. (1963). *Speech correction: Principles and methods.* Englewood Cliffs, NJ: Prentice-Hall.

Wallesch, C.-W. (1990). Repetitive verbal behaviour: Functional and neurological considerations. *Aphasiology, 4*, 133–153.

Washino, K., Kasai, Y., Uchida, Y., & Takeda, K. (1981). Tongue movements during speech in a patient with apraxia of speech. *Current issues in neurolinguistics: A Japanese contribtion* (Supplement to Language Sciences). Tokyo: International Christian University.

Watson, R.T., Fleet, S., Rothi, L.J.G., & Heilman, K. (1986). Apraxia and the supplementary motor area. Archives of Neurology, 43, 787–792.

Wernicke, C. (1874). *Der aphasische Symptomenkomplex.* Breslau: Cohn & Weigert.

Wernicke, C. (1885). Die neueren Arbeiten über Aphasie. *Fortschritte der Medizin, 3*, 824–830.

Wertz, R.T., LaPointe, L.L., & Rosenbek, J.C. (1984). *Apraxia of speech in adults: The disorder and its management.* New York: Grune & Stratton.

Wilson, S.A.K. (1908). A contribution to the study of apraxia. *Brain, 31*, 164–216.

Wise, S.P. (1985). The primate premotor cortex: Past, present, and preparatory. *Annual Review of Neuroscience, 8*, 1–19.

Wise, S.P. & Strick, P.L. (1984). Anatomical and physiological organization of the non-primary motor cortex. *Trends in Neuroscience, 7*, 442–446.

12 Subcortical Limb Apraxia

Bruce Crosson

Any behavioural neuroscientist who peruses the literature on limb apraxia and subcortical dysfunction inevitably will arrive at two conclusions: First, the phenomenon is not well studied, with a relatively small number of investigations addressing the issue. Second, there is little consensus regarding the existence of the phenomenon and, therefore, little agreement about the role of subcortical structures in praxis. The literature which does exist on this subject has numerous problems. Foremost among them is that the definition of praxis varies from study to study, and the definition determines what tests of praxis are used. Some authors have heavily weighted their measures with the imitation of nonsense hand postures and movements. However, in this chapter we will apply the more classical definition of praxis as learned, skilled movements (Heilman & Rothi, 1993).

Given the state of the literature, one should immediately ask: What is the purpose of writing a chapter on apraxia accompanying subcortical dysfunction? Paillard (1982) suggested that a great deal could be learned in the process of developing models for praxis, and in keeping with Paillard's observation, the focus of this chapter will be in large part heuristic. The question of what role subcortical structures play in praxis is one which arises naturally from a more general interest in how brain systems produce behaviour. But, one must ask: What conceptual frameworks can we tap to generate hypotheses about subcortical functions in praxis?

Most behavioural neuroscientists would agree that the basal ganglia and thalamus play some role in movement, though the specifics of this role have been debated. Because praxis involves the execution of learned, skilled movements (Heilman & Rothi, 1993), it is of interest to learn how the basal ganglia and thalamus may participate in such movements as opposed to other motor functions. Thus, the literature on subcortical structures and movement might contain some clues useful in developing models of praxis. Examining parallels between language and praxis also may suggest hypotheses regarding how the basal ganglia and thalamus participate in learned skilled movement. Indeed, Ellis and Young's (1988) lexical–semantic model for word processing proved to be a productive starting point for the cognitive neuropsychological model of praxis discussed in Chapter 4 of this volume.

The use of language models to generate hypotheses about praxis assumes some parallels between language and praxis. The parallels important to this discussion can be summarised as follows. Language and praxis are both complex, goal-directed cognitive activities which are dependent on stored representations and processes heavily involving the dominant hemisphere. Stored representations for these activities include both the meaning or purpose of a word or gesture (semantics) and the sensory and motor forms of a word or gesture (which can be referred to as its lexical aspects). For language, syntax involves the ordering and relationships between words in sentences and phrases. The ordering of and relationship between component actions appears to be important for praxis as well (Jeannerod, 1994; Paillard, 1982). Nonetheless, as I shall discuss shortly, there are limitations in the parallels between praxis and language.

Thus, this chapter will begin by asking the following question: What are the possible roles of the basal ganglia and thalamus in praxis? First, general functions of these structures will be discussed. In particular, I will discuss whether these structures perform information processing or regulatory functions. Subsequently, the focus will shift to subcortical models of language functions and movement, and how they might be applied to praxis. Once the possible roles have been established, the small empirical literature on the subject will be reviewed. Finally, a synthesis of the possible roles with the available data will lead to some tentative conclusions about the role of the basal ganglia and thalamus in praxis and suggest directions for future research.

NEUROPHYSIOLOGICAL BASES OF SUBCORTICAL MECHANISMS

In all cognitive activity, including praxis, neural structures must perform information processing functions: the decoding, modification, or generation of information coded in neural signals. Generation of an idea, transformation of an idea into language symbols, and comprehension of spoken language are examples of information processing. Regulatory functions also are involved in cognition. Such regulatory functions may take place over longer periods of time, for example, optimal wakefulness facilitates cognitive activity, or regulatory functions may involve rapid transitions, such as the initiation of a spoken phrase.

Divac, Oberg, and Rosenkilde (1987) have discussed the different types of neural activity underlying information-processing versus regulatory functions. According to these authors, information is coded in the nervous system through temporal-spatial patterns of neural activity. In other words, the pattern of firing in a single neuron and the temporal sequences of firing in ensembles of neurons are the means through which the nervous system encodes information. On the other hand, the quantity of neural activity may serve to regulate, in some fashion, the activity of the target structure. Dopaminergic input to the basal ganglia probably operates on a quantitative basis, having tonic influences on voluntary muscle activity. Patterned and quantitative activity may interact in various ways in the nervous system (Divac et al., 1987).

It is important to consider whether structures like the basal ganglia and thalamus contribute to various cognitive activities through actual information processing, which is likely accomplished through patterned neural activity, or through regulating the activity of other structures, which is likely based on quantitative activity. To complicate matters further, it is possible that both types of activity may be relevant and interact at the subcortical level. The mode of operation of subcortical structures has a profound influence on the role of the structure in cognition.

THEORIES OF SUBCORTICAL FUNCTIONS IN LANGUAGE AND MOVEMENT: IMPLICATIONS FOR PRAXIS

As noted above, similarities between language and praxis justify a look at language theories to generate hypotheses about praxis. Indeed, in Chapter 4 of this volume, as well as elsewhere (Rothi, Ochipa, & Heilman, 1991), Rothi and her colleagues have proposed a model of praxis which not only parallels a lexical–semantic processing model proposed by Ellis and Young (1988) but also demonstrates how lexical–semantic and praxis processing may interact. Thus, a precedent exists for using models of language to generate hypotheses about apraxia, and we will follow this precedent.

Another place to look for analogues to generate hypotheses about praxis is the literature on movement. Since praxis refers to a particular category of movement (i.e. learned, skilled movements), praxis is likely to have at least some aspects in common with other types of movements. For instance, the way in which various programs are combined to produce a movement sequence is relevant to praxis. The differences between praxis and other types of movements may also be instructive. For example, one way in which praxis differs is having stored representations (i.e. a lexicon) to guide movement production (see

Chapter 4), whereas for other types of actions, the sequence of movements have to be formulated *de novo*. Thus, we will include some concepts from the literature on subcortical structures in movement in the ensuing discussion. In the following paragraphs on theories, potential roles for the basal ganglia will be covered first. Subsequently, potential roles for thalamic structures in praxis will be explored.

Hypotheses Regarding the Role of the Basal Ganglia in Language and Movement: Implications for Praxis

Typically, the term basal ganglia has been used to refer collectively to a group of deep brain structures that include the putamen, caudate nucleus, and globus pallidus. These structures are represented in Fig. 1. The internal structure and connections of the putamen and caudate nucleus are similar, and together these structures are referred as the neostriatum. The nucleus accumbens and olfactory tubercle have similar connections to the neostriatum, except that they are more closely related to the limbic system. For these reasons, these latter two structures are sometimes referred to as the archistriatum. The reader wishing more information may consult a basic neuroanatomy text such as Parent (1996). The basal ganglia are involved in cortico-striato-pallido-thalamo-cortical loops. Areas of anterior and posterior cortex project into a specific area of the striatum which in turn projects to the globus pallidus. The globus pallidus projects to the thalamus, which then projects to the anterior cortical region projecting to the striatal portion of the loop. This chapter will assume some knowledge of these loops. The unfamiliar reader may wish to consult works describing them (e.g. Alexander, DeLong, & Strick, 1986; or Strick, Dum, & Picard 1995).

The role of the basal ganglia in language continues to be a controversial subject. No specific aphasia syndromes with basal ganglia lesions can be defined (Crosson, 1992b; Nadeau & Crosson, in press), and the existence of aphasic syndromes caused by basal ganglia dysfunction has been questioned. Nonetheless, examination of theoretical positions regarding the role of the basal ganglia in language may suggest possibilities regarding their role in praxis. Such hypotheses have run the gamut from no significant role for the basal ganglia in language (Alexander, 1992; Alexander, Naeser, & Palumbo, 1987) to a role in regulatory functions (Crosson, 1985; 1992b) to a role in information processing (Wallesch, 1985; Wallesch & Papagno, 1988). We shall review each of these positions and their relevance to praxis. However, there are likely limitations in the parallels that can be drawn between praxis and language. As noted above, models of movement can be another source for hypotheses about the basal ganglia in language.

The importance of the basal ganglia in movement is, of course, generally accepted. However, the specific movement functions of these structures have been debated. Hypotheses have included running preformulated movement

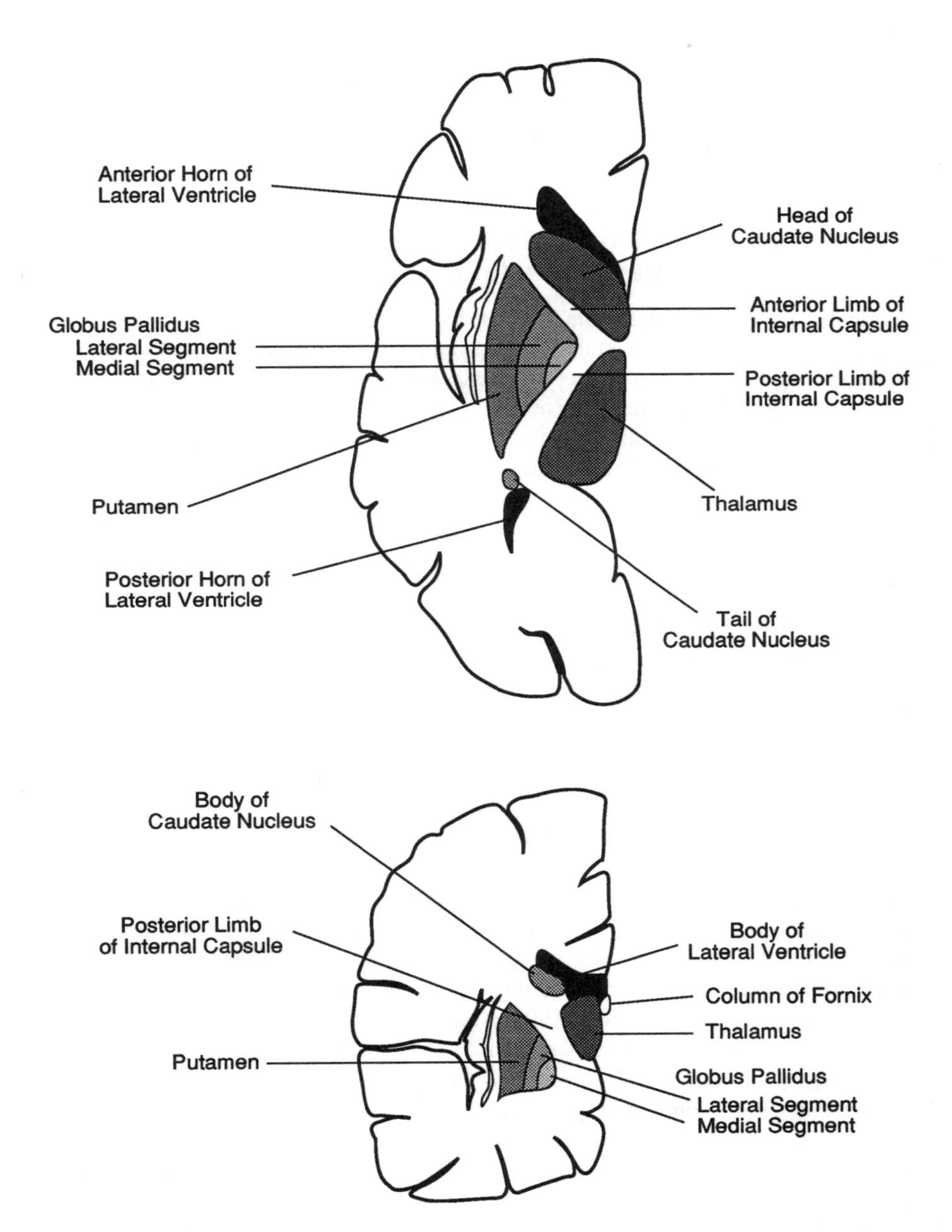

FIG. 1. Sketch of horizontal and coronal sections through the human basal ganglia (from Crosson, 1992, *Subcortical functions in language and memory*. New York: Guilford Press, p.12. Reprinted by permission).

plans (e.g. Wing & Miller, 1984), selecting appropriate movements and inhibiting unwanted movements (e.g. Penney & Young, 1986), and turning off one activity to allow another to operate and combining coactive motor programs (Mink & Thach, 1991). Hypotheses which deal primarily with movement execution may not be so relevant to praxis as those which deal with movement planning. Thus, the review of movement-related hypotheses will be selective.

Hypothesis 1: The Basal Ganglia Play no Direct Role in Language or Praxis. Alexander and his colleagues have strongly advocated little or no role for the basal ganglia in language. In their 1987 paper, these investigators noted that lesions involving the basal ganglia usually involve a good deal of white matter as well, and they attributed the aphasias following nonthalamic subcortical lesions to damage in various white matter structures. Alexander (1992) examined some cases in which lesions were confined to the basal ganglia and/or nearby white matter structures. He divided fluency into those aspects having to do with the motor act of speech (i.e. articulatory agility and melodic line) and those having to do with language proper (variety of grammatical forms and phrase length). Patients with basal ganglia lesions were frequently impaired on speech fluency (i.e. motor execution). Language structure generally was characterised as fluent with variable word finding difficulty and semantic paraphasias, frequent truncated or elided responses, and initiation deficits. Alexander argued that the latter symptoms carried little instrumental specificity, and on the basis of functional imaging (PET) findings, he suggested that the severity of symptoms was related to loss of facilitation in the perisylvian language cortex.

Nadeau and Crosson (in press) also reached the conclusion that the basal ganglia are minimally involved in language, but explained the existence of language symptoms after nonthalamic subcortical lesion in a different way. These authors noted the variability in nonthalamic subcortical infarctions. Whether one examines language symptoms after the involvement of specific subcortical structures (Crosson, 1992b) or after infarcts specifically limited to the putamen, caudate head, and anterior limb of the internal capsule (Nadeau & Crosson, in press), no specific syndromes have consistently been found. In fact, most conceivable combinations of fluency, comprehension, and repetition deficits were present. Nadeau and Crosson reasoned that this variability was related to the vascular dynamics. Such stratio-capsular lesions might be the result of blockage of the initial segment of the middle cerebral artery. Sufficient viability of various cortical areas to prevent cystic infarction detectable by CT or MRI scan may be maintained by end-to-end anastomosies between the middle cerebral artery circulation on the one hand and the posterior and/or anterior cerebral artery circulation on the other. However, the viability of these anastomoses may vary, and some cortical areas in the middle cerebral artery territory may suffer from either ischemic neuronal drop out or insufficient circulation to maintain normal tissue function. The variability in location of functional anastomosies between patients would account for the variability in symptoms. Like Alexander (1992), Nadeau and Crosson (in press) suggested the role of the basal ganglia in language was at best indirect.

To summarise, Alexander (1992) and Nadeau and Crosson (in press) both have suggested little direct role for the basal ganglia in language processes. Alexander was careful to separate aspects of language fluency which are intact with basal ganglia and white matter lesions from speech fluency which is impaired in these cases. From these conclusions, we can generate our first hypothesis regarding the basal ganglia in praxis: Short of actual motor execution, the basal ganglia play no direct role in praxis.

Hypothesis 2: The Basal Ganglia Control the Release of Action Segments. In addressing motor functions of the basal ganglia, Wing & Miller (1984) suggested that the basal ganglia played a role in activating preplanned movements. Marsden (1984) hypothesised that the basal ganglia automatically ran a sequence of motor programs that constituted a motor plan. Similarly, Crosson (1992b) proposed that the basal ganglia signaled anterior cortical areas to release a language segment for motor programming after the segment had been monitored for semantic accuracy. According to this view, language was formulated and then put into a buffer in the more anterior perisylvian regions of the left hemisphere and then monitored for semantic accuracy via cortico-thalamo-cortical pathways. Once semantic accuracy has been verified, convergence of cortical inputs to the striatum would trigger a series of events leading the basal ganglia (via the cortico-striato-pallido-thalamo-cortical loop) to signal the cortex to release the segment in the buffer for motor programming.

A similar role for the basal ganglia in praxis can be proposed. Some evidence suggests that the basal ganglia may be involved directly or indirectly in triggering movement onset in sequences of movements. For example, Marsden (1987) found that when Parkinsonian patients tried to perform two movements in rapid succession, the onset of the second movement was prolonged compared with normal controls. Marsden hypothesised that the basal ganglia used a readout of the current movement parameters from the cortex to set up the premotor cortex to select the parameters of the next movement in the sequence. Onset time is one such parameter.

This conclusion leads us to our second hypothesis: The basal ganglia trigger release of cortically assembled action segments (e.g. language and praxis segments) for motor programming. Although such a function could be performed after information processing occurs at the level of the striatum, it does not require that the basal ganglia actually be involved in the information processing requirements of such language or praxis segments. Rolls and Johnstone (1992) discussed evidence that the striatum receives from the cortex a signal resulting from cortical processing but is not privy to all the information which the cortex must process. Crosson (1992a) had suggested that the striatum could act as a pattern-to-quantity converter by processing only the strength of signals it receives from divergent cortical areas comprising the language network. According to this view, release of a segment for motor programming happens when adequate signals converge within the basal ganglia.

The notion that striatal output was quantitative in nature was based upon the assumption that electrical stimulation of a structure will disrupt patterned neural activity but could mimic increases in quantitative neural activity (Divac et al., 1987). Evidence suggests that language segments are released when the dominant striatum (caudate head) is stimulated (Van Buren, 1963; 1966; Van Buren, Li, & Ojemann, 1966), or when the thalamic nuclei to which the basal ganglia (globus pallidus) project are stimulated (Schaltenbrand, 1965; 1975). While these studies have not been replicated, the elicitation of language segments from cortical stimulation has never been reported, probably because the patterned activity necessary to produce language segments cannot be mimicked by stimulation. The absence of evoked language with cortical stimulation makes it difficult to dismiss the elicitation of language segments with subcortical stimulation. It should be noted that the proposed role for the basal ganglia involves a function necessary to support extended language or praxis sequences, but one might argue that the proposed role does not directly involve language or praxis "calculations."

Hypothesis 3: The Basal Ganglia are Involved in Combining Coactive and Sequential Movements During Praxis. Marsden (1987) cited evidence that the basal ganglia are involved in more than the onset timing for sequential movements. In addition to abnormally prolonged onset for the second movement in a sequence, both simultaneous and sequential movements in Parkinsonian patients are performed more slowly than when the same movement is performed alone, suggesting that the basal ganglia have some significant role in coactive or sequential programs. In addition to controlling onset times between movements, the size of the initial agonist burst for sequential or coactive movements may be affected by the readout of the current motor activity. Indeed, many of Marsden's (1987) Parkinsonian patients found it difficult if not impossible to learn to perform two simple movements in sequence. Similarly, based on their own observations of learned movements in monkeys, Mink and Thach (1991) have suggested that the basal ganglia are involved in switching off one activity to allow another to operate and in blending coactive motor programs.

This leads us to our third hypothesis: The basal ganglia are involved in combining the coactive and sequential movements necessary for praxis. For example, in hammering a nail, the hammer (real or imagined) must be gripped while the arm is swinging the hammer toward the target. The present hypothesis suggests that basal ganglia are involved in combining the motor program for gripping the hammer and the motor program for swinging the hammer toward the nail. If Marsden's (1987) observations can be applied to praxis, then we might expect lesser impairment if these portions of the movement program were performed separately.

Hypothesis 4: The Basal Ganglia are Involved in Selection of Lexical or Praxis Alternatives. Wallesch and his colleagues (Brunner, Kornhuber, Seemuller, Suger, & Wallesch, 1982; Wallesch, 1985; Wallesch & Papagno, 1988) have proposed an alternative model of language production which suggests the striatum does possess information processing capacities. Wallesch has hypothesised that multiple lexical alternatives to express an idea are generated in the posterior perisylvian cortex and conveyed to the anterior perisylvian cortex and the striatum in parallel modules. The function of the basal ganglia is to monitor the lexical alternatives and choose the best suited given various constraints. Because the striatum receives input from a variety of cortices, including limbic cortex, the basal ganglia are in a position to integrate various situational and motivational factors in selecting lexical alternatives. Once the best alternative is chosen, it is "selected" via a cortico-striato-pallido-thalamo-cortical loop which acts to dampen inappropriate alternatives. In some respects, Wallesch's lexical selection hypothesis fits with Alexander's (1992) data on nonthalamic subcortical lesions in which some patients experienced substantial naming problems.

The idea that the basal ganglia are involved in selection of specific actions to be performed has also appeared in the movement literature. Penney and Young (1986) suggested that the pathway from the cortex through the striatum, medial globus pallidus, and thalamus to the premotor (supplementary motor) cortex is involved in facilitating desired movements. On the other hand, the pathway from the cortex through the striatum, lateral globus pallidus, subthalamic nucleus, medial globus pallidus, and thalamus to the premotor (supplementary motor) cortex acts to suppress unwanted movements. This theme continues in subsequent literature; for example, Mitchell, Jackson, Sambrook, and Crossman's (1989) 2-deoxyglucose study of experimental chorea seemed to confirm the role of the latter pathway in suppressing unwanted movement. Gerfen (1992) has explored in some detail the histochemical mechanisms of the basal ganglia by which behaviour may be "facilitated or disfacilitated." Although the pathway through the subthalamic nucleus is probably not involved in cognition, Wallesch and Papagno (1988) suggested an alternative means through which less desirable lexical items are suppressed.

Finally, the idea that the basal ganglia play a role in selecting the most suitable motor programs for praxis, based on the intended goal, has previously appeared in the literature. Denny-Brown and Yanagisawa (1976) hypothesised the basal ganglia could accumulate samples of cortically projected activity and facilitate some actions while suppressing others. Paillard (1982) suggested that the basal ganglia select the best action strategy from various alternatives based on the specified goal and the contextual constraints. According to this model, the frontal cortex processes information relevant to goal specification while the parietal cortex processes somatosensory information and environmental cues relevant to the goal. Both channel information to the basal ganglia where selection occurs. Subsequently, sensory and motor structures are pretuned to enable execution of the act, and the outcome of the act is compared to the expected consequences.

This leads to our fourth hypothesis: The basal ganglia are involved in the selection of language and praxis patterns which are most appropriate to the external circumstances and internal motivation. As noted above, this theoretical position suggests that the basal ganglia are involved in information processing for the purpose of choosing the most appropriate alternative. For language, the level at which selection occurs is explicit: it is the lexical level. In attempting to apply this model to praxis, the level at which the basal ganglia might operate is not intuitively clear. Could the basal ganglia select from whole movement patterns such hammering a nail versus sawing a board? Or, could the basal ganglia select from possible variations within a specific pattern, such as how to hold a hammer and how to move the hand and arm in space given the external constraints?

Hypotheses Regarding the Role of the Thalamus in Language and Movement: Implications for Praxis

The position of the thalamus relative to the basal ganglia is shown in Fig. 1. The thalamus can be divided into several nuclei, which are depicted in Fig. 2. The reader unfamiliar with thalamic anatomy may wish to consult a basic neuroanatomy text such as that of Parent (1996) or even Jones' (1985) work about the thalamus. For praxis, the most relevant nuclei include: the ventral lateral nucleus which is a part of the basal ganglia motor loop, the centromedian/parafascicularis complex which may selectively engage cortical mechanisms, and the pulvinar which projects to cortical regions involved in praxis.

Theoretically, the role of the thalamus in language is more secure than the role of the basal ganglia in language. After reviewing the literature on thalamic aphasia, Crosson (1984) concluded that four characteristics generally described thalamic aphasia: (1) predominance of semantic over other types of paraphasia; (2) paraphasia sometimes deteriorating into jargon; (3) language comprehension better than output would normally indicate; and (4) minimally impaired or unimpaired repetition. These conclusions were based more on cases of hemorrhagic lesion than infarction, and cases of dominant thalamic infarction reported since 1984 (see Crosson, 1992b) indicate that the syndrome is more variable than originally concluded. Nonetheless, one attribute does consistently characterise thalamic aphasia: cases of dominant thalamic lesion with aphasia very rarely show repetition which is more than minimally impaired. The reader will

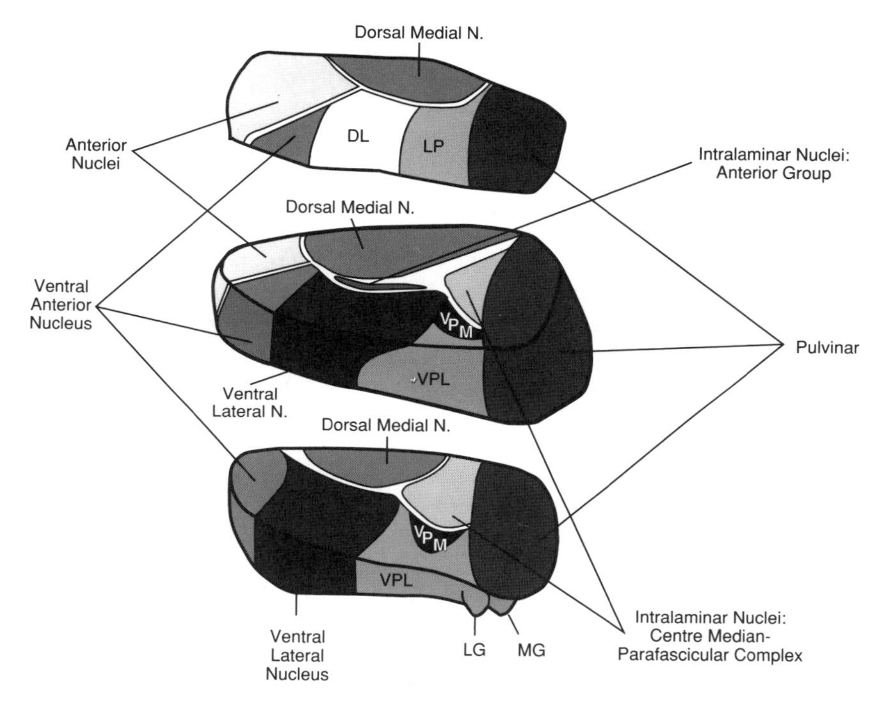

FIG. 2. Sketch of the nuclei of the human thalamus with two horizontal cuts (from Crosson, 1992, *Subcortical functions in language and memory*. New York: Guilford Press, p.24. Reprinted by permission).

recall that such consistency is absent in lesions which include basal ganglia structures, and it suggests that the thalamus plays some role in language. Further, it is not at all uncommon to see vascular lesions confined to the thalamus, a different state of affairs than for basal ganglia lesions which frequently include surrounding white matter.

The consistency of at least some aspects of thalamic aphasia in addition to the fact that isolated thalamic lesions occur has led many students of subcortical structures in language to conclude that the thalamus plays a direct, or at least indirect, role in language. Proposed roles for the thalamus in language have ranged from integrating language functions (e.g. Penfield & Roberts, 1959) to creating an optimal level of arousal for language (e.g. Horenstein, Chung, & Brenner, 1978; McFarling, Rothi, & Heilman, 1982) or directing activation important for language (e.g. Cooper, Riklan, Stellar, Waltz, Levita, Ribera, & Zimmerman, 1968; Samra, Riklan, Levita, Zimmerman, Waltz, Bergmann, & Cooper, 1969). The following paragraphs will focus on two more recent derivations of these positions: (1) the involvement of the thalamus in pre-execution feedback loops; and (2) the involvement of the thalamus in selective engagement of cortical processors. A third hypothesis regarding declarative stores also will be explored.

Hypothesis 5: Thalamic Mechanisms are Involved in Pre-execution Feedback for Language and Praxis. Crosson (1985) had proposed a feedback mechanism which monitored language segments for semantic accuracy before releasing them into an output stream. If such a feedback mechanism existed, he reasoned that damage to the feedback mechanism would lead to semantic paraphasias, perhaps severe enough to deteriorate into jargon, and comprehension less impaired than output. Repetition could be accomplished through intact phonological mechanisms. This pattern, of course, is frequently seen in thalamic aphasia. The proposed feedback loop was supposed to monitor frontally formulated language segments for semantic accuracy. A later modification of the theory (Crosson, 1992b) suggested that the feedback could be accomplished through the pulvinar, a thalamic nucleus which projects both to posterior perisylvian cortex and to frontal cortex (Asanuma, Andersen, & Cowan, 1985; Goldman-Rakic & Porrino, 1985). Bechtereva (1987) reviewed evidence from Russian recording studies which indicated that there are neuronal populations in the human thalamus, as well as in the cortex and basal ganglia, which are specifically responsive to errors in task performance. Evidence also exists that cells in the monkey pulvinar show movement-related changes in firing rate well in advance of the earliest changes in motor cortex (Cudeiro, Gonzalez, Perez, Alonso, & Acuna, 1989). This evidence leaves open the possibility that thalamic nuclei, including the pulvinar, play a role in monitoring aspects of movements before they are actually intitiated.

These constructs and obversations lead us to our fifth hypothesis: Thalamic nuclei are involved in monitoring segments of language and praxis for accuracy before they are executed. At some level, such a feedback would have to involve the transfer and conversion of information; therefore, it would have to be dependent on patterned neuronal activity.

We have discussed some problems applying Wallesch and Papagno's (1988) lexical selection model of basal ganglia function to praxis. It was uncertain whether to apply the model at the level of the entire gesture (e.g. pounding a nail) or to different aspects of the movement sequence (e.g. how to hold the hammer, how to swing the arm, relationship to target object). These same concerns apply to the present monitoring hypothesis. At what level(s) is (are) monitoring taking place?

There are some problems with the semantic monitoring hypothesis. Although patients with thalamic aphasia usually have a preponderance of semantic substitutions at the lexical level during discourse or naming (Crosson, 1992b), some evidence suggests this is not the case in all circumstances. Crosson, Parker, Warren, Kepes, Kim, and Tulley (1986) found the preponderance of errors shifted to lexical items phonologically related to the target when their thalamic aphasia patient had to read aloud. This finding raises the possibility that the problem is not semantic per se, but an underspecification of the attributes needed to access the correct lexical item, whether access is attempted from a semantic or a phonological route. Such an underspecification suggests degradation of the signal used to access lexical items in one of three places: (1) at the origin of the signal (i.e. in the semantic or phonological systems); (2) at the reception site of the signal (i.e. at the lexicon itself); or (3) in the connections that carry the signal to the lexicon (i.e. the white matter). Nadeau and Crosson (in press) have offered an alternative interpretation of thalamic participation in language which might account for lexical access problems.

Hypothesis 6: Thalamic Mechanisms Participate in Selective Engagement of Cortical Processors Necessary for Language Output and Praxis. The hypothesis of Nadeau and Crosson (in press) was based on observations about the nucleus reticularis thalami made by McCormick and Feeser (1990); Steriade, Domich, and Oakson (1986); Watson, Valenstein, and Heilman (1981); Yingling and Skinner (1977) and others. The nucleus reticularis is a thin shell of grey matter surrounding most of the ventral, lateral, and anterior surfaces of the thalamus (Jones, 1985). It receives inputs from thalamic nuclei, the cerebral cortex (including the frontal lobes), and the midbrain reticular formation. It sends inhibitory gamma amino butyric acid (GABA) fibres to thalamic nuclei. Further detail regarding the anatomy and physiology of the nucleus reticularis and related structures is too complicated to address in the limited space available for this chapter. However, the main features of the Nadeau and Crosson (in press) construct are as follows.

Fronto-thalamic axons pierce the anterior portion of the nucleus reticularis. At this point, collaterals from these axons form synapses with dendrites from neurons that reside in more posterior portions of the nucleus reticularis. These nucleus reticularis neurons give off GABAergic fibres which, in turn, synapse with local GABAergic inhibitory neurons in various thalamic nuclei, including the centro-median/parafascicularis complex. Thalamic neurons from the centro-median/parafascicularis complex send fibres to various cortical regions. When an organism intends to perform a particular cognitive activity, this frontal nucleus reticularis-thalamic mechanism is used to selectively engage areas of the cortex necessary to perform the desired cognitive activity. Frontal neurons excite nucleus reticularis neurons which inhibit local inhibitory interneurons in the thalamus, freeing the thalamic neurons to engage the appropriate cortical mechanisms. Note that there must be some spatial relationships which are maintained throughout such a system in order for frontal neurons to selectively engage other cortical areas needed for an activity.

This conjecture leads us to our sixth hypothesis: Thalamic mechanisms are involved in a system by which the frontal cortex selectively engages other cortical areas necessary for language or praxis tasks. Selective engagement does not imply information processing per se, and therefore could be accomplished by quantitative neuronal activity. Deficient selective engagement would cause a decrease in signal-to-noise ratios in the needed cortical systems, resulting in decreased efficiency. This concept could explain the degradation in lexical selection seen in many cases of thalamic aphasia (Nadeau and Crosson, in press), and could also explain why the type of lexical error appears to depend upon the way in which the system is accessed. Degradation of a signal carrying phonological information would cause phonological errors in lexical selection, and degradation of a signal carrying semantic information would cause semantic errors.

Again, the level at which such a hypothesis could be applied to praxis must be determined. Would degradation of signals be likely to affect the whole gesture or the component movements of the gesture? If praxis involves tightly bound, invariant action sequences, then one would expect misselection of whole gestures from the gestural output lexicon. However, if one views praxis gestures as a more general program which governs the selection of more specific components, then one might expect errors to be made at the component level.

Hypothesis 7: The Thalamus is Involved in Aspects of Praxis which Depend on a Readout from Declarative Memory. Recent conceptualisations (e.g. Squire, 1987, 1992; Squire & Zola-Morgan, 1991) divide memory into declarative and nondeclarative processes. Declarative memory consists of knowlege of facts and events which are available to conscious recollection, and therefore can be declared. Nondeclarative memory refers to a conglomeration of processes in which behavioural change is demonstrated after exposure to stimuli or repeated performance of an activity, but is not available to conscious recollection unless a concomitant declarative memory has been established. Nondeclarative memory includes skills and habits, classical conditioning, and priming.

Superficially, since praxis has been defined as learned, skilled movement, it would seem to fall under the rubric of motor skill, which involves nondeclarative memory. However, Nadeau, Roeltgen, Sevush, Ballinger, and Watson (1994) have noted that aspects of declarative memory are most likely involved in praxis. Foremost among this declarative knowledge are concepts of what types of tasks are accomplished by each individual tool. It has been suggested that this knowledge, at least to some extent, is a part of a separate semantic system for actions (see Chapters 4 and 5). Further, it has been suggested that there is a gestural output lexicon which contains movement formulae for the production of learned, skilled actions (again, see Chapter 4). To the extent such formulae can be consciously evoked for the purpose of demonstrating tool use, they can be considered declarative. Nadeau et al. (1994) reasoned that when praxis is tested in the absence of a tool, declarative knowledge is more likely to be needed. For example, if the knowledge of the purpose for which a tool is used cannot be accessed, then the apraxic patient will have difficulty demonstrating how one would use the tool. On the other hand, when a tool is actually in such a patient's hand, procedural memory for the actions involved in using the tool are more likely to be evoked.

Nadeau et al. (1994) cited evidence that some thalamic nuclei might be more involved in declarative than procedural memory. For example, most patients with thalamic aphasia appear to have difficulty in lexical–semantic functions, leaving grammar and phonology more intact (e.g. Raymer, Moberg, Crosson, Nadeau, & Rothi, 1992). Lexical–semantic functions can be considered declarative in nature while phonological and grammatical processes are procedures that are usually used without much conscious thought. Also, alcoholic Korsakoff's patients whose lesions include the dorsal medial thalamus have difficulty establishing declarative memories but are able to establish procedural memory (Cohen & Squire, 1980; Martone, Butters, Payne, Becker, & Sax, 1984; Talland, 1965). Nadeau et al. hypothesised that patients with thalamic lesion would have difficulty on praxis tasks when they had to rely on declarative memories for the learned, skilled movements.

This analysis leads us to our seventh hypothesis: The thalamus is involved in aspects of praxis that require a readout from declarative memory. In particular, Nadeau et al. (1994) noted that gesturing tool use in the absence of the actual tool requires access to knowledge about tool use. These authors have argued on the basis of this distinction that patients with thalamic lesions should have difficulty with the recognition of gestures because this involves access to the same semantic knowledge. Thus, Nadeau et al. do take a specific stand that access to gestures as a whole, not just parts of gestures, should be impaired in cases of thalamic apraxia.

RECAP OF HYPOTHESES REGARDING SUBCORTICAL FUNCTIONS IN PRAXIS

We have explored the possibility that subcortical structures might be involved in limb praxis either by quantitative or by patterned neural activity. Quantitative activity would involve either tonic or phasic regulation of target structures but would have little informational value. Patterned activity would be involved in generating, altering, or transferring information and would be capable of passing information on to target structures (Divac et al., 1987). Empirically, it is difficult to determine whether the basal ganglia and thalamus are involved in praxis by quantitative or by patterned neural activity. If techniques for resolving this issue could be developed, then we would have some idea which of our seven hypotheses were correct. Some hypotheses would require patterned activity in subcortical structures to accomplish information-processing functions, and others could be accomplished on the basis of quantitative activity.

The four hypotheses we developed regarding the participation of the basal ganglia in praxis range from the concept of no significant participation, to the release of already formulated responses into an output stream, to coordination of coactive and sequential motor programs, to selection of the most appropriate item from the gestural output lexicon. If Marsden (1987) is correct, then the basal ganglia could be simultaneously involved in the release of responses into an output stream and in the coordination of coactive motor programs. The last of these hypotheses, regarding the selection of the most appropriate activity, requires information processing, and therefore patterned neuronal activity. Once this selection process is completed the basal ganglia might or might not be involved in triggering its implementation. Thus, the last three hypotheses regarding basal ganglia function are not necessarily mutually exclusive.

The three hypotheses regarding the role of the thalamus have included participation in a preresponse feedback loop to monitor the motor plan for accuracy, selective engagement of the necessary cortical structures for processing praxis, and involvement in the delarative aspects of praxis. The feedback hypothesis would require patterned activity while selective engagement most likely would involve quantitative activity. The declarative stores hypothesis does not necessarily require one or the other. It is possible that multiple thalamic systems exist. For example, the selective engagement hypothesis suggests involvement of the intralaminar nuclei, and it is possible other thalamic nuclei are involved in different functions supporting praxis. Thus, these hypotheses are not necessarily exclusive of one another.

The literature on subcortical functions in praxis is small and sometimes confusing. Nonetheless, examination of the data should help refine our hypotheses. Now we turn to that literature.

STUDIES OF SUBCORTICAL STRUCTURES IN PRAXIS

As noted at the beginning of this chapter, the literature addressing apraxia with subcortical dysfunction is sparse, and consensus regarding the definition of praxis is lacking. As in the theoretical discussion, it is convenient to divide the topic by anatomy into studies covering the basal ganglia plus subcortical white matter and those addressing the thalamus. Papers dealing with basal ganglia functions in praxis can be further divided into investigations of vascular lesion and investigations of degenerative diseases affecting the basal ganglia. Studies of thalamic functions in praxis come primarily from vascular lesion.

The Basal Ganglia and Subcortical White Matter and Apraxia

Vascular Lesion Studies. As already noted, the vascular lesion literature regarding the participation of the basal ganglia in praxis is problematic. Not all studies have even defined praxis in the same way which, of course, can lead investigators to draw different conclusions about the existence and severity of apraxia. Frequently, studies of subcortical lesions have either not assessed for praxis or have not reported findings in any detail. From a localisation standpoint, lesions of the basal ganglia are usually accompanied by damage to the surrounding white matter, making it difficult to distinguish the effects of gray matter versus white matter lesions. Finally, the intricacies of vascular lesions may make it difficult to say with great confidence that the lesion seen on the scan is the sole cause of behavioural and cognitive dysfunction. Acutely in cases of haemorrhage, pressure effects from a haematoma under dynamic tension can be expected to produce effects distant from the lesion site. Unfortunately, such pressure effects can lead to reduction in blood flow resulting in ischemic neuronal dropout, which is not detectable on CT or MRI scanning and which is likely to affect patients on a chronic basis. In cases of infarction, there can be areas of decreased blood flow which are not sufficient to cause cystic infarction but are sufficient to keep the tissue from functioning normally or to cause ischemic neuronal dropout without cystic infarction (Heiss, 1992; Lassen, Skyhoj-Olsen, Hajgaard, & Shriver, 1983; Skyhoj-Olsen, Bruhn, & Oberg, 1986). With these cautions in mind, we can explore the literature on apraxia with subcortical lesion. The discussion is organised chronologically.

Basso, Luzzatti, and Spinnler (1980) studied 123 right-handed cases of left-hemisphere stroke. Based on an examination devised by De Renzi, Pieczuro, and Vignolo (1968), patients were asked to imitate 10 intransitive (i.e. symbolic) gestures with the left hand. Lesion location was obtained from CT scans within 10 days of praxis testing. Patients were divided into four groups based on chronicity (≤90 days vs. >90 days) and praxis findings (apraxic vs. nonapraxic). There was a greater proportion of patients with deep lesions in the nonapraxic group acutely (≤ 90 days) and, apraxia, as measured in this study, was rare (four of 26, or 15%) in patients with purely deep lesions. Problems with the measure of praxis in this study should be noted. First, only intransitive movements were used in the assessment, and evidence suggests that intransitive movements are less likely to be disrupted than transitive movements (Goodglass & Kaplan, 1963). Second, although some patients fail gestural imitation with intact gesture to command (Heilman & Rothi, 1993), imitation of limb gestures is generally less difficult than performing the movement to command (Goodglass & Kaplan, 1983). Thus, these authors' measure of praxis not only avoided transitive actions but also may have been less sensitive to apraxia because it studied imitation rather than pantomime to command. This insensitivity is important because the test may have failed to diagnose milder forms of apraxia and milder forms of apraxia tend to be seen with subcortical (deep) lesions.

Alexander and LoVerme (1980) described six cases of aphasia after dominant intracerebral haemorrhage centering on the putamen. All patients were right handed and all experienced some degree of hemiparesis. Four of these six patients were tested for apraxia, and three showed mild and one moderate apraxia. The authors noted that apraxic errors consisted exclusively of using body part as object[1], e.g. using the hand as if it were a pair of scissors rather than holding scissors.

Agostoni, Coletti, Orlando, and Tredici (1983) used the same test of intransitive gestures along with a test of object usage in seven patients with purely deep lesions (haemorrhage or infarct). In three cases, the lesion was in the right hemisphere, and in the remainder, the lesion was in the left hemisphere. Handedness was not mentioned. In four cases, the lesion appeared to involve the putamen and surrounding white matter; in two cases, the lesion appeared to involve primarily the thalamus or the thalamus and posterior limb or the internal capsule; and in one case there was substantial involvement of the thalamus, the putamen, and the intervening white matter. Five of seven patients had contralateral motor symptoms. Contrary to the report of Basso et al. (1980), all patients were reported to have impaired production of intransitive gestures, though the impairment was mild. Since the same intransitive gestures and the same cut-off score for apraxia was used in both the Basso et al. and Agostoni et al. studies, content and scoring criteria cannot account for the difference between studies. The Agostoni et al. study does not say if gestures were requested verbally or by imitation, but since they cited the original De Renzi et al. (1968) article, it is presumed they followed the origninal procedure using imitation.

Kertesz and Ferro (1984) studied 177 cases of ischemic left-hemisphere stroke, all of whom were right handed. Many patients were studied in both the acute and chronic phases, though some were only studied in one phase or the other. The test of praxis involved five items for buccofacial praxis, five items for intransitive limb praxis, five items for transitive limb praxis, and five items for complex praxis. Of interest were nine patients with small lesions (< 10% of the

---

[1]In the following discussion when studies assessing transitive praxis are reviewed, the term "tool" will be reserved for items which are used to gain a mechanical advantage, consistent with the definition of this term in other chapters of this book. When movements include manipulation of an item that does not provide a mechanical advantage (e.g. flipping a coin), the term "object" will be used. Contrary to its definition in other chapters, I will also use the term "object" as a supraordinate category under which tools may be subsumed. Thus, when praxis is tested both for actions involving items providing a mechanical advantage and for actions involving items providing no mechanical advantage, the term "object" will be used. In some instances, it cannot be determined if actions involving items providing no mechanical advantage were included in praxis testing, and the term "object" will be applied.

total hemisphere volume) who had moderate or severe apraxia. In general, these patients also had severe nonfluent aphasia and right hemiplegia. Seven of these patients had subcortical lesions, mostly frontal and periventricular in location. However, involvement of the basal ganglia was not prominent. The authors concluded that apraxia was due to involvement of white matter pathways: fronto-parietal pathways, the occipito-frontal fasciculus, and the callosal fibres connecting left and right premotor areas.

Basso and Della Sala (1986) used a test developed by De Renzi, Motti, & Nichelli (1980) to measure praxis in a right-handed aphasic patient with right hemiparesis who had a lesion of the left caudate head and anterior limb of the internal capsule. Half of the items of this test involve simple and complex intransitive gestures with the fingers and hand; the other half of the items involve the imitation of simple and complex finger and hand postures or movements which have no meaning. On this test, the patient was severely apraxic, with greater problems on complex movements. The patient also had some difficulty demonstrating object use. It should be noted that mixing intransitive gestures with nonmeaningful movements to develop a single score is problematic. The intransitive gestures have stored movement patterns (i.e. an output lexicon) which the nonmeaningful movements do not have. Nonetheless, since this patient had difficulty demonstrating the use of actual objects, he was certainly severely apraxic.

De Renzi, Faglioni, Scarpa, and Crisi (1986) found limb apraxia in five of 14 patients with lesions confined to the left basal ganglia or thalamus. Three of these lesions had significant involvement of the lenticular nucleus and surrounding white matter. Patients were relatively acute; all three were right handed. Aphasia was present to some degree in all three cases, and right-sided motor impairment ranged from mild to severe. Both imitation of movement and object pantomime were impaired in all patients. For one patient, pantomimes were reported as generally recognisable, but the fingers and hand were improperly positioned or body part was used as object. A second patient reportedly either failed to initiate movements or performed "totally unrelated movements."

Rothi, Kooistra, Heilman, and Mack (1988) reported three cases of limb (ideomotor) apraxia with subcortical lesions. All patients were right handed, and two were aphasic. The two aphasic cases were severely apraxic, and greater detail on these patients was given in an unpublished manuscript (Kooistra, Rothi, Mack, & Heilman, 1991). In both cases, the lenticular nucleus, portions of the caudate nucleus, and the internal capsule were included in the lesion. In one case, the lesion extended into the subinsular white matter and a small area of the ventral corona radiata. One patient had decreased right hemi-body strength, and the other had a right hemiparesis. Patients pantomimed the use of tools with their left hand and arm for 30 items. Responses were videotaped and scored in accordance with a system developed by Rothi, Mack, Verfaellie, Brown, &

Heilman (1985) (see Chapter 6). One apraxic patient errored on 73% of the pantomimes, while the other errored on 93% of the items. The average error rate of six normal controls was about 4%, and no normal errored on more than 10% of the items. Neither of the subcortical lesion patients made intact pantomimes that were unrelated to the target action, nor did they make unrecognisable movement. They never made sequencing or timing errors or demonstrated significant delays in production. Rather, they tended to make perserverative errors, or spatial errors. The latter included errors in movement trajectory, errors in the relationship between the limb or imagined tool and the imagined object to receive the action, and errors in the position of the fingers and hand relative to the imagined tool. These subcortical lesion patients generally made errors similar to those of nine patients with cortical lesions and apraxia. However, one subcortical patient made numerous perseverative errors, which their nine apraxic patients with cortical lesions did not do. Perseverative errors were defined as production of all or part of a previously produced pantomime.

Sanguineti, Agostoni, Aiello, Apale, Bogliun, and Tagliabue (1989) reported a right-handed woman who had apraxia associated with an infarct of the subcortical white matter in the fronto-temporal region, superior and lateral to the head of the caudate nucleus and putamen. Whereas her initial aphasia cleared rapidly, her apraxia for both intransitive and nonmeaningful gestures (De Renzi et al., 1980) remained severe.

Della Sala, Basso, Laiacona, and Papagno (1992) studied 35 cases of deep left-hemisphere lesion (31 ischemic, four haemorrhagic). Praxis was tested using the method of De Renzi et al. (1980) that involves imitating intransitive movements of the fingers and hand as well as an equal number of nonmeaningful hand and finger postures. Using the cut-off score on the praxis test of this study, seven of 35 patients had apraxia. Five of these seven cases had lesions predominantly in the white matter, one included the lenticular nucleus and surrounding white matter, and one included the thalamus and surrounding white matter. All seven cases had right-sided motor dysfunction. The incidence of lesions impinging upon the lenticular nucleus and/or caudate nucleus was higher in the nonapraxic group. The authors attributed apraxia to lesions of the periventricular white matter, especially occipito-frontal fibres. As noted above, imitation of intransitive gestures may underestimate apraxia, and the relationship between the ability to perform nonmeaningful movements and praxis is uncertain.

To summarise, apraxia can be found in nonthalamic subcortical lesions. In general, studies have focused on left-hemisphere lesions, though apraxia has occasionally been reported in right-hemisphere lesions (Agostoni et al., 1983). Most frequently, apraxia is accompanied by some form of aphasia, which may be fluent or nonfluent, and a hemiparesis. Apraxia in these cases is typically described as mild. There is some indication that apraxia may be less common in lesions of the basal ganglia and subcortical white matter than in cortical lesions (Basso et al., 1980); however, this data must be viewed with some caution because the incidence of apraxia may have been underestimated. In two large studies including subcortical lesions (Kertesz & Ferro, 1984; Della Sala et al., 1992), apraxia was found primarily in cases where subcortical white matter, but not the basal ganglia, was damaged. However, cases of apraxia after lesions including portions of the dominant basal ganglia exist; indeed, apraxia was particularly severe in two such cases reported by Kooistra et al. (1991) and in one reported by Basso and Della Sala (1986). Many errors made by apraxic patients with cortical lesions are also made by apraxic patients with subcortical lesions. Nonetheless, one of Kooistra and colleagues' patients had numerous perseverative errors which did not occur in their patients with cortical lesions. Kooistra et al. also reported that patients with subcortical apraxia did not substitute intact but unrelated gestures for the target action, though De Renzi et al. (1986) did report this phenomenon in one of their cases and Shelton and Knopman (1991) reported it occasionally in Huntington's disease patients. Unfortunately, as mentioned above, lack of consistency in definitions and in measures of apraxia make it difficult to compare findings between studies.

Degenerative Diseases of the Basal Ganglia and Apraxia. Although the term "subcortical dementia" had been used earlier (Mandell & Albert, 1990), it was the almost simultaneous articles of Albert, Feldman, and Willis (1974) and McHugh and Folstein (1975) that revived modern interest in the concept. Since that time, the hallmarks of subcortical dementia have been considered to be memory dysfunction, impairment in the manipulation of acquired knowledge, and perhaps slowness of mentation and personality change *in the absence of aphasia, apraxia, or agnosia* (Mandell & Albert, 1990). Parkinson's disease, Huntington's disease, and progressive supranuclear palsy were considered prototypical for subcortical dementia. While popular, this concept has not gone unchallenged. Of note was an extensive critique of the concept by Brown and Marsden (1988). In addition to challenging the concept on an anatomic basis, two other criticisms are of interest. First, Brown and Marsden questioned whether many of the differences between "cortical" and "subcortical" dementias were differences of degree as opposed to differences of type. They cited the literature on naming, word fluency, and vocabulary, noting that deficits in all these functions can be found in Parkinson's and Huntington's disease patients as well as in dementia of the Alzheimer's type. A second criticism revolved around whether cognitive deficits were similar enough in the different varieties of "subcortical dementias" to justify calling them a syndrome. Ideally, Brown and Marsden reasoned, patients with equivalent levels of dementia would be studied and patterns of deficit could be reviewed to determine if different types of deficits emerged. Shortly after Brown and Marsden's (1988) critique, Huber,

Shuttleworth, and Freidenberg (1989) published a study of Parkinson's and Alzheimer's patients matched for level of dementia and provided evidence for some differences in pattern.

These controversies should be kept in mind as we review the evidence regarding apraxia in degenerative diseases involving the basal ganglia. With respect to apraxia, the question of differences of degree versus differences in type is a particularly critical concept. We shall review the limited evidence concerning praxis in Parkinson's disease, Huntington's disease, and cortical-basal ganglionic degeneration.

Parkinson's disease is a process resulting in the loss of cells in the midbrain that supply dopamine, serotonin, and norepinephrine to other areas of the brain. Primary among these are the dopamine containing neurons of pars compacta of the substantia nigra which are the main source of dopamine for the caudate nucleus and putamen. Because of the disruption of these neostriatal structures, Parkinson's disease has been considered one model of basal ganglia dysfunction. It should be remembered that in Parkinsonian patients with dementia, but without histologic evidence of Alzheimer's disease, there are some changes in the cortex relative to normal controls. These changes are not as great as in demented Parkinson's patients with histologic evidence of Alzheimer's disease (de la Monte, Wells, Hedley-Whyte, & Growdon, 1989). Although different interpretations regarding the importance of these cortical changes can be made, it is uncertain what role they might play in cognition.

Goldenberg, Wimmer, Auff, and Schnaberth (1986) compared praxis in 42 patients with idiopathic Parkinson's disease to findings in 38 age-matched neurological patients with no evidence of cerebral damage. Apraxia was tested with nonmeaningful finger and hand postures and movement sequences. Instransitive (symbolic) gestures and transitive gestures (pantomimes of object use) were also administered. Parkinson's patients were impaired relative to controls on the nonmeaningful finger and hand postures and movement sequences; however, there were no differences between patients and controls on intransitive or transitive gestures. Thus, by our definition of apraxia, there was no evidence for apraxia in these Parkinson's patients. Further, strong correlations between the finger postures and movement sequences on the one hand and visual-spatial skills on the other suggest that the imitation of nonmeaningful postures and movement sequences was due to an inability to process them visually.

Huber et al. (1989) compared 10 patients with Parkinson's disease and dementia and nine patients with Alzheimer's disease, matched for severity of dementia, with normal controls on a number of cognitive measures. On the test of praxis, individuals were asked to pantomime use of five common tools, and items were apparently scored as correct or incorrect. The Alzheimer's patients scored significantly below normal controls and the Parkinson's disease patients did not score significantly below controls; however, there was no significant difference between Alzheimer's and Parkinson's patients. The authors concluded that there was evidence of apraxia for the Alzheimer's patients but not the Parkinson's patients, but this is most certainly an overinterpretation of the data. The controls' mean was a perfect score of 5. The mean score of the Alzheimer's group was 4.6 ($SD = 0.4$) while the mean of the Parkinson's group was 4.7 ($SD = 0.2$). Given the closeness of these means to each other and to the perfect score of 5, and given the larger standard deviation for the Alzheimer's patients, the authors would have done better to have emphasised the lack of differences between the patient groups. In both patient groups, a majority had to obtain a perfect score, and it would have to be concluded that in neither group was the evidence for more than a mild apraxia strong.

Grossman, Carvell, Gollomp, Stern, Vernon, and Hurtig (1991) tested praxis and sentence comprehension in a group of 22 nondemented idiopathic Parkinson's patients and 11 neurologically intact individuals. On praxis testing, the groups were asked to pantomime the use of a familiar tool, and were also asked to mirror meaningless gestures demonstrated by the examiner. Contrary to the findings of Goldenberg et al. (1986), patients were impaired relative to controls on transitive as well as nonmeaningful gestures. Using a discriminant analysis, 63.6% of the Parkinson's patients were worse than controls on these tasks, but it is difficult to separate out the contribution of transitive gestures to this latter finding. A majority of patients' errors were substituting body part as object. Authors suggested that while a frank apraxia was not evident in these Parkinson's disease patients, impairments in praxis were evident when tested under controlled conditions.

Huntington's disease is an autosomal dominant disease which involves degeneration of the neostriatum. In beginning stages, the degeneration is particularly severe in the head of the caudate nucleus, and Huntington's disease has also been used as a model of basal ganglia dysfunction. However, as with Parkinson's disease, cortical changes are not absent. Indeed, the frontal lobes are the first cortical structure to demonstrate atrophy in Huntington's disease, and the atrophy moves to more posterior areas of the cortex as the disease progresses (Sax, O'Donnell, Butters, Menzer, Montgomery, & Kayne, 1983).

Shelton and Knopman (1991) compared the performance of nine Huntington's disease patients, six Alzheimer's disease patients, and 20 normals on a test of praxis. The test included 10 transitive, four intransitive, five buccofacial, and one axial item. Three of nine patients with Huntington's disease (33%), and one of six patients with Alzheimer's disease (17%) were rated as apraxic on clinical examination. On the test of praxis, the Huntington's patients made significantly more errors (on 26% of items) than normal controls, but the Alzheimer's disease patients (errors on 12.5% of items) did not make more errors than controls. For Huntington's disease patients, the most common error was in hand posture (33%), followed by body part as object (26%), errors where the movement produced was a recognisable gesture inappropriate to the target item (17%), errors in movement trajectory (i.e. spatial errors, 11%), and various other errors making up the remainder. Heilman, Rothi, and Valenstein (1982) had shown that apraxic patients with posterior parietal lesions or fluent aphasia are impaired in recognition of gestures compared to apraxic patients with anterior lesions or nonfluent aphasia. Only one of Shelton and Knopman's (1991) Huntington's patients made any errors picking the correct gestures from three choices, and he had made no errors on praxis production. Thus, Huntington's patients seemed to fit more the anterior type of apraxia. Given that the major output of basal ganglia loops is to anterior cortex (Alexander, DeLong, & Strick, 1986) and given that frontal cortex is the first cortex to atrophy in Huntington's disease (Sax et al., 1983), this finding is not surprising.

Lenti and Bianchini (1993) used Shelton and Knopman's (1991) praxis test items to test a patient with childhood onset of Huntington's disease. There were 15 errors. Eight of these errors fell into one of the following categories: recognisable but inappropriate gestures, use of body part as object, and incorrect hand posture.

Cortical basal ganglionic degeneration is a disease in which signs of both basal ganglia and cortical dysfunction are present. These signs frequently are decidedly unilateral and include: akinesia and rigidity, postural-action tremor, limb dystonia, hyperreflexia, postural instability, sensory loss, focal reflex myoclonus, alien hand phenomenon, and apraxia (Doody & Jankovic, 1992; Riley, Lang, Lewis, Resch, Ashby, Hornykiewicz, & Black, 1990). Although common, apraxia is not inevitable. Riley et al. (1990) reported apraxia in 11 of 15 of their own cases and 9 of 13 cases reviewed in the literature. This apraxia may exist in the absence of significant cognitive deficit. Mesial frontal or parietal lobes may be involved (Riley et al., 1990). Eidelberg, Dhawan, Moeller et al. (1991) found lower global metabolism and significant metabolic asymmetries in the thalamus, mesial temporal lobe, and inferior parietal lobe contralateral to the involved side of the body in cortical basal ganglionic degeneration compared with normal controls or Parkinson's disease. If this disease process represents degeneration of a system which may be involved in praxis (i.e. a system including frontal cortex, parietal cortex, and the basal ganglia), it may be relevant to our discussion. However, the appearance of apraxia could be primarily a function of what cortical areas are involved.

In summary, cortical dementias have traditionally have been thought of as including aphasia, apraxia, and agnosia while an absence of these phenomena characterise subcortical dementias (Mandell & Albert, 1990). However, in order to insure that differences between dementias is one of pattern and not merely one of degree, Brown and Marsden (1988) suggested that differences between cortical and subcortical dementia should be investigated on the basis of patient groups equated for level of dementia. There is some evidence that, when equated for level of dementia, Parkinson's and Alzheimer's disease patients do not differ greatly in praxis (Huber et al., 1989); otherwise, Brown and Marsden's critique as applied to apraxia is unanswered. Data are equivocal with respect to the existence of praxis deficits in Parkinson's disease (Goldenberg et al., 1986; Huber et al. 1989 vs. Grossman et al., 1991). There is some evidence of apraxia in Huntington's disease (Lenti & Bianchini, 1993; Shelton & Knopman, 1991). The apraxic errors found in Huntington's disease by Shelton and Knopman (1991) are also found in apraxia after cortical lesion. Apraxia is a common feature of cortical basal ganglionic degeneration. Nonetheless, there are varying degrees of cortical involvement in all these disorders which may have a direct bearing on the level of apraxia.

THE THALAMUS AND APRAXIA

Data on the thalamus and apraxia come from cases of vascular lesion. Infarcts of the thalamus, unlike those involving the basal ganglia, often are limited to the thalamus. Thus, the complication of accompanying white matter damage is less of a problem for thalamic lesion cases than it is for basal ganglia lesion cases. Detail on reported cases has sometimes been scant, even to the point of not stating how praxis was tested. However, a few studies have reported details of apraxia.

In one of three cases with dominant thalamic haemorrhage and aphasia, Cappa and Vignolo (1979) described marked apraxia with difficulty initiating limb movements and difficulty demonstrating the use of actual objects. Alexander and LoVerme (1980) reported nine cases of aphasia after dominant intracerebral haemorrhage centring on the thalamus. Mild apraxia was found in two of the nine cases, and remitted in one. These findings in thalamic lesions were in contrast to those from basal ganglia lesions, where all patients tested showed some apraxia. Errors for limb praxis were reported to be exclusively use of body part as object. Graff-Radford, Eslinger, Damasio, and Yamada (1984) reported apraxia in three out of three cases with lesions in the polar artery territory of the dominant thalamus[2], but not in a case of right polar artery territory lesion. All three patients were also aphasic. Neither the method for testing apraxia nor the clinical manifestations were further described. It should be noted that the polar artery territory contains nuclei to which the basal ganglia project (ventral lateral, ventral anterior), which could have some bearing on the involvement of the basal ganglia in apraxia. It also supplies a portion of the nucleus reticularis important in Nadeau and Crosson's (in press) theory of selective engagement in language.

According to schematic representations of lesions, three of Agostoni and colleagues' (1983) seven patients (two right, one left) had lesions which

[2]Normally, the polar artery distribution includes the ventral anterior nucleus, the ventral anterior portions of the nucleus reticularis, and parts of the ventral lateral, dorsal medial, and anterior nuclei (Crosson, 1992; Nadeau & Crosson, in press).

significantly impinged on the thalamus; all had difficulty imitating intransitive movements. De Renzi et al. (1986) reported two cases of haemorrhagic lesions centred on the dominant thalamus and posterior limb of the internal capsule. The first case had impaired pantomime of object use and impaired object use, in addition to impaired imitation of intransitive and nonmeaningful gestures. The second was impaired on imitation of intransitive and nonmeaningful gestures. Della Sala et al. (1992) showed one out of four cases of thalamic lesion to have difficulty imitating intransitive and nonmeaningful movements. The same caveat as mentioned in the last section regarding the insensitivity of imitation and intransitive actions also applies here. In their nine cases of small left-hemisphere lesion with moderate or severe apraxia, Kertesz and Ferro (1984) found damage to the subcortical white matter but not the thalamus or basal ganglia.

Nadeau and colleagues (1994) studied apraxia in a case of a rather large dominant thalamic infarction with extensive involvement of the ventral lateral, ventral posterior lateral, and lateral posterior nuclei and with some involvement of the pulvinar. The patient had a mild right hemiparesis and aphasia of a type typical for dominant thalamic lesion. Testing for praxis was done with the left hand and included a number of transitive movements on which the patient demonstrated a severe apraxia. The patient did not make many unrecognisable responses nor did he tend to substitute an intact gesture for the wrong tool. Errors were primarily temporal (missequencing; increasing or decreasing duration of movements; delay in initiation) or spatial (increased or decreased amplitude; incorrect relation between hand and imagined tool; incorrect movement trajectory; body part as tool; alteration of characteristic movement). Similar to Heilman and colleagues' (1982) patients with fluent aphasia or inferior parietal lesions, Nadeau and colleagues' patient could not select the correct gesture from several presented by the examiner. As noted earlier, Nadeau et al. attributed the difficulty to a lack of access to declarative memories necessary to perform the pantomime in the absence of tool use.

Shuren, Maher, and Heilman (1994) described a case of apraxia in a case of left medial occipital, inferior temporal, and pulvinar infarct. Since medial occipital and inferior temporal infarcts do not usually produce apraxia, the authors attributed this deficit to the pulvinar damage. The patient had anomic speech; there was a mild increase of reflexes on the right. Transitive and intransitive gestures were tested with both the right and left hands. The patient showed a severe apraxia for transitive gestures for both hands, while intransitive gestures were noticeably less impaired. Errors were primarily spatial (increased or decreased amplitude; incorrect relation between hand and imagined tool; incorrect movement trajectory; body part as tool; alteration of characteristic movement), though a few temporal errors were also made. Although performance improved considerably when the patient was given tools, she was still severely apraxic under this condition. The patient's ability to recognise the correct gesture was below but close to the performance of normals, indicating that her gesture recognition was considerably better than her gesture performance. Motor skill learning (rotary pursuit) was impaired. The authors noted connections of the pulvinar to cortical areas important for praxis and suggested the pulvinar participates in selection of the movement representation, the translation of the representation into spatio-temporal coordinates, and the mediation of feedback for gesture monitoring.

In summary, apraxia with thalamic lesion has been reported and may be severe. Indications from Alexander and LoVerme's data suggest that apraxia may be more rare in dominant thalamic than dominant basal ganglia lesions. Nonetheless, Graff-Radford et al. (1984) found apraxia in all three cases of polar artery lesion. Error types are similar to those seen in cortical apraxia, though Shuren and colleagues' (1994) patient showed a heavier predominance of spatial errors. Impairment of gesture recognition may or may not be as severe as the inability to produce gestures (Nadeau et al., 1994; Shuren et al., 1994), as is true of apraxic patients with cortical lesions, depending on location. The case study of Shuren et al. suggests that the pulvinar, in addition to thalamic nuclei connected to the basal ganglia, may be involved in thalamic apraxia.

SYNTHESIS

Now, having generated alternative hypotheses regarding the role of the basal ganglia and thalamus in limb apraxia and having reviewed the limited empirical data, we shall review the hypotheses in light of the data. In the process of doing so, we will highlight unanswered questions which will suggest future research possibilities.

Our first hypothesis regarding the basal ganglia in limb praxis was that they had no direct role. This thesis is supported by some of the data. Basso et al. (1980) found a greater proportion of patients with deep lesions in their nonapraxic group acutely. Kertesz and Ferro's (1984) cases with small lesions and moderate to severe apraxia demonstrated subcortical lesions which involved primarily white matter pathways and largely avoided subcortical gray structures. Two studies (Goldenberg et al., 1986; Huber et al., 1989) have provided little evidence of apraxia in Parkinson's disease, suggesting no involvement of the basal ganglia in praxis (as we have defined it). On the other hand, findings of other studies do suggest disturbance of praxis in Parkinson's disease (Grossman et al., 1991) and in some Huntington's disease patients (Lenti & Bianchini, 1993; Shelton & Knopman, 1991). Thus, no clear conclusion regarding disturbance of praxis can be reached from examination of the literature on progressive subcortical diseases.

If the basal ganglia are not directly involved in praxis, one must explain those cases in which lesions are centred in the basal ganglia and apraxia is present. Two possible explanations present themselves. First, as previously noted, basal ganglia lesions rarely present without damage to surrounding white matter. Similar to Alexander and colleagues' (1987) conclusions regarding non-thalamic subcortical aphasia, Kertesz and Ferro concluded that it could be the interruption of posterior to anterior pathways that caused apraxia with deep lesions. Second, similar to Nadeau and Crosson's (in press) hypothesis regarding nonthalamic subcortical aphasia, temporary ischemic dysfunction or ischemic neuronal dropout may occur in the cortex without being detected on CT or MRI scan. The resulting cortical dysfunction might be responsible for apraxia.

Although the evidence for white matter involvement does seem strong, participation of the basal ganglia cannot be ruled out at the present time. Regarding Kertesz and Ferro's (1984) white matter hypothesis, some of the interrupted pathways will affect cortical input to the basal ganglia. One way to investigate this hypothesis would be to use metabolic or cerebral blood flow imaging. It has been shown that alterations of glucose metabolism occur primarily in target structures of the affected areas (Sokoloff, 1991), and at least one study suggests this principle can be applied to the study of the basal ganglia (Mitchell et al., 1989). If this is true, then large decreases in cortical metabolism with deep white matter lesions and apraxia would favour the cortico-cortical hypothesis. However, large decreases in basal ganglia metabolism in such cases would favour deafferentiation of the basal ganglia.

However, if cortical hypometabolism occurs, one would have to distinguish between two possible causes. As just mentioned deafferentiation from damage to cortico-cortical fibres is one possible cause. A second possible cause might be ischemia sufficient to cause tissue malfunction but insufficient to cause cystic infarction. If lesions are carefully mapped, one should be able to predict which cortical targets might be deafferentiated. When areas not predicted to be involved show decreased metabolism, one might suspect the ischemic as opposed to the deafferentiation hypothesis. According to Nadeau and Crosson (in press) such ischemia might be due to blockage of the M1 segment of the middle cerebral artery. If thrombosis were the cause of such a blockage, then cerebral angiography early after lesion would also be useful in detecting such blockage.

Since basal ganglia participation in praxis cannot be ruled out by the current data, we must review the status of the other hypotheses. Our second hypothesis was that the basal ganglia were involved in the release of praxis segments for motor programming. Such a role in language was proposed (Crosson, 1992b) in part on the basis of the elicitation of irrelevant language by stimulation of the caudate head (Van Buren, 1963; 1966; Van Buren et al., 1966) or thalamic nuclei related to the basal ganglia (Schaltenbrand, 1965; 1975). One possible way such a deficit might be manifested in apraxia is by errors in initiation, timing, or sequencing. Such errors were notably absent in the cases described by Kooistra et al. (1991), in spite of the fact that the scoring system would have picked them up. Further, Shelton and Knopman (1991) did not report such errors in patients with Huntington's disease. It is possible that simple movements in which one motion is repeated (e.g. hammering a nail or using a screw driver) are less sensitive to deficits in timing because multiple programs do not have to be coordinated; the same program is run over and over. In Parkinson's disease patients, Marsden (1987) found that additional deficits are found when multiple programs must be run simultaneously or in sequence. It might be that complex activities, such as lighting a candle with a match and blowing out the match where a series of actions is necessary but no action is repeated, might be a better means of eliciting problems with sequencing and timing. Thus, a number of such items should be included in testing apraxias after basal ganglia lesions.

It is also likely that basic neuroscience research regarding the neural functions of the basal ganglia can contribute to exploring the release-of-praxis segments hypothesis. As already noted, this function could be performed on the basis of quantitative neuronal activity, and does not necessarily require intensive information processing. A greater knowledge of basal ganglia neurophysiology, regarding the quantitative versus patterned capacities, would be helpful in evaluating this hypothesis. Distinguishing these two types of neuronal activity may be quite difficult.

The third hypothesis was that the basal ganglia were involved in combining coactive and sequential movements. If this theory is correct, what would apraxia after basal ganglia lesions look like? One could conceive of transitive praxis as involving multiple movement programs: one for governing how an imagined tool is held in the hand and one or more programs governing the movement trajectory with respect to an imagined object that is operated upon. An inability to combine such programs might involve missequencing the movements or derailment of one or both programs. Although their patients did not make sequencing errors, Kooistra and colleagues' (1991) patients made errors in movement trajectory, errors with respect to the imagined object, and errors in the position of the hand relative to the imagined tools. Thus, the basal ganglia could combine coactive and sequential programs. This analysis would imply that the programs exist but cannot be properly combined. In this case, different errors might be made performing the same movement on different occasions. For this reason, examining the consistency of errors for the same movement on different occasions would be useful. Shuren et al. (1994) did examine consistency of errors across hands in their case of apraxia with posterior cerebral artery and pulvinar infarction and noted that the same error was not generally made across hands.

The fourth and final hypothesis regarding the basal ganglia is that they are involved in selection of alternatives from an action lexicon. Paillard (1982) and Denny-Brown and Yanagisawa (1976) have previously suggested the basal ganglia are involved in selection of action during praxis. If one considers a lexical item for apraxia as the whole gesture, then the substitution of one whole gesture for another would be seen. Jeannerod (1994) cited some evidence that movements may be processed at this level. However, Kooistra and colleagues'

(1991) patients with basal ganglia lesions did not make this type of error, although it has been reported in one of De Renzi and colleagues' (1986) patients and occasionally in Shelton and Knopman's (1991) Huntington's disease patients. Nonetheless, the lack of predominance of this type of error suggests that if the basal ganglia participate in lexical selection it must be at a more molecular level. Perhaps praxis items are represented at more than one level: the whole gesture and the components necessary to execute the gesture. At the whole gesture level, praxis items probably are less constrained than is an individual lexical item in language. For example, if a jeweler and a heavy equipment mechanic are both asked to demonstrate the use of a screw driver, the relationship of the fingers and hand to the imagined tool and the resulting movement might be quite different. Yet, the basic function performed by the tool is the same, and each of these individuals would be able easily to perform the action used by the other. Further, the position of the object to receive the action of the tool would also make a difference in movement trajectory. If selection at this more molecular level is affected, the data would be consistent with the spatial errors common in Kooistra and colleagues' (1991) patients and Shelton and Knopman's Huntington's patients. Yet, if selection at this level is affected, then one might expect a hand posture for an inappropriate tool to be used. For example, a patient might hold an imagined hammer like a needle for sewing. It is unclear if this type of error is made. If errors are merely the deterioration of the correct program for holding the tool or the movement trajectory, then an explanation other than lexical selection would have to be sought.

Again, basic neuroscience research might be useful in addressing the lexical hypothesis. If the basal ganglia are the instrument of lexical selection, this implies that these structures must be able to monitor the different alternatives. This would require processing of patterned neuronal information. Definitive proof that the basal ganglia are capable of processing patterned input would be needed to support the hypothesis.

Now, turning our attention to the thalamus, our fifth hypothesis stated that thalamic nuclei are involved in feedback mechanisms for praxis. Regarding thalamic apraxia, current data do not seem to speak to this hypothesis, though Shuren et al. (1994) have suggested that the thalamus is involved in gesture monitoring. In order to perform gesture monitoring, a patient would need to be able to observe their intended or actual response, detect errors, initiate an attempt to correct, and monitor this attempt. One way to address this hypothesis would be to note whether patients make attempts to correct initial errors and whether such attempts are correct. For example, Crosson et al. (1986) noted that their patient made attempts to correct lexical errors in language, but he was seldom successful. Thus, he was able to detect errors and initiate an attempt at correction, but somehow the information necessary to correct the error was inaccessible. Similar analysis would be useful in cases of subcortical apraxia.

Our sixth hypothesis was that the thalamus is involved in selective engagement of cortical mechanisms necessary for praxis. A failure of selective engagement would result in decreased efficiency in the cortical processing and information processing necessary to suport praxis. If the principles of parallel distributed processing (Rumelhart & McClelland et al., 1986) are applied to the neural nets necessary for praxis, then the outcome of decreased efficiency would be manifested by graceful degradation of praxis. In graceful degradation, the most redundantly represented information is the best preserved. Indeed, Raymer et al. (1992) found fewer errors on high- than low-frequency words in cases of anomia after dominant thalamic lesion. It might be of use to develop some frequency ratings for tool use and apply them to praxis to see if gestures for more common as opposed to less common tools are better preserved after dominant thalamic lesion.

Careful mapping of lesions in thalamic apraxia would be necessary to evaluate the selective engagement hypothesis. Nadeau and Crosson's (in press) analysis was that lesions of the anterior portion of the nucleus reticularis thalami, in which fronto-thalamic fibres synapse, could lead to impairment of selective engagement. This portion of the nucleus reticularis is in the polar artery distribution. It is of interest that all three cases of dominant polar artery lesion reported by Graff-Radford et al. (1984) had some apraxia. In light of the lower incidence of thalamic apraxia reported by Alexander and LoVerme (1980), i.e. two of nine cases, this lesion location may be important. It would be consistent with the selective engagement hypothesis and involvement of the anterior nucleus reticularis but, as noted before, this location is also consistent with the involvement of the basal ganglia in praxis since the polar artery distribution includes the ventral anterior and ventral lateral nuclei to which the globus pallidus projects.

Our seventh hypothesis was that the thalamus is involved in declarative memory processes which are necessary for pantomime production in the absence of tools. According to this hypothesis information stored in long-term declarative memory about tools and their use is used to help produce pantomimes when the actual tool is absent. When the tool is present, patients may rely on procedural memory to demonstrate tool use. If this hypothesis is correct and reflects the primary function of the thalamus in praxis, then patients with thalamic apraxia should demonstrate impaired pantomime of tool use and possibly impaired tool selection (Nadeau et al., 1994), but they should demonstrate errorless performance when given an actual tool and a command because they can rely on procedural knowledge to perform the action.

The case of Shuren et al. (1994) actually addresses this hypothesis. Their patient was severely apraxic with both hands on pantomime, and although the patient improved on gesture with tools in hand, she still was correct on less than 50% of the items. Further, there was evidence that procedural (motor skill) learning was impaired. These data do not rule out that declarative knowledge was important for pantomime in their patient; however, they do indicate that more than access to declarative stores was amiss. Since there were other differences between the cases of Nadeau and colleagues (1994) and Shuren and colleagues (1994) (e.g. poor gesture recognition in the former), more patients need to be tested on pantomime of tool use both with tool present and tool absent. Further, since even pantomime with tool present requires some declarative knowledge if the object to be acted upon is not present, actual tool use with both tool and object of tool use present should be tried.

Finally, it is possible the thalamus participates in multiple subsystems relevant to praxis; therefore, localisation in cases of thalamic lesion must be carefully undertaken. The findings of Shuren et al. (1994) and of Nadeau et al. (1994) may be instructive in this regard, since careful localisation was done. The case of Shuren et al. appeared to involve primarily the pulvinar, whereas the case of Nadeau et al. (1994) involved the ventral lateral, ventral posterior lateral, and lateral posterior nuclei in addition to a small portion of the pulvinar. Thus, variation in the thalamic nuclei affected may account for some of the differences in these two cases. In this regard, it should not be assumed that the thalamus is functionally monolithic.

CONCLUSIONS

We have examined seven hypotheses regarding the participation of subcortical structures in praxis and have drawn tentative conclusions regarding how the limited data support each hypothesis. The state of the current data will not allow for any definitive models to be advanced at this time. Clarification of the issues will await further research. The value of continuing study of subcortical structures in praxis is that it allows us to examine how these structures participate in an activity which involves an extensive interface between cognitive and motor functions. Principles uncovered in this investigation will not only tell us more about praxis, they may be applicable to other forms of cognition or movement. A better understanding of subcortical participation in praxis will put us one step closer to understanding the fundamentals of how brain systems produce complex behaviours.

REFERENCES

Agostoni, C., Coletti, A., Orlando, G., & Tredici, G. (1983). Apraxia in deep cerebral lesions. *Journal of Neurology, Neurosurgery, and Psychiatry*, *46*, 804–808.

Albert, M. L., Feldman, R. G., & Willis, A. L. (1974). The "subcortical dementia" of progressive supranuclear palsy. *Journal of Neurology, Neurosurgery, and Psychiatry*, *37*, 121–130.

Alexander, G. E., DeLong, M. R., & Strick, P. L. (1986) Parallel organization of functionally segregated circuits linking basal ganglia and cortex. *Annual Review of Neuroscience*, *9*, 357–381.

Alexander, M. P. (1992). Speech and language deficits after subcortical lesions of the left hemisphere: A clinical, CT, PET study. In G. Vallar, S. F. Cappa, & C.-W. Wallesch (Eds.), *Neuropsychological disorders associated with subcortical lesions* (pp. 455–477). New York: Oxford University Press.

Alexander, M. P. & LoVerme, S. R. (1980). Aphasia after left hemispheric intracerebral hemorrhage. *Neurology, 30,* 1193–1202.

Alexander, M. P., Naeser, M., & Palumbo, C. (1987). Correlations of subcortical CT lesion sites and aphasia profiles. *Brain, 110,* 961–991.

Asanuma, C., Andersen, R. A., & Cowan, W. M. (1985). The thalamic relations of the caudal inferior parietal lobule and the lateral prefrontal cortex in monkeys: Divergent cortical projections from cell clusters in the medial pulvinar nucleus. *Journal of Comparative Neurology, 241,* 357–381.

Basso, A. & Della Sala, S. (1986). Ideomotor apraxia arising from a purely deep lesion. *Journal of Neurology, Neurosurgery, and Psychiatry, 49,* 458.

Basso, A., Luzzatti, C., & Spinnler, H. (1980). Is ideomotor apraxia the outcome of damage of well-defined regions of the left hemisphere? *Journal of Neurology, Neurosurgery, and Psychiatry, 43,* 118–126.

Bechtereva, N. P. (1987). Some general physiological principles of the human brain functioning. *International Journal of Psychophysiology, 5,* 235–251.

Brown, R. G. & Marsden, C. D. (1988). 'Subcortical dementia': The neuropsychological evidence. *Neuroscience, 25,* 363–387.

Brunner, R. J., Kornhuber, H. H., Seemuller, E., Suger, G., & Wallesch, C. W. (1982). Basal ganglia participation in language pathology. *Brain and Language, 16,* 281–299.

Cappa, S. F. & Vignolo, L. A. (1979). "Transcortical" features of aphasia following left thalamic hemorrhage. *Cortex, 15,* 121–130.

Cohen, N. J. & Squire, L. R. (1980). Preserved learning and retention of pattern-analyzing skill in amnesia: Dissociation of knowing how and knowing that. *Science, 210,* 207–210.

Cooper, I. S., Riklan, M., Stellar, S., Waltz, J. M., Levita, E., Ribera, V. A., & Zimmerman, J. (1968). A multidisciplinary investigation of neurosurgical rehabilitation in bilateral parkinsonism. *Journal of the American Geriatrics Society, 16,* 1177–1306.

Crosson, B. (1984). Role of the dominant thalamus in language: A review. *Psychological Bulletin, 96,* 491–517.

Crosson, B. (1985). Subcortical functions in language: A working model. *Brain and Language, 25,* 257–292.

Crosson, B. (1992a). Is the striatum involved in language? In G. Vallar, S. F. Cappa, & C.-W. Wallesch (Eds.), *Neuropsychological disorders associated with subcortical lesions* (pp. 268–293). New York: Oxford University Press.

Crosson, B. (1992b). *Subcortical functions in language and memory.* New York: Guilford Press.

Crosson, B., Parker, J. C., Warren, R. L., Kepes, J. J., Kim, A. K., & Tulley, R. C. (1986). A case of thalamic aphasia with post mortem verification. *Brain and Language, 29,* 301–314.

Cudeiro, J., Gonzalez, F., Perez, R., Alonso, J. M., & Acuna, C. (1989). Does the pulvinar-LP complex contribute to motor programming? *Brain Research, 484,* 367–370.

de la Monte, S., Wells, S. E., Hedley-Whyte, T., & Growdon, J. H. (1989). Neuropathological distinction between Parkinson's dementia and Parkinson's plus Alzheimer's disease. *Annals of Neurology, 26,* 309–320.

Della Sala, S., Basso, A., Laiacona, M., & Papagno, C. (1992). Subcortical localization of ideomotor apraxia: A review and an experimental study. In G. Vallar, S. F. Cappa, & C.-W. Wallesch (Eds.), *Neuropsychological disorders associated with subcortical lesions* (pp. 357–380). New York: Oxford University Press.

De Renzi, E., Faglioni, P., Scarpa, M., & Crisi, G. (1986). Limb apraxia in patients with damage confined to the left basal ganglia and thalamus. *Journal of Neurology, Neurosurgery, and Psychiatry, 49,* 1030–1038.

De Renzi, E., Motti, F., & Nichelli, P. (1980). Imitating gestures: A quantitative approach to ideomotor apraxia. *Archives of Neurology, 37,* 6–10.

De Renzi, E., Pieczuro, A., & Vignolo, L. A. (1968). Ideational apraxia: A quantitative study. *Neuropsychologia, 6,* 41–52.

Denny-Brown, D. & Yanagisawa, N. (1976). The role of basal ganglia in the initiation of movement. In M. D. Yahr (Ed.), *The basal ganglia* (pp. 115–148). New York: Raven Press.

Divac, I., Oberg, G. E., & Rosenkilde, C. E. (1987). Patterned neural activity: Implications for neurology and neuropharmacology. In J. S. Schneider & T. I. Lidsky (Eds.) *Basal ganglia and behavior: Sensory aspects of motor functioning* (pp. 61–67). Lewiston, NY: Hans Huber Publishers.

Doody, R. S. & Jankovic, J. (1992). The alien hand and related signs. *Journal of Neurology, Neurosurgery, and Psychiatry, 55,* 806–810.

Eidelberg, D., Dhawan, V., Moeller, J. R. Sidtis, J. J., Ginos, J. Z., Strother, S. C., Cederbaum, J., Greene, P., Fahn, S., Powers, J. M. et al. (1991). The metabolic landscape of cortico-basal ganglionic degeneration: Regional asymmetries studied with positron emission tomography. *Journal of Neurology, Neurosurgery, and Psychiatry, 54,* 856–852.

Ellis, A. W. & Young, A. W. (1988). *Human cognitive neuropsychology.* Hove, UK: Lawrence Erlbaum Associates Inc.

Gerfen, C. R. (1992). The neostriatal mosaic: Multiple levels of compartmental organization in the basal ganglia. *Annual Review of Neuroscience, 15,* 285–320.

Goldenberg, G., Wimmer, A., Auff, E., & Schnaberth, G. (1986). Impairment of motor planning in patients with Parkinson's disease: Evidence from ideomotor apraxia testing. *Journal of Neurology, Neurosurgery, and Psychiatry, 49,* 1266–1272.

Goldman-Rakic, P. S. & Porrino, L. J. (1985). The primate mediodorsal MD) nucleus and its projection to the frontal lobe. *The Journal of Comparative Neurology, 242,* 535–560.

Goodglass, H. & Kaplan, E. (1963). Disturbance of gesture and pantomime in aphasia. *Brain, 86,* 703–720.

Goodglass, H. & Kaplan, E. (1983). *The assessment of aphasia and related disorders* (2nd Ed.). Philadelphia: Lea & Febiger.

Graff-Radford, N. R., Eslinger, P. J., Damasio, A. R., & Yamada, T. (1984). Nonhemorrhagic infarction of the thalamus: Behavioral, anatomic, and physiologic correlates. *Neurology, 34,* 14–23.

Grossman, M., Carvell, S., Gollomp, S., Stern, M. B., Vernon, G., & Hurtig, H. I. (1991). Sentence comprehension and praxis deficits in Parkinson's disease. *Neurology, 41,* 1620–1626.

Heilman, K. M. & Rothi, L. J. G. (1993). Apraxia. In K. M. Heilman & E. Valenstein (Eds.), *Clinical neuropsychology.* New York: Oxford University Press.

Heilman, K. M., Rothi, L. J., & Valenstein, E. (1982). Two forms of ideomotor apraxia. *Neurology, 32,* 342–346.

Heiss, W.-D. (1992). Experimental evidence of ischemic thresholds and functional recovery. *Stroke, 23,* 1668–1672.

Horenstein, S., Chung, G., & Brenner, S. (1978). Aphasia in two verified cases of left thalamic hemorrhage. *Annals of Neurology, 4,* 177.

Huber, S. J., Shuttleworth, E. C., & Freidenberg, D. L. (1989). Neuropsychological differences between the dementias of Alzheimer's and Parkinson's diseases. *Archives of Neurology, 46,* 1287–1291.

Jeannerod, M. (1994). The representing brain: Neural correlates of motor intention and imagery. *Behavioral and Brain Sciences, 17,* 187–245.

Jones, E. G. (1985). *The thalamus*. New York: Plenum Press.
Kertesz, A. & Ferro, J. M. (1984). Lesion size and location in ideomotor apraxia. *Brain, 107*, 921–933.
Kooistra, C. A., Rothi, L. J. G., Mack, L., & Heilman, K. M. (1991). *Subcortical aphasia.* Unpublished manuscript.
Lassen N. A., Skyhoj-Olsen, T., Hajgaard, K., & Shriver, E. (1983). Incomplete infarction: A CT negative irreversible ischemic brain lesion. *Journal of Cerebral Blood Flow and Metabolism, 3* (Supplement 1), S602–S603.
Lenti, C. & Bianchini, E. (1993). Neuropsychological and neuroradiological study of a case of early-onset Huntington's chorea. *Developmental Medicine and Child Neurology, 35*, 1007–1014.
Mandell, A. M. & Albert, M. L. (1990). History of subcortical dementia. In J. L. Cummings, *Subcortical dementia* (pp. 17–30). New York: Oxford University Press.
Marsden, C. D. (1984). Function of the basal ganglia as revealed by cognitive and motor disorders in Parkinson's disease. *Canadian Journal of the Neurological Sciences, 11*, 129–135.
Marsden, C. D. (1987). What do the basal ganglia tell premotor cortical areas? In *Ciba Foundation Symposium 132, Motor areas of the cerebral cortex* (pp. 282–300). New York: Wiley.
Martone, M., Butters, N., Payne, M., Becker, J. T., & Sax, D. S. (1984). Dissociations between skill learning and verbal recognition in amnesia and dementia. *Archives of Neurology, 41*, 965–970.
McCormick, D. A. & Feeser, H. R. (1990). Functional implications of burst firing and single spike activity in lateral geniculate relay neurons. *Neuroscience, 39*, 103–113.
McFarling, D., Rothi, L. J., & Heilman, K. M. (1982). Transcortical aphasia from ischaemic infarcts of the thalamus: A report of two cases. *Journal of Neurology, Neurosurgery, and Psychiatry, 45*, 107–112.
McHugh, P. R. & Folstein, M. F. (1975). Psychiatric syndromes of Huntington's chorea: A clinical and pharmacologic study. In D. F. Benson & D. Blumer (Eds.), *Psychiatric aspects of neurologic disease* (pp. 267–285). New York: Grune & Stratton.
Mink, J. W. & Thach, W. T. (1991). Basal ganglia motor control. III. Pallidal ablation: Normal reaction time, muscle cocontraction, and slow movement. *Journal of Neurophysiology, 65*, 330–351.
Mitchell, I. J., Jackson, A., Sambrook, M. A., & Crossman, A. R. (1989). The role of the subthalamic nucleus in experimental chorea. *Brain, 112*, 1533–1548.
Nadeau, S. E. & Crosson, B. (in press). Subcortical aphasia. *Brain and Language.*
Nadeau, S. E., Roeltgen, D. P., Sevush, S., Ballinger, W. E., & Watson, R. T. (1994). Apraxia due to a pathologically documented thalamic infarction. *Neurology, 44*, 2133–2137.
Paillard, J. (1982). Apraxia and the neurophysiology of motor control. *Philosophical Transactions of the Royal Society of London B, 298*, 111–134.
Parent, A. (1996). *Carpenter's human neuroanatomy* (9th ed.). Baltimore: Williams & Wilkins.
Penfield, W. & Roberts, L. (1959). *Speech and brain mechanisms*. Princeton, NJ: Princeton University Press.
Penney, J. B. Jr. & Young, A. B. (1986). Striatal inhomogeneities and basal ganglia function. *Movement Disorders, 1*, 3–15.
Raymer, A. M., Moberg, P. J., Crosson, B., Nadeau, S. E., & Rothi, L. J. G. (1992). Lexical deficits in two cases of thalamic lesion. *Journal of Clinical and Experimental Neuropsychology, 14*, 33–34.
Riley, D. E., Lang, A. E., Lewis, A., Resch, L., Ashby, P., Hornykiewicz, O., & Black, S. (1990). Cortical-basal ganglionic degeneration. *Neurology, 40*, 1203–12.

Rolls, E. T. & Johnstone, S. (1992). Neurophysiological analysis of striatal function. In G. Vallar, S. F. Cappa, & C.-W. Wallesch (Eds.), *Neuropsychological disorders associated with subcortical lesions* (pp. 61–97). New York: Oxford University Press.

Rothi, L. J. G. & Heilman, K. M. (1985). Ideomotor apraxia: Gestural learning and memory. In E. A. Roy (Ed.), *Neuropsychological studies in apraxia and related disorders* (pp. 65–74). New York: Oxford University Press.

Rothi, L. J. G., Kooistra, C., Heilman, K. M., & Mack, L. (1988). Subcortical ideomotor apraxia. *Journal of Clinical and Experimental Neuropsychology, 10*, 48.

Rothi, L. J. G., Mack, L., Verfaellie, M., Brown, P., & Heilman, K. M. (1985). Ideomotor apraxia: Error pattern analysis. *Aphasiology, 2*, 381–388.

Rothi, L. J. G., Ochipa, C., & Heilman, K. M. (1991). A cognitive neuropsychological model of limb praxis. *Cognitive Neuropsychology, 8*, 443–458.

Rumelhart, D. E., McClelland, J. L., & the PDP Research Group. (1986). *Parallel Distributed Processing: Explorations in the microstructure of cognition* (Vol). Cambridge, MA: MIT Press.

Samra, K., Riklan, M., Levita, E., Zimmerman, J., Waltz, J. M., Bergmann, L., & Cooper, I. S. (1969). Language and speech correlates of anatomically verified lesions in thalamic surgery for parkinsonism. *Journal of Speech and Hearing Research, 12*, 510–540.

Sanguineti, I., Agostoni, E., Aiello, U., Apale, P., Bogliun, G., & Tagliabue, M. (1989). Aphasia and apraxia caused by ischemic damage to the white substance of the dominant hemisphere. *Italian Journal of Neurological Sciences, 10*, 97–100.

Sax, D. S., O'Donnell, B., Butters, N., Menzer, L., Montgomery, K., & Kayne, H. L. (1983). Computed tomographic, neurologic, and neuropsychological correlates of Huntington's disease. *International Journal of Neuroscience, 18*, 21–36.

Schaltenbrand, G. (1965). The effects of stereotactic electrical stimulation in the depth of the brain. *Brain, 88*, 835–840.

Schaltenbrand, G. (1975). The effects on speech and language of stereotactical stimulation in the thalamus and corpus callosum. *Brain and Language, 2*, 70–77.

Shelton, P. A. & Knopman, D. S. (1991). Ideomotor apraxia in Huntington's disease. *Archives of Neurology, 48*, 35–41.

Shuren, J. E., Maher, L. M., & Heilman, K. M. (1994). The role of the pulvinar in ideomotor praxis. *Journal of Neurology, Neurosurgery, and Psychiatry, 57*, 1282–1283.

Skyhoj-Olsen, T., Bruhn, P., & Oberg, R. G. E. (1986). Cortical hypoperfusion as a possible cause of "subcortical aphasia". *Brain, 109*, 393–410.

Sokoloff, L. (1991). Brain energy metabolism: Cell body or synapse (Discussion). In D. J. Chadwick & J. Whelan (Eds.), *Exploring functional anatomy with positron tomography* (pp. 43–51). New York: Wiley.

Squire, L. R. (1987). *Memory and brain.* New York: Oxford University Press.

Squire, L. R. (1992). Declarative and non-declarative memory: Multiple brain systems supporting learning and memory. *Journal of Cognitive Neuroscience, 4*, 232–243.

Squire, L. R. & Zola-Morgan, S. (1991). The medial temporal lobe memory system. *Science, 253*, 1380–1386.

Steriade, M., Domich, L., & Oakson, G. (1986). Reticularis thalami neurons revisited: Activity changes during shifts in states of vigilance. *The Journal of Neuroscience, 6*, 68–81.

Strick, P. L., Dum, R. P., & Picard, N. (1995). Macro-organization of circuits connecting the basal ganglia with motor cortical areas. In J. C. Houk, J. L. Davis, & D. G. Beiser (Eds.), *Models of information processing in the basal ganglia* (pp. 117–130). Cambridge, MA: MIT Press.

Talland, G. (1965). *Deranged memory.* New York: Academic Press.

Van Buren, J. M. (1963). Confusion and disturbance of speech from stimulation in the vicinity of the head of the caudate nucleus. *Journal of Neurosurgery, 20*, 148–157.

Van Buren, J. M. (1966). Evidence regarding a more precise localization of the posterior frontal-caudate arrest response in man. *Journal of Neurosurgery, 24*, 416–417.

Van Buren, J. M., Li, C. L., & Ojemann, G. A. (1966). The fronto-striatal arrest response in man. *Electroencephalography and Clinical Neurophysiology, 21*, 114–130.

Wallesch, C.-W. (1985). Two syndromes of aphasia occurring with ischemic lesions involving the left basal ganglia. *Brain and Language, 25*, 357–361.

Wallesch, C.-W. & Papagno, C. (1988). Subcortical aphasia. In F.C. Rose, R. Whurr, & M. A. Wyke (Eds.), *Aphasia* (pp. 256–287). London: Whurr Publishers.

Watson, R. T., Valenstein, E., & Heilman, K. M. (1981). Thalamic neglect: Possible of the medial thalamus and nucleus reticularis in behavior. *Archives of Neurology, 38*, 501–506.

Wing, A. & Miller, E. (1984). Basal ganglia lesions and psychological analyses of the control of voluntary movement. In *Ciba Foundation Symposium 107, Functions of the basal ganglia* (pp. 242–257). Summit, NJ: Ciba.

Yingling, C. D. & Skinner, J. E. (1977). Gating of thalamic input to cerebral cortex by nucleus reticularis thalami. In J. E. Desmedt (Ed.), *Attention, voluntary contraction and event-related cerebral potentials* (pp. 70–96). New York: S. Karger.

13 Developmental Dyspraxia

Mary K. Morris

HISTORY AND DEFINITIONAL ISSUES

The development of skilled movement is a primary means by which the child gains mastery over the environment. Abnormalities in this domain have been referred to as "clumsiness" (Gubbay, 1975), "developmental dyspraxia" (Ayres, 1972; Cermak, 1985; Denckla, 1984), "developmental apraxia and agnosia" (Walton, Ellis, & Court, 1962), "apractognosia" (Brain, 1961) , and "developmental coordination disorder" (DSM-IV; APA, 1994). Although developmental disorders of motor function are both common and significantly impact growth in other cognitive and social–emotional domains, they have received much less attention than developmental disorders more closely related to academic skills (i.e. language, reading, and arithmetic disability).

This chapter will review attempts to describe a childhood syndrome characterised by impaired execution of skilled movement and present current approaches to the assessment and treatment of these deficits. This literature reflects the divergent pulls of acquired and developmental models and is thus difficult to integrate. Two dominant approaches have been employed. The first has focused on "the clumsy child" and has attempted to characterise the range of associated motor and perceptual deficits, broadly defined, that are associated with this problem. This approach has also led to the development of widely used intervention techniques for which efficacy remains largely unsubstantiated. The second is more tightly tied to adult models of acquired apraxia and has emphasised the assessment of gestural representation. This chapter will primarily review those studies that emphasise the assessment of gesture in children, in both normal and clinical samples. Some studies that reflect the first approach have been included where they suggest potentially fruitful lines of inquiry relevant to questions of underlying mechanism or nosology. Literature relevant to potentially related disorders such as constructional dyspraxia, dysgraphia, and developmental verbal apraxia will not be systematically reviewed. The final section of this chapter will attempt to integrate findings across diverse studies, as well as identify unresolved questions for future research.

Descriptions of subtle motor disabilities in children, not associated with cerebral palsy, date back to the early 1900s when Collier identified a group of children with "congenital maladroitness" (Ford, 1966). Although Samuel Orton is most frequently cited with regard to his pioneering work in developmental disorders of communication, he included abnormal clumsiness or "developmental apraxia" in his description of the six most prevalent developmental disorders (Orton, 1937). Despite these early discussions, almost 30 years passed before case series of "clumsy children" were published (Gubbay, Ellis, Walton, & Court, 1965; Walton et al., 1962), containing early attempts to define this syndrome more precisely.

Gubbay (1975) defined developmental dyspraxia in a manner consistent with traditional definitions of acquired apraxia in adults, emphasising the impairment of skilled voluntary movement in the absence of primary deficits in other sensory, motor or cognitive functions. "... The clumsy child is to be regarded as one who is mentally normal, without bodily deformity, and whose physical strength, sensation, and co-ordination are virtually normal by the standards of routine conventional neurological assessment, but whose ability to perform skilled, purposive movement is impaired" (p. 39). The exclusionary features of this definition have proven difficult to apply in the developmental domain. Attempts to define the extent of permissible cognitive impairment must take into account the increased prevalence of comorbid learning disabilities and attention deficit hyperactivity disorder in this population. Exclusion of sensory impairment is complicated by the interdependent relationship between perceptual and motor development, leading many clinicians to view sensory deficits as an integral and possibly aetiological component of the syndrome (Ayres, 1972). Finally, even in the motor domain, it is difficult to draw a boundary between the child whose motor dysfunction falls at the mildest end of the cerebral palsy spectrum and the most severely affected dyspraxic child who may exhibit subtle motor findings on neurological examination (Denckla & Roeltgen, 1992).

Problems associated with the use of exclusionary criteria in defining developmental dyspraxia reflect the significant differences between acquired and developmental disorders. While apraxia secondary to a left-hemisphere lesion in a premorbidly normal adult represents the loss of previously acquired skilled actions, and possibly the ability to add new motor skills to one's repertoire (Kimura, 1977), developmental dyspraxia represents the failure to acquire these skills in a normal fashion. Like many other developmental disorders, developmental dyspraxia has a presumed, but as yet unidentified neuroanatomic substrate, which may reflect more diffuse anatomic and/or functional abnormality than is associated with the identifiable lesion underlying acquired apraxia. Although our understanding of normal praxis development is quite limited, it seems safe to presume that it is contingent on the integrity of many other functions outside of the motor domain including attention, perception, memory and, in the case of representational gesture, conceptual and linguistic abilities. Thus, deficits in other cognitive domains are more likely to be present in a developmental disorder. These difficulties in applying definitions and models based on acquired disorders in adults to developmental disorders in children are not unique to apraxia and have led to calls for theoretical and empirical efforts that incorporate a developmental perspective (Fletcher & Taylor, 1984). Developmental dyspraxia must ultimately be defined in a manner that appreciates that the causes, concomitants, and consequences of a failure to acquire skilled motor behaviours are likely to differ from those associated with the loss of previously established abilities.

Current nosological systems have made little progress since Gubbay's original definition was published. The DSM-IV (APA, 1994) includes a developmental disorder, consistent with clinical descriptions of developmental dyspraxia, referred to as developmental coordination disorder and defined as follows: "Performance in daily activities that require motor coordination is substantially below that expected given the person's chronological age and measured intelligence. This may be manifested by marked delays in achieving motor milestones (e.g. walking, crawling, sitting), dropping things, 'clumsiness', poor performance in sports, or poor handwriting" (p. 54). The DSM-IV criteria also require that these difficulties must significantly interfere with academic achievement or activities of daily living, must not be attributable to a general medical condition (e.g. cerebral palsy, hemiplegia, or muscular dystrophy), and must be discrepant from the an expected level of motor competence based on intellectual ability.

Consistent with these definitions, many studies of developmental dyspraxia have cast a wide net in defining disorders of skilled motor behaviour, including both gross and fine motor abilities. In contrast, studies of acquired apraxia in adults have focused almost exclusively on gestural ability, particularly representational gestures (those that represent a familiar activity) performed either with the limbs, hands, or oral musculature. The distinctions made in the adult literature between different subtypes of apraxia (i.e. ideomotor, ideational, limb kinetic) have rarely been incorporated into studies of dyspraxic children. Clearly, a "lumping" approach has dominated the developmental dyspraxia literature to date; despite the gains in knowledge associated with "splitting" approaches in the study of other developmental disabilities, particularly learning disabilities, which have dominated the past two decades (Denckla & Roeltgen, 1992).

CLINICAL DESCRIPTION

The child with developmental dyspraxia is most commonly described by parents and teachers as clumsy or awkward, frequently bumping into things, dropping things, tripping, and falling, sustaining more than his or her fair share of bumps and bruises. The following specific characteristics are typically included in clinical descriptions of the syndrome (Cermak, 1985; Denckla, 1984; Henderson & Hall, 1982; Miller, 1986): (1) problems in gross motor skills (e.g. walking, running, riding a bicycle); (2) problems in fine motor skills (e.g. constructional, graphomotor skills); (3) delays in learning activities of daily living (e.g. feeding, dressing, grooming); (4) poor visual and tactile perceptual abilities; (5) impairments of body schema; (6) right–left disorientation; (7) Performance IQ lower than Verbal IQ; (8) soft neurological signs (e.g. overflow, mild choreiform movements); (9) anomalous patterns of manual preference; and (10) behavioural/emotional difficulties typically viewed as secondary to repeated episodes of failure, teasing, and social rejection by peers with consequent low self esteem and avoidant behaviour when faced with situations in which motor skill is essential.

Estimates of the incidence of developmental dyspraxia in school age children range from 5 to 8% (DSM-IV, APA, 1994; Gubbay, 1975). Like many of the developmental disorders, the disorder appears to be more common in males, with a reported ratio of 2–4 to 1 (Gubbay, 1975). Differential diagnosis is complicated by the increased prevalence of several other developmental disabilities in children with dyspraxic symptoms. These include developmental articulatory deficits (i.e. verbal apraxia), developmental learning disabilities, and attention deficit hyperactivity disorder.

The aetiology of developmental dyspraxia is also unknown. A family history of similar problems is frequently reported, raising the possibility of genetic contributions. As with most developmental disabilities, there is an increased incidence of pre- and perinatal complications, estimated at 50% in Gubbay's sample. Analysis of the CT scans of 51 children with developmental apraxia and agnosia revealed an increased probability of nonspecific CT findings, including ventricular dilatation and cortical sulcal prominence (Knuckey, Apsimon, & Gubbay, 1983). Despite the paucity of empirical data, several aetiological theories have been proposed including: (1) alteration of typical cerebral dominance (Reuben & Bakwin, 1968); (2) dysfunction of the left parietal lobe based on the lesion location known to be associated with acquired apraxia (Orton, 1937); (3) cerebellar dysfunction (Lesny, 1980); (4) failure of integration (inter- or intrahemispheric) reminiscent of Geschwind's (1975) disconnection model of apraxia and suggestive of white matter abnormality (Denckla & Roeltgen, 1992; Gordon, 1979); and (5) dysfunction of the dorsal column medial lemniscal system, based on its role in tactile perception (Ayres, 1972).

ASSESSMENT

The assessment of dyspraxic children begins with a detailed developmental history documenting early motor behaviours (e.g. sucking, swallowing, grasping and reaching), traditional motor milestones (e.g. sitting and walking), and attainment of early self-help skills such as feeding and dressing. Beginning with the preschool period, both gross motor skills (e.g. running, throwing, and catching) and fine motor skills (e.g. drawing and using scissors) should be discussed. The standard paediatric neurological examination typically begins with an exclusionary focus, to rule out the presence of known neurological disorders that affect motor function. The addition of a more comprehensive neurodevelopmental motor evaluation allows the examiner to detect two types of subtle neuromotor signs: (1) those findings, often described as "soft" signs, which are subtle suggestions of clear neurological abnormalities (e.g. mild asymmetries of tone, reflex, arm swing or facial expression, choreiform movements, or dysmetria); and (2) failure to perform motor tasks with age-expected proficiency, reflecting not only inability to perform, but also failure to perform tasks with expected speed, smoothness, duration, and precision (i.e. without overflow movement). Denckla and Roeltgen (1992) review these two types of motor findings and discuss their relevance in the assessment of developmental dyspraxia. They stress their usefulness in understanding the child's functional difficulties (e.g. the presence of choreiform movements may explain the child's difficulty on tasks that require motor steadiness, such as writing or pouring liquids), as well as in guiding remediation efforts.

Test batteries that assess behaviours thought to be affected by or causally related to developmental dyspraxia have frequently been used with this population (Henderson, 1987). Several batteries focus on the assessment of motor ability (Bruininks, 1978; Gubbay, 1975; Stott, Moyes, & Henderson, 1984) and include a heterogenous set of motor tasks reflecting ecologically relevant actions, such as catching, throwing, hopping, drawing, and cutting. They are not based on either underlying models of motor function or aetiological models of motor dysfunction and are typically descriptive and normative in emphasis. Performance across a range of tasks is often combined to create composite scores that reflect "motor age," "gross motor ability," or "fine motor skill," based on the questionable assumption that these represent unitary domains. Scoring is quantitative and often based on a pass–fail system; thus, qualitative information about the nature of impaired performance and the child's use of compensatory strategies is not included. Finally, these batteries usually do not include measures of either representational or nonrepresentational gesture although some measures of actual tool use (e.g. drawing, scissors) may be included.

Other test batteries used to assess dyspraxic children include both sensory-perceptual and motor tasks. These batteries are typically based on theories that causally link perceptual deficits to motor impairments. The Southern California Sensory Integration Tests–Revised (Ayres, 1980) are the most frequently cited example of this approach, based on Ayres' aetiological model linking deficits in sensory integration to dyspraxia. This battery does include a measure of nonrepresentational gestural ability, the Imitation of Postures Test, in which the child imitates hand and arm postures modelled by the examiner. Ayres proposes impaired performance on this task in conjunction with evidence of tactile integration deficits as diagnostic criteria of developmental dyspraxia. Although strong empirical support for theories linking sensory-perceptual dysfunction with dyspraxia is lacking, a careful assessment of sensory-perceptual function is an essential component of the evaluation. Denckla (1984) has stressed the contribution of spatial ability to many of the behaviours that are so difficult for the dyspraxic child, particularly in the athletic domain, and emphasises the importance of carefully assessing both domains to avoid inappropriately targeted remediation.

The assessment of gestural ability in dyspraxic children represents a separate area of focus, influenced significantly by the adult literature. While tasks that assess the ability to imitate nonrepresentational gestures have occasionally been included in sensorimotor batteries as described, tasks assessing the ability to perform representational gestures to verbal command or imitation have been borrowed from the adult examination for apraxia (De Renzi, 1985; Goodglass & Kaplan, 1963). Studies of adults have revealed the impact of several task-related variables on practic ability; these variables are just beginning to be explored in children, as will be discussed in detail below. These variables include: (1) transitive vs. intransitive gestures, i.e. gestures directed toward the manipulation of objects vs. gestures that are used to express concepts or feelings; (2) familiar vs. novel gestures; (3) limb vs. orofacial musculature; (4) proximal vs. distal musculature; (5) single vs. sequential movements; (6) differing elicitation conditions, i.e. verbal command vs. imitation vs. actual tool use; and (7) place of action, i.e. peripersonal vs. extrapersonal space. Although some recent studies have begun to assess gestural representation abilities in children who perform poorly on motor and sensorimotor test batteries, the relationship between these two types of motor performance remains largely unexplored.

TREATMENT

Remediation procedures that are designed to address both primary motor dysfunction and underlying sensory-perceptual impairments are widely available. These can be generally divided into three categories: (1) those that treat the motor deficit directly, through drill or practice; (2) those that attempt to develop compensatory techniques or ways to circumvent the motor deficit; and (3) those that attempt to correct the hypothesised cause of the deficit, e.g. providing exercises to improve tactile and vestibular processing (Ayres, 1972). Unfortunately, empirical data regarding the efficacy of intervention techniques with dyspraxic children is almost nonexistent. Treatment efficacy research has been greatly hindered by the absence of clear diagnostic criteria for this disorder. Nonetheless, clinicians can provide considerable benefit to children and their families as part of the diagnostic process. Many parents arrive frustrated by the inconsistency of their child's daily functioning and may mistakenly attribute their child's careless, awkward, and messy behaviours, to laziness or disobedience. This attribution is understandable given that many dyspraxic children are capable of near normal performance in supportive contexts, for brief intervals, when they exert maximal effort (Denckla, 1984). Many dyspraxsic children are equally puzzled and frustrated by their difficulties and may be concerned that they reflect inadequate effort or stupidity. The increased incidence of poor self-image, avoidant behaviours, and depression is not surprising given these attributions. The diagnostician may provide considerable therapeutic benefit by providing an alternative explanation for the problem behaviours, by helping the child and parents distinguish those behaviours most amenable to skill building from those for which compensation may be more appropriate, and by helping the family identify assets that can serve as a foundation for esteem building.

EMPIRICAL STUDIES OF PRAXIS IN NORMAL CHILDREN

Berges and Lezine (1965) presented a series of nonrepresentational gestures involving arm, hand, and fingers to a large series of children between the ages of three and eight. Children were asked to imitate the examiner's posture and items were graded in difficulty from simple unimanual or symmetrical bimanual postures through complex arm–hand combinations that required an increased appreciation of the spatial relationship between body parts. Progressive improvement in the child's ability to imitate nonrepresentational gestures was observed across this age range, with consistently accurate performance across all difficulty levels attained by age eight.

The most comprehensive study to date of the development of representational gestural ability was performed by Kaplan (1968). Twenty-four boys, in three age groups (four, eight, and 12 years) were asked to perform 16 representational transitive gestures to verbal command under several conditions: (1) pretending to use a tool on themselves; (2) pretending to use a tool on another person or an actual object (e.g. a doll or a nail); (3) pretending to use a tool on an imaginary other or object; and (4) actual tool use. A detailed scoring system was developed that included a qualitative analysis of incorrect performance. Kaplan was able to differentiate four levels of gestural maturity, based on the accuracy with which the child is able to represent symbolically the implement or tool in pantomiming the action. At the lowest level, the tool is not depicted; the child may simply point to the object that receives the action or directly manipulate the object without a tool. With increasing development, the child begins to include the tool in the pantomimed action, but uses a body part to directly represent the tool (e.g. when asked to pantomime brushing the teeth, the index finger is extended in the horizontal plane and moved up and down on the teeth). This level appears to be analogous to the body part as tool (BPT) error frequently observed in apraxic adults. The child next begins to represent the tool as separate from the body but initially makes what Kaplan has referred to as "holding" errors in that the extent of the tool (i.e. its size and shape) is not accurately represented relative to the child's hand or in relation to the object that receives the action. Finally, the child is able to pantomime accurately transitive gestures, representing the agent of action, the tool and the object, as well as the spatial relationships between these components.

A clear developmental progression in gestural maturity was found. The four-year-old children in Kaplan's study make a preponderance of BPT responses. Performance characteristic of the lowest level of gestural maturity (i.e. pointing, direct manipulation) is also observed in this group but decreases dramatically by age eight. Eight-year-olds continue to exhibit BPT responses, but by this age the majority of responses (76%) include an attempt to represent the tool. Finally by age 12, 93% of responses include differentiated representation of the agent, tool, and object.

Two other normative studies with somewhat younger samples were identified. Overton and Jackson (1973) replicated Kaplan's findings in an independent population of children aged from three to eight years, extending the lower age limit for which data is available and including equal numbers of boys and girls in the sample. No sex differences in performance were observed. As might be predicted, three-year-olds exhibited an even higher rate of BPT errors (81%); these errors diminished progressively with age and represented only 26% of the eight year old responses, consistent with Kaplan's data. Kools and Tweedie (1975) assessed both oral and limb praxis in 87 males, aged one to six years, using tasks and scoring procedures adapted from De Renzi, Pieczuro, & Vignolo (1966). Not surprisingly, one-year-old children were unable to perform several tasks, but a progressive increase in accurate performance was observed from two to six years. These authors report ceiling performance on their praxis measures by age six, suggesting earlier attainment of mature gestural representation than the studies previously described. However, differences in scoring may account for this difference. The system used by Kools and Tweedie did not specifically code BPT responses, the most frequent type in this age range in previous studies.

In summary, normative developmental studies have identified a consistent maturational progression in gestural ability that is present for both representational and nonrepresentational gesture. This progression appears to be relatively complete by age eight, although some continued developmental change between eight and 12 was documented in a study of representational gesture that utilised a more detailed qualitative scoring system. However, there has been no attempt to relate these changes in gestural abilities to other cognitive developments that may underly maturational shifts (e.g. improvements in spatial representation, abstract/symbolic thought, or the capacity to integrate multiple representations) and may differentially support representational and nonrepresentational gestural competence.

EMPIRICAL STUDIES OF DEVELOPMENTAL DYSPRAXIA

Unilateral Lesions

Acquired apraxia in adults is more frequent after left-hemisphere lesions, suggesting that the left hemisphere may be specialised for the performance of skilled motor movements of both hands. If the neural substrate for praxis is found in the left hemisphere in a manner similar to language, then early left-hemisphere damage may negatively impact the acquisition of skilled movements to a greater degree than similar damage to the right hemisphere. Unfortunately, practic abilities have received very little attention in the study of children who have sustained early hemispheric lesions. Nass (1983) assessed 11 children, aged seven to twelve years, who had been diagnosed with congenital unilateral brain damage as infants. At the time of assessment, all of the children exhibited mild to moderate hemiparesis. Two tasks were used: (1) unimanual repetitive and successive finger movements (Denckla, 1973); and (2) gestural representation to verbal command with the unimpaired extremity.

Significant group differences were observed on the repetitive finger movements task. Although left- and right-hemisphere damaged children performed equally well, and within normative expectations with the unaffected limb, the left-hemisphere group showed significantly more impairment with the hemiplegic right limb. Nass discussed two possible interpretations of this finding, both consistent with early left hemisphere dominance for the execution of skilled distal movement. Damage to the left-hemisphere may affect both the praxis control system and the crossed cortico-spinal tract mediating distal movement of the right extremity, resulting in more significant impairment than damage only to the crossed cortico-spinal tract in instances of damage within the right hemisphere. It is also possible that, in children with congenital lesions, compensatory ipsilateral contributions to distal movement may be present, with the intact left hemisphere presumably better suited to assume this function than the right, given its specialised role in praxis. However, the left-hemisphere dominance hypothesis is not supported by the finding of equivalent performance with the intact limb. An intact praxis control system in the undamaged left hemisphere of the right hemisphere damaged group should have been associated with superior performance in comparison to the left-hemisphere damaged group.

No differences between children with early left- and right-hemisphere lesions were observed on the gestural representation task. Children in both groups reportedly made frequent BPT errors, although no quantitative data was presented in the report. Nass interprets this finding as age appropriate and consistent with the normative data reported by Kaplan (1968); however, in Kaplan's sample, BPT errors represented less than 25% of the responses of eight-year-olds and were quite rare in 12-year-olds. Nonetheless, both left- and right-hemisphere groups reportedly performed similarly. The scoring system used by Nass categorised responses as pass, fail, or BPT and did not permit precise characterisation of qualitative aspects of movement.

Further studies of praxis development in children who sustain early brain lesions are warranted. The single study reviewed does not provide strong support for an innate left-hemisphere lateralisation of praxis. It is possible that more subtle impairments of movement programming may result from early left-hemisphere damage, in a manner analogous to the higher order linguistic deficits observed in children with early left-hemisphere lesions whose basic language skills are intact. This possibility requires that subsequent investigations of praxis in these unique children utilise comprehensive assessments that encompass the full range of gesture types and elicitation conditions and permit fine-grained analysis of error type.

Developmental Learning Disabilities

Several studies have assessed praxis in children with developmental learning disabilities, a population in which the prevalence of dyspraxia is reportedly increased. Cermak, Coster, and Drake (1980) administered a task that included both representational and nonrepresentational gestures to right-handed boys, aged 9–13. Both learning disabled (LD) and normally achieving boys were assessed. The LD group met the following criteria: (1) Wechsler Intelligence Scale for Children (WISC–R) Full Scale IQ (FSIQ) of at least 85; (2) Performance IQ (PIQ) greater than Verbal IQ (VIQ) by at least 15 points; and (3) reading performance at least two grades below expectation. Thus, the LD group was selected to maximise the likelihood of language/reading disability, consistent with the adult literature suggesting that language and praxis deficits frequently co-occur. Control children obtained WISC–R FSIQ scores of 85 or greater, had no reported history of special education services, and were receiving passing grades in all academic areas.

The gestural task was designed to look at the effect of place of action; both gestures performed in near peripersonal space ("on self") and in extrapersonal space ("away from self") were included. Representational gestures were performed under two elicitation conditions, verbal command and imitation; nonrepresentational gestures were performed to imitation. A scoring system for spatial errors (e.g. location, plane, reversal, right–left confusion, and finger position) was employed for both representational and nonrepresentational gestures. Representational gestures were also scored using a simplified version of Kaplan's system.

The learning disabled group performed more poorly on both representational and nonrepresentational gestures, exhibiting developmentally less mature modes of gestural representation and increased incidence of spatial errors on both types of gestures. Place of action was also a relevant variable for the learning disabled group, but not for the controls. Children with learning disabilities made more errors than age-matched controls on gestures performed in near peripersonal space. This study is significant because it represents an early attempt to investigate praxis in a clinical population of children, using tasks that incorporate variables found to be important in understanding adult apraxia and indicating that these variables may also affect the performance of children.

Some studies have attempted to identify subtypes of dyspraxia in learning disabled children using models described in the adult literature. Conrad, Cermak and Drake (1983) used Luria's (1966) model to develop the Praxis Test for Children, designed to classify children into specific subtypes of dyspraxia based on the nature of their impairment. Luria described four types of apraxia: (1) kinaesthetic apraxia, characterised by problems in placing the limbs in a desired position, presumably due to impaired kinaesthetic awareness; (2) optic-spatial apraxia, characterised by spatial-perceptual disorganisation; (3) symbolic apraxia, reflected in a patient's inability to symbolically represent gestures to command despite preserved purposive behaviour; and (4) dynamic apraxia, characterised by a breakdown in the smooth execution of motor programs. Tasks were selected from the work of Luria (1966), Berges and Lezine (1965), Kaplan (1968), and Ayres (1980) to assess each of these types of apraxic breakdown. Kinaesthetic tasks required the imitation of postures without visual input; for these tasks, the child's hand was shielded from view and placed in a specific posture, which the child was asked to duplicate with the same hand after returning to a neutral posture. Optic-spatial tasks required similar imitation of hand and limb postures with the addition of visual input. Symbolic tasks required the pantomime of transitive gestures. Dynamic tasks required the imitation of unimanual and bimanual motor sequences (e.g. "fist–edge–palm" test, alternate tapping with both hands).

Fifty-eight boys, aged 9–12 (41 learning disabled, 17 age-matched controls) were assessed. Learning disabled children were identified based on enrolment in special education services, had WISC–R FSIQs above 85, and were reading at least one grade level below expectation. The learning disabled group was further subdivided based on VIQ/PIQ differences. Conrad et al. hypothesised that dyspraxic deficits would be correlated with the relationship between VIQ and PIQ. Impairments in verbal processing skills would be associated with poorer performance on tasks thought to assess symbolic and dynamic aspects of praxis, while impairments in nonverbal processing skills would be associated with poorer performance on kineasthetic and optic-spatial aspects of praxis. Learning disabled boys received lower overall scores on the praxis battery relative to age-matched controls. This difference was accounted for by their poorer performance on the optic-spatial and dynamic components; the groups were not significantly different on the kinaesthetic or the symbolic components. However, praxis performance was not related to the VIQ/PIQ relationship.

Although as a group, the learning disabled children performed more poorly on both the optic-spatial and the dynamic test items, Conrad et al. also report that scores on these two subsections of the battery were not significantly correlated. Although the authors do not report individual scores on these measures to permit the identification of patterns of dissociation, the absence of correlation between these two types of test items raises the possibility that apraxia in children may result from independent impairments in either the spatial or the dynamic aspects of skilled movement.

Rothi and Heilman (1985) have included the ability to comprehend pantomimed gestures as an important component in the assessment of apraxia. In their model, the ability to recognise pantomimed gestures distinguishes two groups of apraxic patients, those who have lost the representations necessary to guide the programming of skilled movements and those in whom representations are intact but unable to guide motor programming and execution, due to disconnection or damage to the areas responsible for motor programming. Lennox, Cermak and Koomar (1988) assessed the relationship between nonrepresentational praxis production and gesture comprehension in normal and learning disabled children, aged four to six. A modified version of the gesture comprehension task developed by Rothi and Heilman was used. In this task, the child views a video segment depicting an actor pantomiming the use of a common object and then chooses one of four pictures depicting either the tool being used (e.g. hammer) in level 1, or the object upon which the tool is used (e.g. nail) in level 2. Auditory comprehension and the ability to imitate nonrepresentational gestures were also assessed. The performance of learning disabled children did not differ from age-matched normal controls on the gesture comprehension task. In both learning disabled and normal children, gesture comprehension was significantly correlated with auditory comprehension, but was not correlated with the ability to imitate nonrepresentational gestures. Unfortunately, the ability to perform representational gestures, the type of gesture depicted in the comprehension task, was not assessed. The failure to include this critical comparison leaves the relationship between comprehension and production of representational gestures in children unexplored to date.

Deuel and Doar (1992) identified a subset of 24 children whose overall performance on an apraxia battery fell at least one standard deviation below the group mean of a mixed population (*N*=164) of both normal elementary age children and children referred to a paediatric neurologist for school problems. The majority of these children came from the referred group. They used a battery that included imitation of nonrepresentational gestures, pantomime of representational gestures to command, and actual tool use. Each item was rated on five-point scale from excellent to totally unrecognisable. Individual cases were examined to determine if patterns of performance on these three types of gestural tasks revealed underlying subtypes of developmental dyspraxia. Three children were identified who performed normally on actual tool use, but were impaired on imitation of nonrepresentational and pantomime of representational gestures. This group was described as analogous to adult ideomotor apraxia. Another three children showed the opposite pattern, impaired in actual tool use with normal performance on imitation and pantomime tasks. This group was described as analogous to adult ideational apraxia, as defined by De Renzi et al. (1966). These findings suggest that, as in adults, developmental dyspraxics may perform differently with differing elicitation conditions and degree of environmental contextual support and that these differences may be useful in subclassifying this population.

Studies of learning disabled children have demonstrated an increased incidence of dyspraxia; however, attempts to link dyspraxia to a specific pattern of cognitive deficit have not been successful. Both verbally and nonverbally impaired learning disabled children exhibit problems with praxis, but these two groups could not be differentiated on a battery specifically designed to assess several distinct mechanisms underlying praxis. However, a careful examination of this group of studies does provide some preliminary support for the notion that all dyspraxic children are not alike. In one study, tasks emphasising different processing skills (i.e. optic-spatial vs. dynamic) were not correlated. In another, dissociations were found between different elicitation conditions (i.e. pantomime vs. imitation vs. actual tool use). These findings suggest that it may be possible to subgroup dyspraxic children based on dissociations between different types of praxis or between different components of a hypothethical model of the praxis production system.

Developmental Speech and Language Disorders

Developmental verbal apraxia is defined as an impairment in programming movement sequences necessary for speech production (Darley, Aronson, & Brown, 1975). Attempts to document the relationship between these articulatory impairments and deficits in manual and nonspeech oral-motor skill have led to contradictory conclusions. Aram and Horwitz (1983) assessed 10 children, diagnosed with developmental verbal apraxia, on measures of representational transitive gestures and single and sequential volitional oral movements. While Aram and Horwitz noted that three children with verbal apraxia were also impaired on the manual praxis task, the remaining seven children were not impaired on manual gestures or on single oral movements. Cermak, Ward and Ward (1986) also found that children with "functional articulation disorders" were not impaired on a nonrepresentational gestural task, although they did perform more poorly than controls on a global measure of motor coordination.

In contrast, Dewey, Roy, Square-Storer, and Hayden (1988) found that children with verbal apraxia were impaired on multiple measures of limb and orofacial praxis including single limb gestures to command and imitation, single oral gestures to command and imitation, sequencing of limb and oral gestures, and production of novel movement sequences to command and imitation. However, these findings only characterised those children who were impaired on a sequenced motion rate task involving the sequencing of stop-initiated consonants (e.g. pa, ta, ka). A contrast group of children with speech production disorders, who were not impaired on the sequenced motion rate task, only had difficulty with single oral gestures, rather than pervasive impairment across limb and oral tasks.

Research with brain damaged adults has revealed a significant correlation between aphasia and limb apraxia following damage to the left hemisphere (Kertesz, 1985). There is little information about the relationship between developmental language disorder and limb dyspraxia in children. Archer and Witelson (1988) assessed a variety of manual and oral motor skills in children with developmental dysphasia. The dysphasic children were more impaired than age-matched controls on tests of imitation of hand postures and dynamic movements and production of representational gestures. Differences were not observed in oral movement imitation, in fine motor speed and dexterity (i.e. finger tapping and peg moving), in motor sequencing of novel actions, or in pantomime recognition. Thal, Tobias, and Morrison (1991) reported that children with delayed onset of lexical skills did significantly poorer on several gesture reproduction tasks, including single gestures without contextual support (e.g. pretending to drink from a cup) and sequences of gestures embedded within a familiar script (e.g. pretending to put a teddy bear to bed) than age matched controls.

In summary, the few studies of limb apraxia in children with speech and language disorders have found relationships similar to those previously reported in adults. An increased incidence of limb apraxia was observed in children with language delay and dysphasia, consistent with the relationship between apraxia and aphasia in adults, and with theoretical models which emphasise the dominant role of the left hemisphere in both of these abilities. Studies of the relationship between limb apraxia and apraxia of speech suggest that, while these two disorders frequently co-occur, they can be seen independently and must therefore reflect impairment to at least partially independent praxis systems.

Developmental Motor Deficits

Although learning disabled children are at increased risk as a population to experience problems with skilled motor function, the individuals in the studies described above may or may not have had problems in that domain and, given the broad selection criteria used in most studies, were almost certainly a heterogenous group. The number of empirical studies that have assessed praxis in a population of children selected for the presence of documented developmental motor deficits is quite small. Those studies identified have used diverse measures for selection, encompassing both gross and fine motor behaviours, have often failed to describe adequately the specific selection criteria employed, and have rarely provided data regarding the range of comorbid diagnoses (e.g. developmental learning disability, attention deficit hyperactivity disorder) in their samples. These factors hinder efforts to replicate findings and to integrate information across studies.

Visual and tactile perceptual deficits are frequently cited in clinical descriptions of dyspraxic children and have been identified as causal factors in some aetiological models (Ayres, 1972). Hulme, Biggerstaff, Moran, and McKinlay (1982) assessed the ability of "clumsy" children to make visual, kinaesthetic and cross-modal judgments of length. Children in the "clumsy" group were selected from a group of children receiving physical therapy services. All were described as having "educational difficulty." Age- and sex-matched controls were described by teachers as "performing normally." Following group selection, all children received a battery of five motor tasks, based on those described by Gubbay (1975). These included: (1) rolling a ball under the dominant foot around a series of obstacles; (2) throwing a ball; (3) threading beads; (4) inserting plastic blocks of various shapes into appropriately shaped slots; and (5) skipping. Significant differences between clumsy and control groups, in the expected direction, were observed on all five tasks.

All children performed a line estimation task under three conditions: (1) visual-visual, in which the stimulus line was presented visually followed by a response line, also presented visually, which the experimenter increased or decreased until the child indicated that it was identical in length to the stimulus line; (2) kinaesthetic-kinaesthetic, in which the child pushed a rod along a slot to a stop point without visual feedback, and then attempted to produce another movement of the same length with the stop removed; and (3) cross-modal, in which the above two conditions were combined so that the stimulus line was presented in one modality and the response was made in the other. Clumsy children were significantly less accurate and more variable in their responses across all conditions. These findings support the presence of both visual and tactile perceptual deficits in at least a subgroup of children selected for motor problems.

A series of studies by Dewey and colleagues (Dewey, 1991; Dewey, 1993; Dewey & Kaplan, 1991) have assessed praxis in a group of children with "developmental motor deficits". In these studies, potential child participants have been identified by teacher referral and selected based on performance on a screening battery to document motor deficits. This screening battery included the following tests from the Southern California Sensory Integration Tests (Ayres, 1972): (1) Motor Accuracy—Dominant; (2) Motor Accuracy—Nondominant; (3) Standing Balance—Eyes Open; and (4) Standing Balance—Eyes Closed, as well as the Southern California Postrotary Nystagmus Test (Ayres, 1975). To be included in the motor deficit group, a score one standard deviation below the mean on at least one of these measures was required. It is apparent that the tests in this screening battery assess a wide range of behaviours, mediated by different brain systems (i.e. cortical, cerebellar, brain stem) and do not purely reflect motor function. Exclusionary criteria included a history of known neurologic disorder affecting motor function (e.g. muscular dystrophy, cerebral palsy, acquired brain injury), mental retardation, attention deficit hyperactivity, and/or pharmacologic treatment for behavioural or mood disorders. Normal control children in these studies did not exhibit motor problems by teacher report and were in the normal range of academic achievement, based on classroom grades. Controls were not assessed with the screening battery.

The first study in this series (Dewey, 1991) assessed performance on gesture representation, gesture recognition, and motor sequencing tasks in 32 children with motor deficits and 16 normal controls. In this study, the motor deficit group was further subdivided into those with sensorimotor dysfunction and those with motor problems "that did not appear to be due to sensorimotor dysfunction" (Dewey, 1991). This classification was made based on performance on the screening battery. Unfortunately, no information is provided about how these measures were used to to separate the two groups. Examination of group means across the tasks suggests that the sensorimotor dysfunction group performed more poorly on measures of motor accuracy with both hands and on a measure of postrotary nystagmus. However, it is not known if these group differences are significant.

The gesture representation task included both transitive and intransitive gestures, performed to verbal command and imitation. Actual object use was also assessed; however, no results were reported for this condition. Motor sequencing was assessed with an apparatus similar to that described by Roy (1981). Four knobs that could be manipulated in several ways were mounted vertically on a board. Sequences comprised of these possible actions (i.e. turn, pull, slide, point) were performed to verbal command and to imitation. Gestural recognition was assessed by asking the child to select a picture depicting a named gesture and to point to an object associated with a gesture performed by the examiner.

Group differences were not observed on the gestural recognition task; in fact, ceiling effects were observed in all three groups. Both motor deficit subgroups (i.e. with and without sensorimotor dysfunction) performed more poorly than the normal control group in pantomiming transitive gestures and on the motor sequencing task. The sensorimotor dysfunction group were also significantly impaired on intransitive gestures and had poorer motor sequencing ability than the "pure" motor deficit group. Dewey concludes that motor deficits associated with sensorimotor dysfunction result in more severe problems with praxis and motor sequencing than motor deficits alone.

Dewey and Kaplan (1992) further explored task parameters shown to affect performance in adult apraxia in a larger but similarly selected sample of children. In this study, children in the developmental motor deficit group were directly compared to the control group on selected subtests of the Bruininks–Oseretsky Test of Motor Proficiency (Bruininks, 1978) and found to be significantly poorer on measures of balance, bilateral coordination, and upper limb coordination. Tasks were chosen to assess several potentially relevant dimensions including: (1) representational vs. nonrepresentational gestures; (2) transitive vs. intransitive gestures; (3) single vs. sequential gestures; (4) limb vs. orofacial actions; and (5) elicitation condition (verbal command vs. imitation). Children with motor problems were impaired across all tasks relative to normal controls. Consistent with the adult literature, transitive gestures were more impaired than intransitive, and gestures to verbal command were more impaired than to imitation. The motor deficit group also had significantly lower scores on a measure of receptive vocabulary than the normal controls and receptive language impairment was significantly correlated with practic ability, particularly the ability to pantomime transitive gestures. This finding is consistent with the frequent concordance of apraxia and aphasia in adults with left hemisphere damage and suggests a relationship between left hemisphere dysfunction and developmental dyspraxia.

In a subsequent study, based on the same population as the Dewey and Kaplan (1992) report, Dewey (1993) investigated the relationship between limb and orofacial praxis in children by comparing the types of errors made on both tasks, using an adaptation of the error analysis system developed by Roy and his colleagues (Roy, Square, Adams, & Friesen, 1985). Limb praxis errors were classified into eight categories: correct gesture, delay in initiating movement, added movement, movement errors, posture errors, action errors, location errors, and no response; six of the eight categories were used to classify orofacial praxis. Children with motor deficits made significantly more action errors than controls on both types of limb gestures (transitive and intransitive) and on orofacial gestures. Action errors are defined as "distortions in the dynamic action used to perform gestures" (Dewey, 1993) and may reflect an underlying common mechanism for both limb and orofacial dyspraxia. However, the use of this same scoring system in apraxic adults has revealed a predominance of spatial and postural errors in limb gestures, suggesting a different underlying mechanism in developmental dyspraxia. This finding also contrasts with a recent study of adults that found differential proportions of error types in limb and buccofacial praxis, using a similar scoring system (Raade, Rothi, & Heilman, 1991).

The studies by Dewey and colleagues have demonstrated that children with a variety of different motor abnormalities referable to different neuroanatomic substrates, are impaired on tasks that assess the planning and execution of skilled movements. Although factors that have been shown to affect the performance of apraxic adults on such tasks also appear to be relevant in children, it is likely that many of the children in these studies have coexisting impairments in more basic sensory and motor functions and thus would not meet the exclusionary criteria of the classical definition of apraxia. The relationship between lower and higher order motor impairments in these children is unknown, although it has been hypothesised that the integrity of basic sensorimotor functions may be necessary for the subsequent development of normal praxis (Denckla & Roeltgen, 1992). It is also unclear whether specific types or patterns of lower order deficits could be identified within the broad category of subtle sensorimotor abnormalities that might be associated with specific patterns of impairment in higher order motor skills.

Lundy-Ekman, Ivry, Keele, and Woollacott (1991) hypothesised that clumsiness might reflect qualitatively distinct deficits in dissociable components of movement computation (e.g. timing, sequencing, force). These computations are thought to be performed by different neuroanatomic structures, with the basal ganglia implicated in force control (Stelmach & Worringham, 1988; Wing, 1988) and the cerebellum associated with timing control (Ivry & Keele, 1989). They subdivided a group of "clumsy" children, aged seven–eight, based on the pattern of motor findings on a standard neurodevelopmental examination (Touwen, 1979). Two groups were created, those who exhibited motor abnormalities suggestive of basal ganglia dysfunction (e.g. choreiform and athetoid movements, synkinesis) and those with abnormalities suggestive of cerebellar dysfunction (e.g. dysdiadochokinesis, intention tremor, dysmetria). All children scored at or below the 40th percentile on the Bruininks–Oseretsky battery and had at least one motor finding; children with findings suggestive of both basal ganglia and cerebellar abnormality were excluded. Experimental tasks designed to assess timing and force control were administered. A double dissociation between the two "clumsy" groups and the two tasks was found. Children with cerebellar signs were impaired on measures of timing control and time perception but not on measures of force control; children with basal ganglia signs showed the opposite pattern of impairment. These findings suggest that clumsiness in children can result from different underlying mechanisms. Unfortunately, measures of praxis were not included in this study but one might hypothesise that primary deficits in timing and force control would also negatively impact a child's ability to produce the skilled movements necessary for gesture and tool use. From a developmental perspective, lower order motor abnormalities may be clues to help us identify underlying mechanisms, rather than "noise" that obscures our ability to observe "pure" deficits in higher order function.

Integration of the findings described in this section is complicated by the extreme heterogeneity of subjects, both within and across studies. Children who exhibit difficulty in at least some aspects of skilled motor behaviour can certainly be identified, although one suspects that such a broadly defined group might have as many differences as similarities. No consensus can be found regarding the types and range of motor deficits that define this disorder and how they are to be assessed. The concordance between developmental motor disorders and other developmental disabilities also remains unknown. However, the studies reviewed do suggest that there is a relationship between overall "clumsiness" and impaired gestural ability and have documented this relationship for both limb and orofacial praxis, using both representational and nonrepresentational tasks. Lundy et al. (1991) propose an approach that may lead to greater clarity. Their findings suggest that the application of cognitive neuropsychological models of movement computation may help us determine the critical building blocks of skilled movement. Patterns of association and dissociation on tasks specifically designed to assess these building blocks may ultimately lead to a more meaningful classification system for "clumsy" children.

FUTURE DIRECTIONS

Our understanding of developmental dyspraxia has been hindered by the paucity of normal developmental studies of praxis and its relationship to other developing abilities. Those studies which have been published have utilised cross-sectional approaches and have looked at praxis in isolation. Even within the motor domain, the unfolding of practic ability in relationship to the emergence of other types of motor skill remains largely unexplored. Studies which assess the relationship between praxis and other developing competencies, particularly in the motor, perceptual, and linguistic domains would provide a foundation for attempts to understand anomalous development. Bates and colleagues (Bates, Benigni, Bretherton, Camaioni, & Volterra, 1979) have utilised this approach to explore the relationships between language, tool use, and symbolic gesture during early language acquisition. An alternative approach might be to document the gradual acquisition of specific practic skills over time, using a longitudinal design. For example, Connolly and Dalgleish (1989) made repeated observations of 12 infants during the second year of life and found systematic patterns in the acquisition of spoon use. Finally, the role of manual preference in the normal development of praxis has been virtually ignored. Kaplan (1968) observed diminished manual preference, defined as the hand spontaneously to perform a task, in pantomime of representational gesture relative to actual tool use. Voeller, Rothi, Lombardino, Mack, and Heilman (1987) observed more skilful right-hand performance of pantomimed gestures relative to left in normal children, in contrast to the equal proficiency observed in normal adults. It is possible that this observation reflects preferential access of the right hand to critical left hemisphere regions that mediate praxis, during a period in development when myelination of the anterior corpus callosum is not yet complete.

Unresolved nosological issues have also impeded progress in understanding this disorder. The extent to which studies of developmental dyspraxia, from the narrower perspective of impaired gestural ability and from the broader perpective of developmental "clumsiness", describe the same group of children is unknown. Although clinicians assert that this disorder exists as an independent entity, there is no information regarding its incidence as a "pure" syndrome, without evidence of other developmental disabilities. Although many authors have hypothesised that a heterogenous group of dyspraxics might be divided into more homogenous subtypes, empirical support for proposed classification schemes has been lacking. Future studies of clinical populations must begin to document the presence of comorbid conditions such as developmental language and learning disabilities and attention deficit hyperactivity disorder in their "dyspraxic" samples. A more fruitful approach might be to move from comparing single clinical samples (e.g. developmental motor deficits or learning disabilities) to normal controls on a restricted set of motor tasks, to designs which include multiple comparison groups of children with different developmental disabilities and more comprehensive assessment batteries. These contrasts may help us distinguish between potentially dissociable patterns of motor abnormality characteristic of different developmental disorders. For example, dyslexic children may be impaired on tasks that involve rapid repetitive hand or finger movements (Rudel, 1985); children with attention deficit hyperactivity disorder may exhibit increased overflow movements (Denckla & Rudel, 1978); and both of these patterns may be distinct from gestural ability in a subset of children with pure developmental dyspraxia. Clearly, the development of more sensitive assessment techniques in the motor domain will enhance our ability to detect such differences. Techniques developed for studying the performance of apraxic adults such as qualitative error coding systems (Rothi, Mack, Verfaellie, Brown, & Heilman, 1988) and three dimensional computergraphic techniques (Poizner, Mack, Verfaellie, Gonzalez-Rothi, & Heilman, 1990) could be adapted for developmental populations, to better characterise impaired performance on gestural tasks. More comprehensive neuropsychological batteries should also be included in future subtyping studies since it is not yet clear whether classification schemes based on differences in the motor domain or on patterns of associated cognitive impairments will receive empirical support. Advances in understanding the causes of developmental dyspraxia and in knowing how to intervene effectively are contingent on our ability to successfully address these nosological issues.

REFERENCES

American Psychiatric Association. (1994). *Diagnostic and statistical manual of mental disorders* (4th ed.). Washington, DC: APA.

Aram, D.M. & Horwitz, S.J. (1983). Sequential and non-speech praxic abilities in developmental verbal apraxia. *Developmental Medicine and Child Neurology, 25*, 197–206.

Archer, L.A. & Witelson, S.F. (1988). Manual motor functions in developmental dysphasia. *Journal of Clinical and Experimental Neuropsychology, 10*, 47.

Ayres, A.J. (1972). *Sensory integration and learning disorders.* Los Angeles: Western Psychological Services.

Ayres, A.J. (1980). *Southern California Sensory Integration Tests–Revised.* Los Angeles: Western Psychological Services.

Ayres, A.J. (1975). *Southern California Postrotary Nystagmus Test.* Los Angeles: Western Psychological Services.

Bates, E., Benigni, L., Bretherton, I., Camaioni, L., & Volterra, V. (1979). *The emergence of symbols: Cognition and communication in infancy.* New York: Academic Press.

Berges, J. & Lezine, I. (1965). The imitation of gestures. *Clinics in Developmental Medicine No. 18.* London: Heinemann.

Brain, W.R. (1961). *Speech disorders, aphasia, apraxia and agnosia.* London: Butterworths.

Bruininks, R.H. (1978). *Bruininks–Oseretsky Test of Motor Proficiency.* Circle Pines, MN: American Guidance Service.

Cermak, S. (1985). Developmental dyspraxia. In E.A. Roy (Ed.), *Neuropsychological studies of apraxia and related disorders.* Amsterdam: North Holland.

Cermak, S.A., Coster, W., & Drake, C. (1980). Representational and nonrepresentational gestures in boys with learning disabilities. *American Journal of Occupational Therapy, 34*, 19–26.

Cermak, S.A., Ward, E.A., & Ward, L.M. (1986). The relationship between articulation disorders and motor coordination in children. *American Journal of Occupational Therapy, 40*, 546–550.

Connolly, K. & Dalgleish, M. (1989). The emergence of a tool-using skill in infancy. *Developmental Psychology, 25*, 894–912.

Conrad, K.E., Cermak, S.A., & Drake, C. (1983). Differentiation of praxis among children. *American Journal of Occupational Therapy, 37*, 466–473.

Darley, F.L., Aronson, A.E., & Brown, J.R. (1975). *Motor speech disorders*. Philadelphia: W.B. Saunders.

Denckla, M.B. (1973). Development of speed in repetitive and successive finger movements in normal children. *Developmental Medicine and Child Neurology, 15,* 635–645.

Denckla, M.B. (1984). Developmental dyspraxia: The clumsy child. In M.D. Levine & P. Satz (Eds.), *Middle childhood: Development and dysfunction*. Baltimore: University Park Press.

Denckla, M.B. & Roeltgen, D.P. (1992). Disorders of motor function and control. In I. Rapin & S.J. Segalowitz (Eds.), *Handbook of neuropsychology, Volume 6: Child Neuropsychology*. Amsterdam: Elsevier Science.

Denckla, M.B. & Rudel, R.G. (1978). Anomalies of motor development in hyperactive boys. *Annals of Neurology, 3,* 231–233.

De Renzi, E. (1985). Methods of limb apraxia examination and their bearing on the interpretation of the disorder. In E.A. Roy (Ed.), *Neuropsychological studies of apraxia and related disorders*. Amsterdam: North Holland.

De Renzi, E., Pieczuro, A., & Vignolo, L. (1966). Ideational apraxia: A quantitative study. *Neuropsychologia, 6,* 41–52.

Deuel, R.K. & Doar, B.P. (1992). Developmental manual dyspraxia: A lesson in mind and brain. *Journal of Child Neurology, 7,* 99–103.

Dewey, D. (1991). Praxis and sequencing skills in children with sensorimotor dysfunction. *Developmental Neuropsychology, 7,* 197–206.

Dewey, D. (1993). Error analysis of limb and orofacial praxis in children with developmental motor deficits. *Brain and Cognition, 23,* 203–221.

Dewey, D. & Kaplan, B.J. (1992). Analysis of praxis task demands in the assessment of children with developmental motor deficits. *Developmental Neuropsychology, 8,* 367–379.

Dewey, D., Roy, E.A., Square–Storer, P.A., & Hayden, D. 1988). Limb and oral praxic abilities in children with verbal sequencing deficits. *Developmental Medicine and Child Neurology, 30,* 743–751.

Fletcher J.M. & Taylor, H.G. (1984) Neuropsychological approaches to children: Towards a developmental neuropsychology. *Journal of Clinical Neuropsychology, 6,* 39–56.

Ford, F.R. (1966). *Diseases of the nervous system in infancy, childhood and adolescence* (5th ed.). Springfield: Thomas.

Geschwind, N. (1975). The apraxias: Neural mechanisms of disorders of learned movements. *American Scientist, 63,* 188–195.

Goodglass, H. & Kaplan, E. (1963). Disturbance of gesture and pantomime in aphasia. *Brain, 86,* 703–720.

Gordon, N.S. (1979). The acquisition of motor skills. *Brain and Development, 1,* 3–6.

Gubbay, S.S. (1975). *The clumsy child: A study of developmental apraxic and agnosic ataxia*. London: W.B. Saunders.

Gubbay, S.S., Ellis, E., Walton, J.N., & Court, S.D.M. (1965). Clumsy children: A study of apraxic and agnosic deficits in 21 children. *Brain, 88,* 295–312.

Henderson, S.E. (1987). The assessment of clumsy children: Old and new approaches. *Journal of Child Psychology and Psychiatry, 28,* 511–527.

Henderson, S.E. & Hall, D. (1982). Concomitants of clumsiness in young school children. *Developmental Medicine and Child Neurology, 24,* 448–460.

Hulme, C., Biggerstaff, A., Moran, G., & McKinlay, I. (1982). Visual, kinaesthetic and cross-modal judgements of length by normal and clumsy children. *Developmental Medicine and Child Neurology, 24,* 461–471.

Ivry, R. & Keele, S.W. (1989). Timing functions of the cerebellum. *Journal of Cognitive Neuroscience, 1,* 136–152.

Kaplan, E. (1968). *Gestural representation of implement usage: An organismic-developmental study*. Unpublished PhD dissertation, Clark University.

Kertesz, A. (1985). Apraxia and aphasia: Anatomical and clinical relationships. In E.A. Roy (Ed.), *Neuropsychological studies of apraxia and related disorders*. Amsterdam: North Holland.
Kimura, D. (1977). Acquisition of a motor skill after left hemisphere damage. *Brain, 100,* 527–542.
Kools, J.A. & Tweedie, D. (1975). Development of praxis in children. *Perceptual and Motor Skills, 40,* 11–19.
Knuckey, N.W., Apsimon, T.T., & Gubbay, S.S. (1983). Computerized axial tomography in clumsy children with developmental apraxia and agnosia. *Brain and Development, 5,* 14–19.
Lennox, L., Cermak, S.A., & Koomar, J. (1988). Praxis and gesture comprehension in 4-, 5-, and 6-year-olds. *American Journal of Occupational Therapy, 42,* 99–104.
Lesny, I.A. (1980). Developmental dyspraxia-dysgnosia as a cause of congenital children's clumsiness. *Brain and Development, 2,* 69–71.
Lundy-Ekman, L., Ivry, R., Keele, S., & Woollacott, M. (1991). Timing and force control deficits in clumsy children. *Journal of Cognitive Neuroscience, 3,* 367–376.
Luria, A.R. (1966). *Higher cortical functions in man.* New York: Basic Books.
Miller, N. (1986). *Dyspraxia and its management.* Rockville, MD: Aspen Publishers.
Nass, R. (1983). Ontogenesis of hemispheric specialization: Apraxia associated with congenital left hemisphere lesions. *Perceptual and Motor Skills, 57,* 775–782.
Orton, S.T. (1937). *Reading, writing, and speech problems in children.* New York: Norton.
Overton, W.F. & Jackson, J.P. (1973). The representation of imagined objects in action sequences: A developmental study. *Child Development, 44,* 309–314.
Poizner, H., Mack, L., Verfaellie, M., Gonzalez-Rothi, L.J., & Heilman, K.M. (1990). Three-dimensional computergraphic analysis of apraxia. *Brain, 113,* 85–101.
Raade, A.S., Rothi, L.G., & Heilman, K.M. (1991). The relationship between buccofacial and limb apraxia. *Brain and Cognition, 16,* 130–146.
Reuben, R.N. & Bakwin, H. (1968). Developmental clumsiness. *Pediatric Clinics of North America, 15,* 601–610.
Rothi, L.J.G. & Heilman, K.M. (1985). Ideomotor apraxia: Gesture discrimination, comprehension and memory. In E.A. Roy (Ed.), *Neuropsychological studies of apraxia and related disorders.* Amsterdam: North Holland.
Rothi, L.J.G., Mack, L., Verfaellie, M., Brown, P., & Heilman, K.M. (1988). Ideomotor apraxia: Error pattern analysis. *Aphasiology, 2,* 381–388.
Roy, E.A. (1981). Action sequencing and lateralized cerebral damage: Evidence for asymmetries in control. In J. Long & A. Baddeley (Eds.), *Attention and performance IX.* Hillsdale, NJ: Lawrence Erlbaum Associates Inc.
Roy, E.A., Square, P.A., Adams, S., & Friesen, H. (1985). Error/movement notation systems in apraxia. *Semiotic Inquiry, 5,* 402–414.
Rudel, R.G. (1985). Definition of dyslexia: Language and motor deficits. In F.H. Duffy & N. Geschwind (Eds.), *Dyslexia: A neuroscientific approach to clinical evaluation.* Boston: Little, Brown and Company.
Stelmach, G.E. & Worringham, C.J. (1988). The preparation and production of isometric force in Parkinson's disease. *Neuropsychologia, 26,* 93–103.
Stott, D.H., Moyes, F.A., & Henderson, S.E. (1984). *The Henderson revision of the Test of Motor Impairment.* San Antonio: Psychological Corporation.
Thal, D., Tobias, S., & Morrison, D. (1991). Language and gesture in late talkers: A 1-year follow-up. *Journal of Speech and Hearing Research, 34,* 604–612.
Touwen, B.C.L. (1979). Examination of the child with minor neurological dysfunction. *Clinics in Developmental Medicine No. 71.* London: Heinemann.
Voeller, K.K.S., Rothi, L.J.G., Lombardino, L., Mack, L., & Heilman, K.M. (1987). Developmental dyspraxia of the right hand in normal right handed children. *Neurology, 37* (Suppl. 1), 178.

Walton, J.N., Ellis, E., & Court, S.D.M. (1962). Clumsy children: A study of developmental apraxia and agnosia. *Brain, 85,* 602–613.

Wing, A.M. (1988). A comparison of the rate of pinch grip force increases and decreases in Parkinsonian bradykinesia. *Neuropsychologia, 26,* 479–482.

14 Naturalistic Action

Myrna F. Schwartz and Laurel J. Buxbaum

The topic of this chapter is naturalistic action, by which we mean movement in the service of commonplace, practical goals like food preparation and consumption. Naturalistic action involves sequences of movements that are well established through practice, for example, cutting food with fork and knife, spearing food onto the fork, and bringing the fork to the mouth. As such, it represents a class of skilled action. It is also movement carried out with and upon objects[1], and hence it utilises knowledge about the function and usage of objects (e.g. the spearing function of the fork). Finally, since naturalistic action is organised by goal hierarchies that structure behaviour over reasonably long time frames, it requires planning, attention, and working memory—functions served, at least in part, by the frontal lobes. Given these attributes, it is somewhat surprising that naturalistic action and naturalistic action disorders do not feature prominently in neuropsychological theories of skilled action, conceptual (semantic) memory, or frontal lobe functions.[2]

There are several probable reasons why naturalistic action has attracted so little attention in the neuropsychological literature. Traditionally, the neuro-

[1]The apraxia literature often distinguishes between "objects" and the tools, instruments, or appliances that are used to operate on objects. Here and elsewhere in the chapter, we use the term "object" for all manipulable entities, including tools, etc. Exceptions should be evident from the context.

[2]One particular class of naturalistic action disorders—dressing apraxia—has attracted considerable attention for what it reveals about the spatial and spatial-motor requirements of real world action. Dressing apraxia will not be discussed in this chapter.

psychological examination has taken place at the bedside and in the laboratory, neither of which provides adequate opportunity to assess naturalistic action. Perhaps because of this, naturalistic action disorders are unlikely to come to the attention of investigators unless they are dramatic; and dramatic breakdown of naturalistic action is thought to occur only in conditions in which the damage to cerebral structures is diffuse and/or extensive (e.g. late stage Alzheimer's disease; severe traumatic brain injury). Thus, there is widespread belief that focal lesions do not seriously disrupt naturalistic action, especially when performed in the natural setting. But if, as we suggest, naturalistic action draws on such functions as sequential movement control, semantic memory, and goal-based planning, it might be expected that focal lesions compromising any one of these functions should have identifiable effects on performance. Throughout this chapter, we will be examining the small body of studies that bears on this question. But for now, let us assume that it is true that focal cerebral lesions do not seriously impact naturalistic action performed in context, and let us ask why this should be so.

REDUNDANCY IN THE COGNITIVE SUBSTRATE FOR NATURALISTIC ACTION.

The most likely explanation is that there is a great deal of redundancy in the neural systems responsible for naturalistic action, that is, many routes to normal, or at least adequate, performance. For example, the impact of a disorder of perceptual object recognition (e.g. apperceptive agnosia) on naturalistic action could be mitigated by familiarity with the task and the context, coupled with greater reliance on unimpaired modalities and search of the perceptual array for clues to the appropriate action. In essence, defective "bottom-up" input processing would be compensated for by increased reliance on executive functions and stored knowledge (conceptual-semantic memory). This would change the mode of production from automatic to controlled (attention demanding), but except in extreme cases, would not have a serious impact on performance.

Conversely, disorders that compromise executive function or semantic memory might be mitigated in their impact on naturalistic action tasks by the availability of bottom-up input from perception, i.e. the information about the affordances of objects which is perceived more or less directly (e.g. Gibson, 1977; Roy, 1982; 1983). The use of bottom-up information is exemplified in a recent case study of an associative agnosic with poor recognition of the functional and associative attributes of objects (Sirigu, Duhamel, & Poncet, 1991). The patient was asked to describe how he would use various objects presented to sight and to demonstrate correct usage with the object in hand. The patient's descriptions and object manipulations invariably respected the mechanical affordances of the object, although not necessarily its conventional function. For example, shown a nail clipper, he incorrectly identified its function as something that can attach several sheets of paper together, but then went on to add "You turn the piece on the top and tip it back," which indeed describes the manner of usage, and accompanied this comment with an accurate pantomime (Sirigu et al., 1991, p. 2566). In a similar vein, patients with severe executive dysfunction may show "utilisation behaviour" in which they impulsively grasp and manipulate objects and combinations of objects, despite being discouraged from doing so by task demands or explicit instructions (e.g. Lhermitte, 1983; Shallice, Burgess, Schon, & Baxter, 1989). These examples illustrate that neither full semantic specification of an object's properties, nor a fully elaborated intention to use the object purposively, is necessary for object use which instantiates previous learning. A redundant, direct route to action is available through perception (Buxbaum, Schwartz & Carew, in press).

Redundancy in the neuropsychological organisation of naturalistic action also helps explain why patients with ideomotor apraxia (IM), who cannot execute gestures accurately to command and imitation, nevertheless succeed quite well in carrying out real-life activities that incorporate these same gestures. Gesturing to command and imitation depend on the motor system's access to intact gesture–motor, or space–time engrams, localised in left inferior parietal cortex (Heilman, Rothi, & Valenstein, 1982; Liepmann, 1905; Rothi, Ochipa, & Heilman, 1991). But naturalistic action that involves actual objects and that is performed in natural contexts may bypass these engrams entirely, in favour of motor systems specialised for overlearned, automatic movements (e.g. the supplementary motor system or basal ganglia; Marsden, 1982; Paillard, 1982) or right-hemisphere structures brought into play by the concrete nature of this class of action (Liepmann, 1905; Rapcsak, Ochipa, Beeson, & Rubens, 1993). Alternatively, since the sight and feel of the objects of naturalistic action provide information that is relevant to the spatio-temporal programming of actions, the presence of these objects in naturalistic action tasks could support and reinforce defective engrams or defective communication between these engrams and anterior motor systems.

Work by Poizner and colleagues provides empirical support for the latter thesis. Using three-dimensional movement analysis (Poizner, Mack, Verfaeillie, Rothi, & Heilman, 1990), Clark et. al (1994) studied the "slicing" gesture as it was performed by three ideomotor apraxics using their apractic left hands. The gesture was performed to verbal command with and without contextual cues available. Four conditions were studied: (1) no cues (verbal command only); (2) object (bread) present; (3) tool (knife) held in hand; and (4) acting with tool on object (i.e. actually slicing the bread with the knife). The key finding was that the apraxics showed similar deficits in spatial-temporal control of the hand *across all conditions*, that is, adding contextual cues did not eliminate the deficit, as would be expected if an alternative motor control system were being recruited. On the other hand, several kinematic measures showed improvement with added context. For example, movement amplitude and maximum wrist velocity, deficient in the command, tool, and object conditions, improved to normal in the tool + object condition. Also in this condition, the trajectory of the slicing gesture improved from a frontal plane, parallel to the bread, to a near sagittal plane. This allowed the bread to be successfully cut, albeit in small, triangle-shaped slices. Overall, these findings suggest that the space–time engrams implicated in IM are also called upon by real-world tasks; however, the presence of real objects to be acted upon provides an additional, somewhat redundant source of information to that which is encoded in the engrams, thereby mitigating the impact of ideomotor apraxia in real life naturalistic action.

IDEATIONAL APRAXIA VIEWED AS A DISORDER OF NATURALISTIC ACTION.

In the preceding section we have suggested how redundancy in the neuropsychological organisation of naturalistic action might account for its resilience in the face of damage to specific processing systems. Yet disturbances of naturalistic action do occur in real life, or at least in laboratory tests that approximate natural conditions. Indeed, some neuropsychologists take this to be the defining characteristic of ideational apraxia (IA). What follows is a review of the IA literature as it bears on the neuropsychological basis of naturalistic action and the necessary and sufficient determinants of naturalistic action disorders.

Evidence from Praxis Testing

In the traditional assessment for limb apraxia in left brain-damaged patients, IA is revealed by defective object manipulation in the nonparalysed left hand. In the clearest cases, gesturing to imitation is well preserved (Liepmann, 1905), but in cases where IA and IM coexist, the production of gestures to imitation and command is also defective. In these cases, the presence of IA is signalled by an absence of improvement in the object manipulation condition.

Ochipa, Rothi, and Heilman (1992) have recently argued that IA and IM can also be differentiated on the basis of the types of error that are made in the production of transitive gestures. Whereas ideomotor apraxics make production (i.e. spatial and temporal) errors, ideational apraxics make content errors, for example, executing a combing gesture when asked to pantomime the use of a toothbrush. Ochipa et al. claim that such errors are indicative of a deficit in the praxis conceptual system, that is, the knowledge base that supports action involving objects (Roy & Square, 1985). This underscores a running theme in the IA literature, which is that focal lesions of the left hemisphere can disrupt the learned association between objects and their manner of usage. While this is not central to all accounts of IA, it certainly bears on the neuropsychology of naturalistic action disorders and we will return to it throughout this paper.

The presence of IA in praxis testing does not necessarily entail problems in real world action with objects. In the group study conducted by De Renzi,

Pieczuro, and Vignolo (1968), 45 of 160 left-hemisphere patients performed abnormally on an object manipulation test, but informal observation revealed that naturalistic actions like eating and grooming were rarely affected. On the other hand, Ochipa, Rothi, and Heilman (1989) reported an unusual case of ideational apraxia following right hemisphere infarct in which errors did arise in the natural setting. The discrepancy may have been due to this patient's atypical brain organisation—the patient, who was left handed, exhibited aphasia after the right-hemisphere stroke. But there are other possibilities as well. On praxis testing with the nonparalysed, nondominant right hand, the Ochipa et al. patient made content errors when gesturing tool use to command and showed no improvement when given the tool in hand. Further testing with the same tools revealed adequate naming and name comprehension, but poor knowledge of tool function and tool–object association. Thus, the evidence points to impaired conceptual knowledge at the root of his IA, which may not have been the case in the patients studied by De Renzi et al. In addition, Ochipa et al. observed the patient in settings in which foils had been placed, and they report that at least some of the patient's errors involved these foils (e.g. eating with a toothbrush that had been placed on the dinner tray). It may be that the presence of the foils encouraged errors, or that errors in the natural setting are sufficiently infrequent that systematic, videotaped observation such as that conducted by Ochipa et al. is required to observe them. As we will see, most claims about real-world consequences of IA, like those of De Renzi et al., are based on more informal observation.

Evidence from Serial Action/Multiple Objects Tests

Some left hemisphere patients who perform well on praxis tests with single objects break down when assessed on serial actions that involve multiple objects. These serial action/multiple object probes utilise familiar, naturalistic action tasks, such as lighting a candle. The following is from Liepmann (cited in Brown, 1988, p. 17):

> The simplest test for ideational apraxia is the task of lighting a candle or using it to seal. All the necessary articles, a candle and a matchbox, are laid in front of the patient. Now, for example, the patient takes the matchbox and brings the whole thing up to the wick (the last act is anticipated and carried out with the whole box, instead of one match). Or he takes the box from the case, removes a match, and brings it up to the wick unlighted (a partial act: the lighting is omitted). Then, for example, he takes the candle in his hand and strikes it on the frictional surface of the matchbox (the proper movement [striking] with the wrong object). Or he strikes the whole matchbox against the light (proper movement performed on the wrong object with the wrong object).

Liepmann emphasised the effect of complexity: "The simplest motions are always successful and performed well. Even somewhat more complicated ones, such as lighting a match, are generally successful and done well. The disorder only becomes severe when a sequence of motions with different objects is to be made" (cited in Brown, 1988, p. 18).

Because of this, and because patients' errors often disrupt the sequential organisation of the task, the disorder has been described as a sequencing impairment (e.g. Paillard, 1982). Liepmann's writings encouraged such a view. However, his preferred account was that the gesture–motor, or, in his terms, space–time, engrams were highly refractory to retrieval in IA. Depending on the severity of the deficit, then, the problem might be manifested in simple acts (e.g. lighting a match) or more complicated acts involving the organisation of multiple space–time engrams. The latter lends itself to the view that what is disrupted in IA is the *plan* of action (Hécaen, 1981) or its "ideational outline" (Liepmann, cited in Brown, 1988, p. 15) or "conceptual organisation" (Poeck, 1983). All of these descriptors imply a disturbance in the sequential or temporal organisation of behavioural structures larger than the simple gesture.

Poeck and Lehmkuhl (1980) used a serial action/multiple objects test involving activities like hanging a picture and making coffee to diagnose IA in a left-handed patient with right haemorrhagic infarct. (Hereafter, we refer to this type of test as the "multiple objects test", or MOT.) The patient made numerous errors on the MOT, for example, errors of anticipation (e.g. hammering a nail into the wall with a picture already hanging from it), tool substitution (e.g. stirring water with an electrical plug), tool omission (e.g. pouring rather than spooning coffee into a cup, with spoon nearby) and omission of essential steps (e.g. stirring a mug of coffee without pouring water into it). The authors argued that the patient's obvious grasp of the situation and her ability to recognise and name the objects she misused ruled out symbolic deficits and dementia as causal factors.

In a subsequent study, Lehmkuhl & Poeck (1981) found that patients who failed the MOT also had difficulty arranging a series of photographs to convey the sequential order of MOT-type tasks. Five left-hemisphere patients with aphasia and ideational apraxia diagnosed by the MOT performed this serial action photos test, along with two other picture arrangement tests depicting event sequences that did not involve the manipulation of objects (e.g. playing soccer, shopping). The most interesting comparison was between the five IA patients and 30 aphasic patients without IA. These groups performed at the same level on the two control tasks; however, on the experimental test, the IA patients made more errors than any of the patients in the non-IA group. The authors concluded that in contrast to IM, IA is not characterised by defective motor programming, but rather by disturbance of the conceptual organisation of actions involving objects.

Central to Poeck's account is that ideational apraxia emerges only in tasks that bear on the serial organisation of actions involving multiple objects.

Moreover, he claims that this deficit is manifested whenever these conditions are met, which includes most tasks of everyday living. Thus, he writes that ideational apraxics "are conspicuous in everyday behavior because they have problems with preparing a meal, even eating breakfast or doing some professional activity they had been used to doing for years" (Poeck, 1985, p. 104).

Poeck's work grows out of the tradition which links IA to the serial organising properties of the left hemisphere action systems. There is a second tradition, mentioned earlier, which originates in Morlaas' (1928) conception of IA as "utilisation agnosia"—a recognition deficit specific to the functions of objects. According to Morlaas, utilisation agnosia is manifest in both single object and multiple object tests. The reason that some patients perform well on the former but poorly on the latter is simply that the MOT is more sensitive to this deficit.

In accord with this suggestion, De Renzi & Lucchelli (1988) predicted that performance on an MOT would correlate highly with single object manipulation and that errors on the MOT would favour object misuse (as opposed to the perseverations and other sequential problems touted by Poeck, 1983, 1985). They tested this prediction in 20 left-hemisphere patients selected because clinical evaluation indicated that they were likely to experience difficulty acting on objects. Patients performed a MOT consisting of five naturalistic action tasks (e.g. candle lighting, preparing a letter for mailing). In addition, they performed a single object manipulation test, a movement imitation test, and pantomime to sight of object. MOT scores were significant correlated only with the object manipulation test ($r = .85$), which supports the thesis of a common underlying deficit. MOT performance was coded according to an error taxonomy that included perplexity (delays and hesitations), clumsiness (poor control of skilled hand movements), omission of steps essential to the task, mislocation of action (e.g. stamp placed on back of envelope), misuse of objects (tool and object substitutions) and sequence errors (anticipations). The errors that occurred most frequently were omissions, misuse of objects, and mislocation of actions, all of which, the authors argue, are consistent with utilisation agnosia. While this is true, it is also worth noting that there may have been fewer opportunities for sequence errors than there were for omissions or substitutions.

De Renzi and Lucchelli (1988) suggest that what Morlaas called utilisation agnosia might actually be a type of conceptual-semantic memory disorder: (p. 1181, 1183)

> The patient who puts the stamp inside the envelope . . . or strikes the candle instead of the match does not fail because he/she chooses a defective innervation pattern to implement the movement, but because the sight and handling of the object do not evoke the specific action which should be associated with it. . . . The cognitive deficit is specific and concerns the ability to gain access to the semantic repository where the multiple features defining an object are stored, among which there is the way it must be used.

The case study of Ochipa et al. (1989) which we discussed earlier, provides direct support for this view. However, whereas Ochipa et al. recorded spontaneous errors in real world tasks, De Renzi & Lucchelli report that eating, washing, cooking and the like were unimpaired in all but the most severe of their patients. Clearly, more systematic evidence is needed on this question, in particular, evidence that takes into account the nature and severity of the underlying deficit and the cognitive and executive demands of different naturalistic action tasks.

To summarise this discussion of IA as a disorder of naturalistic action production: laboratory-based assessments like the MOT clearly reveal a vulnerability to naturalistic action errors in patients with dominant hemisphere lesions. What is less clear is whether this condition carries over to real-world settings; Poeck's view is that it invariably does, De Renzi and colleagues' view is that it generally does not. Also unclear is the nature of the deficit that gives rise to errors on the MOT. One class of accounts focuses on objects—their functions and usage. Another focuses on the sequential plan of action. There exists evidence to support both accounts, suggesting the possibility that naturalistic action disorders in the context of the MOT can arise in more than one way. This is consistent with our earlier discussion of the multifaceted character of the neuropsychological substrate for naturalistic action. On the other hand, the vulnerability to errors on the MOT implies that redundancy in the system may not be sufficient to overcome local deficits when naturalistic action tasks are performed outside the everyday, real-world context.

ALZHEIMER'S DISEASE AND NATURALISTIC ACTION DISTURBANCE

In all the studies reviewed above, IA is viewed as a dominant hemisphere syndrome. But from earliest reports, "ideational apraxia" has referred also to the disorder of naturalistic action in dementia of the Alzheimer type (DAT). And whereas there is dispute about the real-life consequences of IA in left-hemisphere patients, there is little doubt that IA in DAT manifests in impaired performance of everyday tasks.

In the past, the presence of IA in DAT has been viewed as a nonspecific manifestation of global mental deterioration. However, with growing evidence for focal deficits of all types in this population (Schwartz, 1990), researchers are asking whether the presence of IA signals pathological involvement of left-hemisphere praxis systems (e.g. Lucchelli, Lopez, Faglioni, & Boller, 1993; Rapcsak, Croswell, & Rubens, 1989) or elements of the praxis conceptual system (Ochipa et al., 1992). In view of evidence that working memory is particularly affected in this population (see Morris & Baddeley, 1988, for review), one might also ask how this and other executive impairments contribute to their naturalistic action disorder. We will speculate further on this in the sections below.

FRONTAL APRAXIA AND NATURALISTIC ACTION DISTURBANCE

Extensive frontal lobe injury can produce a behavioural disorder resembling IA. Luria (1966) designated this condition "frontal apraxia." Of a patient asked to light a candle, he writes: "Having lit the candle, he would put it into his mouth, or perform the habitual movements of smoking a cigarette with it, or break it and throw it away (as he usually did with a match)" (pp. 237-238). For Luria, the immediate cause of frontal apraxia was breakdown of the sequential organisation of behaviour, this being a consequence of a severe disorder of executive function: "Actions in relation to objects require a series of successive links, and these must be in the proper order. It is this sequence that is disturbed in patients with the lesion of the frontal lobes; an action that has become firmly established by previous experience disintegrates into a series of isolated fragments" (p. 237).

Schwartz, Reed, Montgomery, Palmer, and Mayer (1991) studied a patient, H.H., who developed frontal apraxia following rupture of a pericallosal artery aneurysm which resulted in right mesial frontal infarction with extension into the corpus callosum. Using an original action coding system, these authors described and quantified the disruption of two naturalistic tasks—toothbrushing and instant coffee preparation—as H.H. performed them daily in the inpatient rehabilitation unit. There was heightened susceptibility to object substitutions and object misuse. At the breakfast tray, for example, H.H. would occasionally pour coffee grinds into his orange juice; spoon butter into his coffee; or use the wrong implements to eat and stir. Such substitution errors occurred sporadically and inconsistently, alongside error types indicative of faulty sequential organisation (omissions, anticipations, perseverations).

Schwartz et al. argue that what is impaired in frontal apraxia is the capacity for online action planning, by which they mean the capacity to retrieve the components of familiar action sequences based on high level goal specification. We will elaborate on this below. However, we wish to point out here that although this account bears some similarity to theories of IA mentioned earlier, there are also important differences. Whereas Liepmann and followers invoke a "plan" whose elements are space–time engrams, Schwartz et al. envision hierarchical assemblies of "schemas" which encode abstract goals as well as specific movement patterns (see also Grafman, 1989; Shallice, 1988).

APRAXIA AND REAL-LIFE NATURALISTIC ACTION DEFICITS

The Schwartz et al. study was the first to describe in detail a disorder of naturalistic action as it is manifested in real-world behaviour. More recently, Foundas et al.(1995) compared the mealtime behaviour of 10 left-hemisphere lesioned ideomotor apraxics (selected on the basis of gesture to command performance) with that of nonapraxic stroke patients and normal controls. In addition to the usual mealtime array, each patient's tray included tool foils (e.g. comb, pencil) arranged in standardised positions. Videotape analysis demonstrated that apraxics performed fewer actions with tools, as well as fewer overall actions, and made more tool–action errors than the other groups.

Disturbances of real-life naturalistic action, like that documented in the Schwartz et al. and Foundas et al. studies, are well known to neurological rehabilitationists; in stroke and brain injury services one frequently encounters patients who can not be relied upon to carry out basic living skills without making errors. The challenge is to clarify how this relates to the various disorders of praxis and how these disorders, especially IA and frontal apraxia, relate to one another.

In a study addressing these questions, Schwartz, Fitzpatrick-DeSalme, and Carew (1995) identified 15 patients from the inpatient stroke and brain injury services at the MossRehab Hospital who had been observed to make errors in everyday activities that could not be ascribed to sensory or motor impairments. Five of these had unilateral brain damage: three right-handers with left CVA; one left-hander with right CVA; and one right-hander with left posterior gunshot wound. These five comprised the unilateral group (UL). The remaining 10 patients were recovering from closed head injury (CHI). The 15 patients, along with 14 matched controls, performed an MOT comprising six tasks: (1) preparing a letter for mailing; (2) preparing toast with butter and jelly; (3) repotting a plant; (4) wrapping a present; (5) preparing a pot of coffee; (6) packing a lunch box. Performance was coded from videotapes, and two scores were assigned: an accomplishment score, which expresses the number of subtasks completed as a proportion of the total, and an error score, which was based on the taxonomy shown in Table 1 (see also Buxbaum, Schwartz, Coslett, & Carew, 1995).

As a group, the patients achieved a lower accomplishment score than controls ($M = 76.9\%$ vs. 98.3%; $t(27) = 3.7, p < .001$) and a higher error score ($M = 13.6$ vs. 2.1; $t(25) = 4.1, p < .01$). This confirms that patients selected for problems in real-world action also make errors on the MOT. The dominant error type for controls was sequence: of the 35 errors scored, 15 were in this category[3]. For the patients, subtask omissions accounted for 30% (61/204) of the errors and sequence errors 24%; action additions and object/location substitutions each accounted for 13%. The remaining error types accounted for only a small proportion of the data.

Looking separately at the data from UL and CHI patients, we found that the distribution of errors collapsed across tasks was very similar for the two groups (Fig. 1). The mean number of errors per patient per task varied significantly across the six tasks ($F = 19.7; p < .0001$), from a low of .93 on the letter task to a

[3]Errors by controls were primarily deviations from what we had determined in a pilot study to be the normative performance of each task.

TABLE 1.
Error Taxonomy used in the MOT study

Type #1:	**Substitution of object or location** e.g., (coffee task) puts coffee can on burner (lunch box) puts thermos lid on juice bottle (present) wraps bottom half of box
Type #2:	**Gesture content error (correct object used with incorrect gesture)** e.g., (present) crumples wrapping paper at ends (vs. folds)
Type #3:	**Gesture motor error (faulty grasp or gesture orientation)** e.g., (plant) incorrect grasp of trowel (toast) incorrect grasp of knife for butter
Type #4:	**Spatial estimation error** e.g., (present) cuts small strip of wrapping paper
Type #5:	**Tool omission** e.g., (toast) spreads jelly with finger
Type #6:	**Subtask omission** e.g., (letter) fails to use stamp (letter) fails to seal envelope (lunch box) does not use juice (lunch box) does not wrap sandwich
Type #7:	**Sequence error (step omission, anticipation, perseveration)** e.g., (coffee) puts pot on burner without pouring water (letter) seals envelope before folding letter (lunch box) seals empty thermos (present) closes box before inserting gift (toast) applies jelly before butter
Type #8:	**Action addition/substitution (action not interpretable as step in task)** e.g., (lunch box) eats cookies (plant) jabs at plant in pot (toast) dips bread in jelly
Type #9:	**Quality Error (inappropriate or inexact quantity)** e.g., (plant) plant too low in soil (present) quality of wrapping (toast) extracts multiple pieces of bread from bag

high of 3.79 on the lunchbox task. More importantly, each task generated a specific error profile. For example, while the rate of errors was the same on the lunch box and coffee tasks, Type I errors (object/location substitutions) accounted for 25% of coffee errors but only 11% of lunch box errors. These task-specific profiles reflect, among other things, differences in number of actions

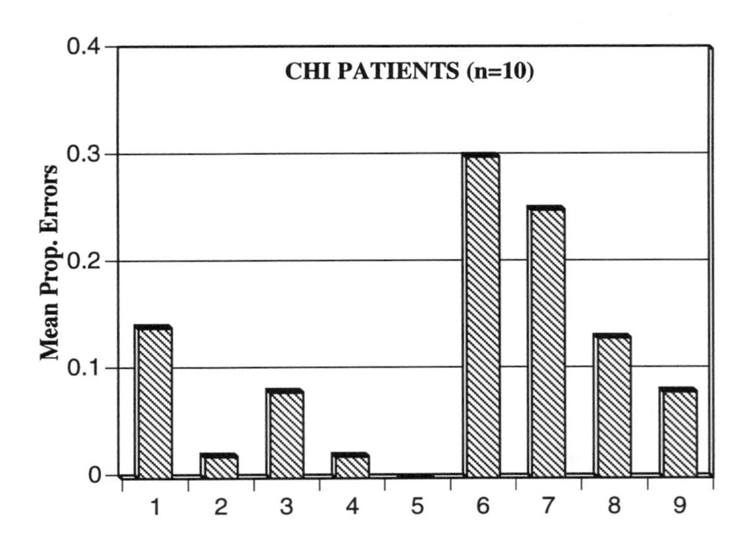

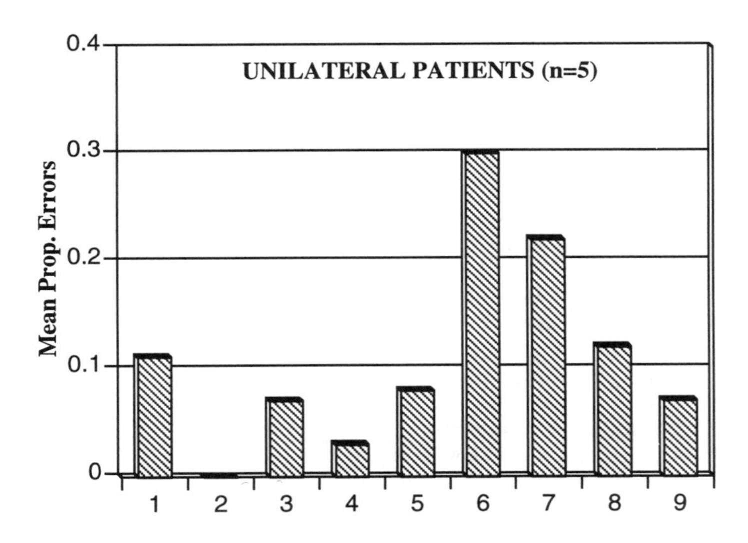

FIG. 1. Distribution of errors by closed head injury (CHI) and unilateral lesioned (UL) patients on the Multiple Objects Test. Type error 1 = object/location substitution; 2 = gesture-content; 3 = gesture-motor; 4 = spatial estimation; 5 = tool omission; 6 = subtask omission; 7 = sequence error; 8 = action addition; 9 = quality.

required, in the degree of sequencing constraints, and in the number of substitutable objects. The important point for present purposes is that the CHI and UL patients responded in a similar fashion to these task factors, i.e. the task-specific error profiles were similar for the UL and CHI patients.

Thus, the comparison of UL and CHI patients reveals considerable similarity, both in overall error profile and in task-specific patterns. One must be cautious in interpreting this finding: the group sizes are small and the patients within each group vary considerably in their neuropsychological findings. Nevertheless, it does underscore the need for further clarification of the role of aetiology and patient factors in the expression of naturalistic action disorders.

The data also revealed that the same patient may appear unimpaired in one task and apraxic in another. This often overlooked effect of task and situational complexity may begin to explain some of the inconsistency in the literature on the real world performance of patients with IA.

All patients in the study were tested for apraxia using a gesture to command test with 20 transitive gestures. Responses were scored for accuracy, with no attempt to differentiate production from content errors, or IM from IA (cf. Ochipa et al., 1992). Two of the UL patients had apraxia in the ipsilesional hand (with hemiparesis in the contralesional hand) and two others had bimanual apraxia. Six of the 10 CHI patients showed apraxia in one or both hands. Poor performance on the MOT was associated with apraxia. Ranking the patients from most to fewest errors on MOT, eight of eight high ranking patients had apraxia while only two of seven low ranking subjects did.

MOT scores also correlated with other tests sensitive to parietal lobe damage, including the Behavioural Inattention Test (Wilson, Cockburn, & Halligan, 1987) ($r_s = .72, p < .01$) and a gesture recognition test ($r_s = .70$). There was also a significant correlation ($p < .05$) with tests that assess conceptual-semantic knowledge, for example a test requiring the matching of objects on the basis of function (r_s –.57), a tool use judgment test in which the patient selects from a set of photographs the one that depicts a correct action–tool match (–.58); and a serial action photos test based on Lehmkuhl and Poeck (1981) (–.65). Among the tests that did *not* correlate with MOT performance ($p > .10$) were several that measure mental control and frontal lobe functions, e.g. Wechsler Memory Scale Attention Index ($r_s = .09$); Trails B (–.43); Tower of London Test (–.09); Modified Wisconsin Perseveration Score (.40). The failure to obtain significant correlations here cannot be due to insensitivity of the measures, since all but the Tower of London Test differentiated patients from controls at $p < .01$.

DO CONCEPTUAL-SEMANTIC MEMORY IMPAIRMENTS UNDERLIE NATURALISTIC ACTION DEFICITS?

Because CHI frequently produces frontal lobe damage, we conceptualised the naturalistic action disorder in CHI patients as frontal apraxia. However, the pattern of correlations we observed, and the similarities between CHI and UL patients, raises the possibility that damage to nonfrontal processing or knowledge systems may also play a role.

Naturalistic action is dependent on various types of knowledge, including knowledge of task organisation, the functions of objects in tasks, the mechanical affordances of objects that serve as tools, and the movement sequences that realize these affordances (Ochipa, Rothi, & Heilman, 1992; Roy & Square, 1985). Accounts of ideational apraxia often make reference to conceptual knowledge, or access to this knowledge. Lehmkuhl & Poeck (1981), drawing on evidence of impaired serial ordering of pictured actions, attributed the deficit to faulty knowledge of task organisation. De Renzi & Lucchelli (1988) and Ochipa et al. (1992) argued that the problem in IA centres on knowledge of object function and usage.

A recent study conducted in our laboratory speaks to the role of conceptual-semantic memory as it bears on naturalistic action production. Schwartz et al. (1995) studied a patient, J.K., who developed a profound disturbance of routine action production subsequent to closed head injury. Analysis revealed a high incidence of errors, especially object–use errors, in videotaped sessions of grooming and breakfast eating. An MRI study showed multiple focal lesions, including lesions in the left occipito-parietal area and right frontal lobe. The most prominent lesions, however, were to the temporal lobes bilaterally. These lesions affected inferior and anterior regions which play a role in object recognition and object knowledge, leading to the hypothesis that associative or utilisation agnosia might be responsible for J.K.'s naturalistic action disorder.

To test this hypothesis, we administered a extensive battery of tests assessing J.K.'s recognition and understanding of objects, including their function and usage. In general, he performed sufficiently well on these tests to rule out visual or associative agnosia. For example, on the Function Matching Test assessing access to functional semantic knowledge from vision (e.g. Coslett & Saffran, 1989; Warrington, 1982), J.K. scored 31/36 correct (86%), which is within the range of normal controls (29–36; $M = 34.2$; $SD = 2.15$; $N = 10$). He did, however, perform below ceiling or control levels on some tests of tool recognition; and on praxis testing he demonstrated gesture production and recognition defects. He was also unable to reliably demonstrate the use of tools in isolation, but this defect was completely overcome when the object of the tool's action was presented along with the tool.

We concluded that J.K. had a combination of IM and IA, the former due to defective gesture–motor engrams and the latter to faulty access to these engrams from perception. We also granted that J.K.'s access to other conceptual-semantic memory stores, while good, may have been less than perfect. Could these deficits, alone or in combination, account for J.K.'s real-life naturalistic action disorder? From the literature reviewed here, one would have to say no. The vast majority of patients with disorders of praxis are not plagued by real-world action errors; nor are patients with impaired conceptual-semantic memory access, including Alzheimer's patients (e.g. Chertkow & Bub, 1990; Schwartz & Chawluk, 1990). In the Ochipa, Rothi, and Heilman (1992) study demonstrating deficits in the praxis conceptual system in Alzheimer patients, naturalistic action performance was not examined, but based on the reports of caregivers, it is unlikely that they experienced serious difficulties in everyday tasks (C. Ochipa, personal communication).

We believe that J.K.'s praxis and memory retrieval deficits had the effect of disrupting the automaticity of his naturalistic action planning, causing it to become more dependent on the executive system that controls attention and working memory (Baddeley, 1986; Hasher & Zacks, 1979; Norman & Shallice, 1980; 1986; Reason, 1990; Shiffrin & Schneider, 1977). If the executive system had been functioning adequately, the impact on naturalistic activities of daily life should have been minimal, as indeed it is in most IM, IA, and semantic memory patients. Thus, to account for J.K.'s flagrant deficit, we postulated additional damage to the executive system, such that the resource demands of the once automatic activities of everyday life could not reliably be met.

This account of a disorder of naturalistic action manifested in real world settings has implications for our understanding of ideational and frontal apraxia. The final section spells out the theoretical rationale for this account, and those broader implications (see also Schwartz, 1995).

TOWARDS A THEORETICAL ACCOUNT OF NATURALISTIC ACTION DISORDERS

Information-processing theories of skilled action production postulate interconnected ensembles of knowledge specialists, or "schemas", which are created through practice and which are selected automatically, that is, with minimal demand for capacity-limited attention and memory resources (e.g. Norman, 1981; Norman & Shallice, 1986; Reason, 1990; Shallice, 1982; 1988). Schemas can be viewed as processes for retrieving information from long-term memory stores. One class of schemas ("goal schemas") draws on memory for the goal structure and sequential organisation of familiar activities (so-called script knowledge, see Schank, 1982; Schank & Abelson, 1977). Another class of schemas ("primitive action schemas") draw on spatio-temporal movement descriptions for gestures and combinations of gestures. Goal schemas prime primitive action schemas, thus readying them for selection when the environmental conditions are right (Cooper, Shallice & Farringdon, 1994; Norman, 1981). Selection is achieved when perceptual-semantic inputs to primitive action schemas match specified input conditions. For example, in toothbrushing, the input conditions for the brush-across-teeth action would specify a pasted toothbrush in the hand; for selection of this action schema to occur, this condition must be met in the environment. This assures that primitive actions are carried out in the correct serial order (i.e. pasting before brushing).

In the automatic mode, multiple goal and action schemas are generally activated in parallel. This makes automatic schema selection fast and efficient but also subject to "slips", as when the wrong object is taken, or an action is performed at the wrong time, or in the wrong context (Reason, 1979; 1984; Reason & Mycielska, 1982). This error tendency is kept in check by executive systems, which come into play when the likelihood of errors is high. For example, in cases where the environmental context biases toward a different, but perhaps more familiar, action, controlled or "supervisory" attention (SA) alters the activation dynamics of the action–schema system in ways that benefit the intended behaviour (Norman & Shallice, 1980; 1986). And when satisfaction of a goal cannot be immediately achieved, say because an object necessary to meet the input conditions is not available, working memory (WM) buffers the goal schema while the requisite object is sought. The involvement of these executive systems, hereafter SA/WM, defines a switch from the automatic to the controlled mode. Novel or taxing tasks are always carried out in the controlled mode.

Within this framework, frontal lobe lesions are thought to impact naturalistic action production by compromising SA/WM, thus allowing automatic action selection to run more or less unchecked (Shallice, 1988; Shallice et al., 1989). In contrast, lesions posterior to the frontal lobes are liable to affect the integrity of the schemas themselves. In particular, we hypothesise that parietal lobe lesions impair primitive action schemas, either by compromising the movement descriptions instantiated in these schemas (IM), or impacting how they are primed and selected (IA). Due to the redundant routes to naturalistic action, discussed earlier, the relevant information may still be conveyed to the effector system, but at a considerable cost in speed and efficiency. Moreover, the disruption effected by this damage to primitive action schemas will have ramifications throughout the system; for example, if the sight and feel of the pasted toothbrush does not trigger the brush-across-teeth action in a timely manner, SA/WM may be required to maintain the tooth brushing goal in an activated state.

The basic idea, then, is that damage to primitive action schemas causes them to cease operating as schemas, that is, fast and efficiently. And this compromises the automaticity of the system as a whole, such that operations and whole tasks that were once automatic now depend on capacity-limited resources for proper execution. Failure to meet the resource demands of a given operation disposes to error. In the example above, failure to maintain the goal in working memory invites the omission of critical steps.

The vulnerability to error created by loss of automaticity in naturalistic action planning is highly sensitive to task demands. Activities that are less familiar, or have a more complex goal structure, or are carried out in a more challenging environment, place a greater burden on SA/WM. Under these conditions, it becomes more likely that a particular operation will lack the resources it needs to be performed effectively. This explains why the MOT reveals the presence of ideational apraxia where simpler tasks (e.g. single object manipulation) or more routine activities (eating, grooming) often do not. It also may explain the finding in Foundas et al. (1995) that ideomotor apraxics made action errors when executing routine activities in an environment in which semantically related foils were present; in the context of reduced automaticity, the extra resources needed to reject the foils may have exceeded capacity. If these suggestions are correct, the notion that IM and IA do not materially affect naturalistic action in the real world will have to be qualified in recognition of the loss of automaticity that ensues, and the vulnerability to errors when the task and/or situational complexity exceeds some threshold.

We have suggested that the local damage to primitive action schemas compromises the automaticity of the system as a whole. This helps explains why, when ideational apraxics commit errors on the MOT, their errors are so diverse. Since all operations have become more resource dependent, all are subject to error when the resource requirements cannot be met. At what point errors arise, and how they are expressed (i.e. as omissions, substitutions, etc.) is primarily determined by the resource demands at each step of the task and what alternative actions the environment affords when the demands are not met.

Vulnerability to action errors in the MOT is not limited to patients with focal left-hemisphere lesions. The CHI patients run in the Schwartz, Fitzpatrick-DeSalme, and Carew (1995) study were indistinguishable from the UL group. While it is true that the most impaired of the CHI patients were also apraxic in one or both hands, we believe that the vulnerability to action errors in this population does not reduce to a single cause. Rather, the diffuse and multifocal damage that attends CHI probably compromises many of the operations on which automatic action selection depends, along with the frontal executive functions needed to compensate effectively for reduced automaticity.

Indeed, we hypothesise that, in extreme cases, this combination of (1) disturbed automatic action selection and (2) impaired SA/WM has a disruptive effect on even the simplest and most routine activities of daily life (hereafter, ADL). We offer this as an explanation for the flagrant ADL disorder which we observed in the closed head-injury patient, J.K., as well as for those patients with ideational apraxia who similarly exhibit difficulties with ADL performed in real-world settings. The presumption is that in such IA patients, the apraxia producing lesion encroaches, as well, on the neural substrate for SA/WM.

We call the foregoing account of naturalistic action disorders the "unified hypothesis" because it accounts for disparate, and often contradictory, observations in the literature, and because it offers a common framework from which to investigate naturalistic action disorders associated with left-hemisphere lesions (ideational apraxia), bilateral lesions involving the frontal lobes (frontal apraxia), and progressive dementing conditions.

We close with a summary of the major premises of the unified hypothesis:

1. Damage to selective components of the action–schema network compromises the automaticity of action selection, bringing about greater dependence on executive systems that allocate attention and working memory.

2. As long as these executive systems (SA/WM) are intact, the shift to the controlled mode will not affect performance of routine activities of daily living. But when these executive systems are compromised as well, the consequences for ADL may indeed be severe.

3. SA/WM damage by itself does not materially affect ADL tasks because of their normally low resource demands. But damage to the action schema network that renders ADL production non automatic exaggerates the consequences of the SA/WM deficit for ADL tasks.

4. The combination of loss of automaticity in action selection and impaired SA/WM is thus responsible for the severe naturalistic action disorder that manifests in errors in real world tasks like eating and grooming. This is most likely to occur in CHI and DAT, which often impact both the anterior substrates for SA/WM as well as posterior substrates for action and object schemas. However, we assume that this combination may occur with large focal lesions, as well.

5. Formal assessments like the MOT, which tax the action system with less familiar tasks and novel contexts, have the capacity to reveal the vulnerability to errors associated with the breakdown of automatic action selection, even with an intact SA/WM.

6. Automatic action selection is a highly integrative activity that rests on the efficient operation of numerous specialised knowledge and processing systems distributed throughout the brain. Accordingly, it is not just damage to primitive action schemas that should impact automatic action selection, but many other types of neuropsychological deficit, as well. Indeed, we interpret the conflicting accounts of ideational apraxia as supporting just such a diversity of mechanism. For example, on the assumption that automatic action selection draws on script knowledge, and semantic memory for objects, damage to these representational systems should predispose to errors. Each of these conditions will cause SA/WM to play a greater role in the programming of naturalistic action than is normal and, depending on the task complexity, and the status of SA/WM, introduce vulnerability to action errors when SA/WM fails. It remains to be seen whether this error tendency is expressed differently in patients with different neuropsychological syndromes, or, alternatively, whether the increased dependence on SA/WM has an homogenising effect that obscures underlying causes. Answering this question will require more refined methods for analysing naturalistic action production than those currently in use.

ACKNOWLEDGMENT

Preparation of this manuscript was supported in part by grant #1R01 NS31824 from the National Institutes of Health/National Institute of Neurological Diseases and Stroke.

REFERENCES

Baddeley, A. D. (1986). *Working memory*. Oxford: Clarendon Press.

Baddeley, A. D. (1993). Working memory or working attention? In A. Baddeley & L. Weiskrantz (Eds.), *Attention: Selection, awareness and control*. Oxford: Clarendon Press.

Brown, J. W. (1988). *Agnosia and apraxia: Selected papers of Liepmann, Lange, and Potzl*. Hillsdale, NJ: Lawrence Erlbaum Associates Inc.

Buxbaum, L. J., Schwartz, M. F., & Carew, T. G. (in press). The role of semantic memory in object use. *Cognitive Neuropsychology*.

Buxbaum, L. J., Schwartz, M. F., Coslett, H. B., & Carew, T. G. (1995). Naturalistic action and praxis in callosal apraxia. *Neurocase, 1*, 3–17.

Chertkow, H. & Bub, D. (1990). Semantic memory loss in Alzheimer-type dementia. In M.F. Schwartz, (Ed.), *Modular deficits in Alzheimer-type dementia*. Cambridge: MIT Press/ Bradford Books.

Clark, M., Merians, A. S., Kothari, A., Poizner, H., Macauley, B., Gonzalez-Rothi, L. J., & Heilman, K. M. (1994). Spatial planning deficits in limb apraxia. *Brain, 117*, 1093–1106.

Cooper, R., Shallice, T. & Farringdon, J. (1994). Symbolic and continuous processes in the automatic selection of actions. Technical report no. UCL–Psy–Adrem–TR11. University College, London, October 1994.

Coslett, H. B. & Saffran, E. M. (1989). Preserved object recognition and reading comprehension in optic aphasia. *Brain, 112*, 1091–1110.

De Renzi, E. & Lucchelli, F. (1988). Ideational apraxia. *Brain, 111*, 1173–1185.

De Renzi, E., Pieczuro, A., & Vignolo, L. A. (1968). Ideational apraxia: A quantitative study. *Neuropsychologia, 6*, 41–52.

Foundas, A. L., Macauley, B. L., Raymer, A. M., Maher, L. M., Heilman, K. M., & Rothi, L. J. G. (1995). Ecological implications of limb apraxia: Evidence from mealtime behavior. *Journal of the International Neuropsychological Society, 1*, 62–66.

Gibson, J. J. (1977). The theory of affordances. In R. Shaw & J. Bransford (Eds.), *Perceiving, acting, and knowing: Toward an ecological psychology*. Hillsdale, NJ: Lawrence Erlbaum Associates Inc.

Grafman, J. (Ed.). (1989). *Plans, actions, and mental sets: Managerial knowledge units in the frontal lobe*. Hillsdale, NJ: Lawrence Erlbaum Associates Inc.

Hasher, L. & Zacks, R. T. (1979). Automatic and effortful processes in memory. *Journal of Experimental Psychology: General, 108*, 356–388.

Hécaen, H. (1981). Apraxias. In S. B. Filskov & T. J. Boll (Eds.), *Handbook of clinical neuropsychology*. New York: Wiley.

Heilman, K. M., Rothi, L. J., & Valenstein, E. (1982). Two forms of ideomotor apraxia. *Neurology, 32*, 342–346.

Lehmkuhl, G. & Poeck, K. (1981). A disturbance in the conceptual organization of action in patients with ideational apraxia. *Cortex, 17*, 153–158.

Lhermitte, F. (1983). Utilization behavior and its relation to lesions of the frontal lobes. *Brain, 106*, 237–255.

Liepmann, H. (1905). *The left hemisphere and action* (Doreen Kimura, 1980, Trans.). London, Ontario: University of Western Ontario.

Lucchelli, F., Lopez, O. L., Faglioni, P., & Boller, F. (1993). Ideomotor and ideational apraxia in Alzheimer's disease. *International Journal of Geriatric Psychiatry, 8*, 413–417.
Luria, A. R. (1966). *Higher cortical functions in man.* New York: Basic Books.
Marsden, C. D. (1982). The mysterious motor function of the basal ganglia: The Robert Wartenberg Lecture. *Neurology, 32*, 514–539.
Morlaas, J. (1928). *Contribution a l'étude de l'apraxie.* Paris: Legrand.
Morris, R. G. & Baddeley, A. D. (1988). Primary and working memory functioning in Alzheimer-type dementia. *Journal of Clinical and Experimental Neuropsychology, 10*, 279–296.
Norman, D. A. (1981). Categorization of action slips. *Psychological Review, 88*, 1–15.
Norman, D. A. & Shallice, T. (1980, 1986). *Attention to action: Willed and automatic control of behavior* (CHIP No. 99). University of California.
Ochipa, C., Rothi, L. J. G., & Heilman, K. M. (1989). Ideational apraxia: A deficit in tool selection and use. *Annals of Neurology, 25*, 190–193.
Ochipa, C., Rothi, L. J. G., & Heilman, K. M. (1992). Conceptual apraxia in Alzheimer's disease. *Brain, 115*, 1061–1072.
Paillard, J. (1982). Apraxia and the neurophysiology of motor control. *Philosophical Transactions of the Royal Society, London, B298*, 111–134.
Poeck, K. (1983). Ideational Apraxia. *Journal of Neurology, 230*, 1–5.
Poeck, K. (1985). Clues to the nature of disruptions to limb praxis. In E. A. Roy (Ed.), *Neuropsychological studies of apraxia and related disorders.* Amsterdam: North-Holland.
Poeck, K. & Lehmkuhl, G. (1980). Ideatory apraxia in a left handed patient with right-sided brain lesion. *Cortex, 16*, 273–284.
Poizner, H., Mack, L., Verfaellie, M., Rothi, L. J. G., & Heilman, K. M. (1990). Three-dimensional computergraphic analysis of apraxia: Neural representations of learned movement. *Brain, 113*, 85–101.
Rapcsak, S. Z., Croswell, S. C., & Rubens, A. B. (1989). Apraxia in Alzheimer's disease. *Neurology, 39*, 664–668.
Rapcsak, S. Z., Ochipa, C., Beeson, P. M., & Rubens, A. B. (1993). Praxis and the right hemisphere. *Brain and Cognition, 23*, 181–202.
Reason, J. T. (Ed.). (1979). Actions not as planned: The price of automatization. In G. Underwood & R. Stevens (Eds.), *Aspects of Consciousness* (vol. 1). London: Academic Press.
Reason, J. T. (1990). *Human error.* London: Cambridge University Press.
Reason, J. T. & Mycielska, K. (1982). *Absent minded? The psychology of mental lapses and everyday errors.* Englewood Cliffs, NJ: Prentice-Hall.
Rothi, L. J. G., Ochipa, C., & Heilman, K. M. (1991). A cognitive neuropsychological model of praxis. *Cognitive Neuropsychology, 8*, 443–458.
Roy, E. A. (1982). Action and performance. In A. W. Ellis (Ed.), *Normality and pathology in cognitive functions.* London: Academic Press.
Roy, E. A. (1983). Neuropsychological perspectives on apraxia and related action disorders. In R. A. Magill (Ed.), *Memory and control of action.* Amsterdam: North-Holland.
Roy, E. A. & Square, P. A. (1985). Common considerations in the study of limb, verbal and oral apraxia. In E. A. Roy (Ed.), *Neuropsychological studies of apraxia and related disorders.* Amsterdam: North-Holland.
Schank, R. C. (1982). *Dynamic memory.* Cambridge: Cambridge University Press.
Schank, R. C. & Abelson, R. (1977). *Scripts, plans, and understanding.* Hillsdale, NJ: Lawrence Erlbaum Associates Inc.
Schwartz, M. F. (1990). *Modular deficits in Alzheimer-type dementia.* Cambridge, MA: MIT Press.
Schwartz, M. F. (1995). Re-examining the role of executive functions in routine action production. *Annals of the New York Academy of Science 769*, 321–335.

Schwartz, M. F. & Chawluk, J. B. (1990). Deterioration of language in progressive aphasia: A case study. In M. F. Schwartz (Ed.), *Modular deficits in Alzheimer-type dementia*. Cambridge, MA: MIT Press.

Schwartz, M. F., Fitzpatrick-DeSalme, E. J., & Carew, T. G. (1995). The Multiple Objects Test for ideational apraxia: Etiology and task effects on error profiles (Abstract). *Journal of the International Neuropsychological Society, 1*, 149.

Schwartz, M. F., Montgomery, M. W., Fitzpatrick-DeSalme, E. J., Ochipa, C., Coslett, H. B., & Mayer, N. H. (1995). Analysis of a disorder of everyday action. *Cognitive Neuropsychology, 12*, 863–892.

Schwartz, M. F., Reed, E. S., Montgomery, M. W., Palmer, C., & Mayer, M. H. (1991). The quantitative description of action disorganization after brain damage: A case study. *Cognitive Neuropsychology, 8*, 381–414.

Shallice, T. (1982). Specific impairments of planning. *Philosophical Transactions of the Royal Society of London, B298*, 199–209.

Shallice, T. (1988). *From neuropsychology to mental structure*. Cambridge: Cambridge University Press.

Shallice, T., Burgess, P. W., Schon, F., & Baxter, D. M. (1989). The origins of utilization behavior. *Brain, 112*, 1587–1598.

Shiffrin, R. M. & Schneider, W. (1977). Controlled and automatic human information processing: II. Perceptual learning, automatic attending, and a general theory. *Psychological Review, 84*, 127–190.

Sirigu, A., Duhamel, J., & Poncet, M. (1991). The role of sensorimotor experience in object recognition. *Brain, 114*, 2555–2573.

Warrington, E. K. (1982). Neuropsychological studies of object recognition. *Philosophical Transactions of the Royal Society of London, B298*, 15–34.

Wilson, B., Cockburn, J., & Halligan, P. (1987). *Behavioural inattention test*. Titchfield, UK: Thames Valley Test Company.

Glossary

action lexicon = Performance knowledge about previously experienced actions that is stored such that it can be called upon and used the next time the action is needed. In this way the individual does not need to reconstruct the production attributes of each action every time it is performed. That is, the action lexicon is an action memory containing spatial-temporal representations of learned purposive skilled movements.

action semantic system = That portion of the semantic system that contains knowledge related to actions and tools such as the function of tools, the objects that tools act upon, the meaning of actions and how actions may be organised with other actions to accomplish a goal.

addressed = When something is previously experienced and a memory is formed for all or a portion of that experience, the memory can be called upon to reconstitute that experience at the discretion of the individual. The process of retrieving stored information is called "addressing" the information.

afferent dysgraphia = A writing disorder caused by the inability to utilise visual and/or kinaesthetic sensory feedback to control the execution of writing movements.

affordances = Clues regarding the nature of a movement provided by the physical attributes of the associated tool.

agnosia = A modality-specific defect in recognition.

agraphia = A term used to refer to various acquired disorders of spelling and writing caused by neurological damage.

allographs = Physically different forms of the same letter (e.g. upper versus locase).

allographic conversion = The stage in the writing process where abstract graphemic representations are assigned specific letter shapes.

allographic store = Long-term memory store of abstract spatial codes representing letter shape information.

allographic disorders = Writing disorders characterised by an inability to activate or select the letter shapes appropriate for the set of abstract graphemic representations generated by central spelling routes.

aphasia = An acquired, neurologic disorder of language.

apraxia = A disorder of learned, voluntary actions resulting from neurologic impairment.

apraxic agraphia = A writing disorder characterised by poor letter formation that cannot be attributed to impaired letter shape knowledge or elementary sensorimotor dysfunction affecting the writing limb.

articulation (speech) = Vocal tract movements which result in speech sound production. Parameters of speech articulation include accuracy in placement and trajectories of movements of the articulators (i.e. lips, jaw, tongue, velum) which results from the timing and force of muscle contractions.

assembled = The occasion when a response must be constructed without advantage of prior experience.

asymbolia = The inability to understand or appreciate symbols.

athetosis = Involuntary, slow, writhing movements of the trunk or limbs.

attention = The process by which the brain selects what stimuli to process and how much these stimuli should be processed.

attention deficit hyperactivity disorder = A persistent pattern of inattention and/or hyperactivity/impulsivity that is present prior to age seven and that is more frequent and severe than is typically observed in individuals at a comparable level of development.

body schema = The mental representation or map of the sensorimotor organisation of the body.

cerebral palsy = A nonprogressive, permanent disorder of motor function secondary to brain injury with onset at birth or shortly thereafter.

closed-loop circuit = A form of control in which the motor program is modified online, during movement using sensory information or sensory feedback (e.g. visual, proprioceptive).

cognitive neuropsychology = The study of the structure of normal cognitive processing systems through investigation of patients with brain disorders.

compensatory treatment = Treatment designed to achieve a behavioural goal (due to impairment of the system's normal means of achieving the desired behaviour) by altering the task strategy.

conceptual apraxia = A disorder of tool–action knowledge resulting in impairments in tasks requiring knowledge of tools, their functions, object relationships, and actions.

conceptual praxis = See action semantics.

degraded memory trace = In the case of apraxia, when an apraxic patient is not able to get into the movement memory system at all because it is damaged and therefore the patient has no representation to compare received gestures to nor do they have a representation that guides the production of previously learned gesture.

developmental dysphasia = A variety of communication disorders of early childhood that are characterised by failure to acquire language normally and at the appropriate age, despite adequate hearing, normal nonverbal intelligence and the absence of major sensorimotor defect or congenital malformation of the vocal tract. They are assumed to reflect dysfunction of brain systems necessary for language.

developmental learning disability = A heterogeneous group of disorders manifested by significant difficulties in the acquisition and use of listening, speaking, reading, writing, reasoning, mathematical ability, or social skills. These disorders are intrinsic to the individual and presumed to be due to central nervous systems dysfunction.

developmental verbal apraxia = An impairment in programming movement sequences necessary for speech production.

dissociation apraxia = A modal-specific defect in gesture production.

dysdiadochokinesia = An inability to perform rapidly alternating movements.

dysmetria = An inability to accurately estimate distances associated with voluntary motor acts.

electropalatography = Instrumental analysis of the patterns and areas of contact of the tongue with the hard palate using electrical current.

elemental motor deficit = A loss of the ability to perform movements that may be induced by weakness, ataxia, abnormal movements (such as athetosis, chorea, ballismus, tremors or dystonia).

excess and equal stress (speech) = Excess stress on usually unstressed parts of speech (e.g. monosyllabic words and unstressed syllables of polysyllabic words).

fundamental frequency (f_0) = The lowest component frequency of the sound generated by the larynx and perceived as the pitch of the voice.

grapheme = A letter or a sequence of letters that corresponds to a single phoneme of spoken language. Spelling requires computing sets of abstract graphemic representations specifying the identity and serial order of the letters that make up a word (or nonword).

graphemic buffer = A working memory system that temporarily stores abstract graphemic representations while they are being converted into specific letter names or letter shapes.

graphic innervatory patterns = Sequences of motor commands to specific effector systems involved in writing. Graphic innervatory patterns specify concrete movement parameters such as absolute stroke size, duration and force.

graphic motor programs = Stored memory representations that guide the execution of the skilled movements of writing. Graphic motor programs specify the sequence, direction and relative size of the strokes necessary to create a given letter, without specifying absolute stroke size or duration.

graphemic output lexicon = The long-term memory store of learned word spellings. Spelling familiar words normally relies on the retrieval of information from the graphemic output lexicon.

handedness = Preference for using one hand (e.g. right) when performing skilled acts (e.g. writing).

hand preference = See handedness.

hemiplegia = Weakness of one side of the body.

ideational apraxia = Definitions vary across researchers and include impairment in the sequencing of tool use, failure to use single tools appropriately, or amnesia of use in that there is a failure to recall the appropriate action when attempting to use a tool.

ideomotor apraxia = A deficit of learned skilled purposive movements that specifically effects the efficiency and accuracy of the production of a movement and cannot be accounted for by elemental sensorimotor, cognitive or attentional deficits.

implement = An object or tool.

innervation = To supply with neural input.

innervatory patterns = A term coined by Liepmann (1905) to describe that portion of the praxis system which allows the representational form of the movement memories or action lexicon to be converted to a form that allows motoric implementation; that is, temporally sequential information regarding those neurons that should be activated or inhibited.

input action lexicon = Stored representations containing information relative to the physical attributes of seen actions.

intention = Voluntary preparation for goal-oriented actions.

intention tremor = Rhythmic purposeless movements that increase during voluntary actions.

intransitive movement = A movement that does not involve the use of a tool or object such as waving goodbye.

kinetic memories = A term used by Liepmann (1905) in reference to "... properties of the senso-motorium" that he defined as "... a functional linkage tak[ing] place between innervations ... which runs its course without intervention from orientation and visual images, through short circuiting."

lexicon = Memory store of previously experienced words that provides a processing advantage in subsequent exposures to those familiar words.

limb kinetic apraxia = A term used by Liepmann to indicate a loss of the ability to make precise, independent finger movements.

memory egress disorder = With respect to the praxis system specifically, the occasion when an apraxic patient is able to get into the movement memory system and retrieve information that would allow them to recognise gestures when they see them but could not use that information to guide their own productions.

micrographia = An extreme diminution of letter size typically observed in the handwriting of patients with Parkinson's disease.

modality = A sensory channel or system of input (i.e. vision).

mode = A channel or system of output (i.e. writing, gesture, etc.).

mode/modality consistency = When all modes and modalities are disturbed or spared to the same degree when compared to one another.

motor program = An abstract representation of intended movements that describes the relationship among the goal of an action, the external event or object it is directed toward, and the organism's interactions with the external event (as per Harrington & Haaland, Chapter 9).

movement formulae = Described by Liepmann (1905) as "... space–time sequences" of movement that give "... general knowledge of the course of the procedure to be realized."

muscular dystrophy = A group of inherited myopathies characterised by progressive muscle degeneration and weakness.

myelination = The process by which the glial cells of the nervous system surround axons providing them with insulation which facilitates neural transmission.

naturalistic action = Action in the service of commonplace, practical goals. Includes routine activities of daily living, such as eating and grooming, as well as less routine activities such as those tested on the Multiple Objects Test for ideational apraxia (e.g. lighting a candle; hanging a picture).

nonlexical action processing = Sublexical or assembled.

object = A thing that receives an action; for example, a nail, a piece of stationery, etc.

open-loop control = A form of control in which movements are carried out based upon a motor program which is not subsequently modified online, during movement by sensory information (e.g. visual, proprioceptive).

optic aphasia = Visual modality specific aphasia in which patients are unable to name viewed objects, but they can provide meaningful information about unnamed objects such as function or appropriate gesture.

orthographic analysis = An "early" visual processing stage in reading where individual letters are identified.

orthographic input lexicon = The long-term memory store of written word forms. This "sight vocabulary" mediates the recognition of visually presented familiar words.

output action lexicon = A memory system containing a code about the physical attributes of a "to be performed" action.

perceptual speech analysis = The description of speech as discerned by the human ear in terms of articulatory pitch, loudness, quality.

phoneme = Group or family of closely related speech sounds all of which have the same distinctive acoustic characteristics in spite of subtle differences; each phoneme corresponds roughly to one of the symbols in the phonetic alphabet; e.g. p=/p/.

phonemic paraphasia = The production of unintended sounds or syllables in the production of a partially recognisable word.

phonetic transcription = Recording in proper order the phonetic symbols which represent the speech sounds heard in a spoken speech sample. Broad transcription involves the use only of standard symbols which represent phonemes. Narrow transcription makes use of additional symbols or modifying notations in an attempt to more nearly code the exact pronunciation of a speaker.

phonology = Study of the sound system of a language including pauses and stress.

postrotary nystagmus = An oculovestibular reflex which occurs in response to rotatory stimulation.

praxis = Skilled movement.

praxis conceptual system = Knowledge related to tools such as their functions, the objects they act upon, their actions and how actions may be organised with other actions to accomplish a complex action sequence.

praxis production system = The sensorimotor component of action knowledge, including the space–time information contained in action programs, and the translation of these programs into action.

processing advantage = The assistance in processing provided by a system that can be called upon to reconstitute previously constructed action programs.

prosody = Physical attributes of speech that signal linguistic qualities such as stress and intonation; includes the fundamental frequency of the voice, the intensity of the voice, and the duration of the individual speech sounds.

purposeful movements = Movements that are intended to yield a result.

restitutive treatment = Treatment designed to recover impaired function within the limits set by the system.

schemas = Generally, specialised knowledge units which instantiate the regularities of past experience. Here, units of skilled (including naturalistic) action programming.

semantic system = The core conceptual system, store of meaningful information for words and objects such as category, functions, and other associated relationships.

skilled movement = Learned actions.

speech analysis =

acoustic: Measurement by analysing properties of the acoustic wave in terms of frequencies of source and resonating cavities, amplitude and timing of speech articulation.

physiological: Measurement of the dynamic functioning of the structure of the vocal tract as used for speech production.

speech errors =

distortion: A sound readily identified as belonging to a phoneme category (e.g. /s/, /t/, /m/ . . .) but not accurate enough to be considered to be produced normally.

substitution: The production of an erroneous phoneme in the place of that which was intended.

substitutive treatment = Treatment aimed at achieving a behavioural goal in a way novel to the system.

syllable = A unit consisting of a vowel which may stand alone or be surrounded by one or more consonants (e.g. I, in, me, men . . .). A syllable is considered to be the basic physiological and acoustic unit of speech and is acoustically realized as a series of pulses of sound energy.

syllable dissociation = A severe form of syllable segregation (see below) which results in the impression that speech is produced syllable by syllable in that there are silent intervals between syllables.

syllable segregation = A speech pattern which results in the temporal separation of syllables sometimes referred to as "scanning" speech.

sympathetic apraxia = Apraxia of the left hand resulting from a left-hemisphere lesion.

synkinesis = Involuntary movement that occurs during and is elicited by a voluntary action.

tool = An implement that provides a mechanical advantage in an action such as a hammer.

transcranial magnetic stimulation = The use of magnetic radiation to stimulate cortical neurons.

transitive movement = Involving tool use.

vicariative treatment = Treatment designed to facilitate a functional reorganisation of the nervous system to achieve a behavioural goal when the normal means of achieving the goal is impaired.

visuokinaesthetic motor engrams = A term coined by Heilman and Rothi (1985) that means the same as Liepmann's "movement formulae."

WADA Test = Selective hemispheric barbituate anaesthesia used preoperatively to determine language dominance and other lateralised functions.

Author Index

Abbs, J. H., 181, 184–185, 191–192, 194
Abelson, R., 283
Abrams, R. A., 116–117
Acuna, C., 218
Adair, J. C., 70
Adamovich, S., 94
Adams, S., 113–114, 143, 262
Adams, S. G., 180, 194–196
Agostoni, C., 224, 226, 231–232
Agostoni, E., 226
Ahern, M. B., 195
Aielo, U., 226
Ajuriaguerra, J. de, 1
Akelaitis, A. J., 10–12
Alajouanine, T., 187–190
Albert, M. L., 196, 227, 230
Alexander, G. E., 163–164, 210, 230
Alexander, M. P., 142, 155–158, 178, 185, 187, 189, 212–213, 215, 224, 231–234, 237
Allport, D. A., 51, 53, 55
Alonzo, J. M., 218
Andersen, R. A., 218
Anderson, K., 93–95
Anderson, K. C., 42
Anderson, S. W., 157–159
Anson, G., 116
Ansquer, J.C., 161
Aoki, Y., 187
Apale, P., 226
Apeldoorn, S., 180, 188–191, 193–195
Apsimon, T. T., 248
Aram, D. M., 258
Archer, L. A., 258
Archibald, Y., 9, 129, 131, 136, 178, 194
Aronson, A. E., 174–176, 184–185, 189–192, 196, 258
Asaba, H., 142
Asanuma, C., 218
Ashby, P., 230
Ashe, A., 118
Aten, J. L., 178
Athenes, S., 115–117, 120, 128–129
Auff, E., 228–229, 231, 233
Ayres, A.J., 245–246, 249–251, 255, 259–260

Baddeley, A. D., 276, 283
Baker, E., 142
Bakwin, H., 248
Ballinger, W. E., 220–221, 232–233, 237–238
Baratz, R., 78
Barbieri, C., 52
Barbizet, J., 161
Barraquer, L. L., 113–114
Barresi, B., 78
Barry, C., 156–157
Barter, D., 78
Basaglia, N., 31, 161

Bass, K., 149, 156–157
Basso, A., 76, 193, 223–227, 232–233
Bastard, V., 39
Bates,E., 264
Baum, S. R., 185
Baxter, D. M., 157–158, 271, 284
Beauvois, M. F., 37–39, 43, 53–54
Bechtereva, N., 218
Becker, J. T., 221
Beeson, P. M., 162, 271
Behigni, L., 264
Behrmann, M., 149, 156–157
Bell, B. D., 36
Bellaire, K., 79, 88
Bellugi, U., 63, 178
Benabio, A. L., 185
Benke, T., 41
Benson, D. F., 185, 187
Benton, A. L., 2, 114
Berges, J., 251, 255
Bergmann, L., 217
Berkinblit, M., 94
Bermudez, A., 41
Berndt, R. S., 54
Bernstein, N., 116
Berthier, M., 160
Bertrand, I., 188
Bianchini, E., 230–231, 233
Biggerstaff, A., 259
Binder, L., 154, 157–158
Black, S., 175, 230
Black, S. E., 149, 156–157
Bloedel, J. R., 163–164
Blumstein, S. E., 177, 180, 185
Bogen, J. E., 11–12, 14
Bogliun, G., 226
Bohm, C., 140
Boldrini, P., 31, 161
Boller, F., 276
Bonvillian, J. D., 78
Borod, J. C., 78
Bowers, D., 21–22
Bracewell, M., 32, 76
Brain, W. R., 245
Brenner, S., 217
Bretherton, I., 264
Brinkman, C., 13, 34
Broca, P., 20, 24, 175–177, 185, 189
Brooks, R., 57
Brooks, V. B., 164–165
Brown, E. R., 129–130
Brown, J. R., 174–176, 184–185, 189–192, 258
Brown, J. W., 78, 273–274
Brown, L., 175, 195
Brown, P., 14, 63–65, 81–83, 101, 113–114, 194, 225–226, 265
Brown, R. G., 227, 230–231
Bruhn, P., 223
Bruininks, R. H., 249, 261
Brunner, R. J., 215
Bub, D., 283
Buckingham, H. W., 180, 188
Burgess, P. W., 271, 284
Burns, B. D., 130
Bushnell,M. C., 21–22
Butters, N., 221, 229–230
Buxbaum, L. J., 271, 278

Caligiuri, M., 180
Calvanio, R., 23, 154–155, 157
Camaioni, L., 264
Camras, L., 181
Cantagallo, A., 31, 161
Canter, G. J., 177, 188, 190, 194
Capitani, E., 76
Caplan, D., 175
Caplan, L. R., 40
Cappa, S. F., 231
Caramazza, A., 51, 53–54, 149, 156–157
Carew, T. G., 271, 278–282, 285
Cartz, L., 158
Carvell, S., 229, 231, 233
Castaigne, P., 187
Cederbaum, J., 230
Cermak, S., 245, 248
Cermak, S. A., 254, 255–256, 258
Charlton, J., 194
Chase, T. N., 57
Chatterjee, A., 64–65
Chawluk, J. B., 283
Cherktow, H., 283
Chiappa, K. H., 23
Christopoulou, C., 78
Chumpelik, D. A., 195–196
Chung, G., 217
Cipolotti, L., 153
Clark, C. J., 179, 194
Clark, M. A., 41, 93–101, 271–272
Clerebaut, N., 78
Coates, R., 186
Cockburn, J., 281
Code, C., 78, 86
Coelho, C. A., 78–79
Cohen, A., 116, 129
Cohen, L. G., 23
Cohen, N. J., 221
Coletti, A., 224, 226, 231–232
Collard, R., 129–130
Collins, M. J., 192, 196
Coltheart, M., 29, 43–44, 51, 53
Connolly, K., 264
Conrad, K. E., 255–256

Consoli, S., 177
Contamin, F., 187
Cooper, I. S., 217
Cooper, R., 283
Coren, S., 20–21, 23–24
Coslett, H. B., 39–40, 54, 157–158, 278, 282
Coster, W., 254, 255
Court, S. D. M., 246
Cowan, W. M., 218
Coyle, J. M., 9, 25–26, 112, 158
Crary, M. A., 155, 157
Crisi, G., 32, 225, 227, 232
Croisile, B., 157, 160
Cros, D., 23
Crossman, A. R., 215, 234
Crosson, B., 210–214, 216–221, 231, 234, 236–237
Croswell, S. C., 57, 276
Cubelli, R., 78, 86
Cudeiro, J., 218

Dalgleish, M., 264
Damasio, A. R., 93, 157–159, 161, 185, 187, 231, 233, 237
Damasio, H., 157–159
Darley, F. L., 174–178, 180, 184–185, 188–194, 258
David, E. D., 153
David, R., 78
Davis, G. A., 86
Davis, K. R., 185–186
Deal, J., 178
De Bastiani, P., 156–157
de Bleser, R., 186
Deck, J., 196
Dee, H. L., 114
Degos, J. D., 161
Dejerine, J., 177, 188
de la Monte, S., 228
Della Sala, S., 225–227, 232
DeLong, M. R., 163–164, 210, 230
Demiati, M., 158
Denckla, M. B., 245–246, 248–251, 262, 264, 272–273, 275–276, 282
Denes, G., 54, 153
Denes, P., 153
Denny-Brown, D., 188, 215, 235
De Renzi, E., 2, 14–15, 32, 41, 52, 61, 72, 75, 86, 112, 114, 131, 141–143, 223–227, 232, 250, 252–253, 257
Derouesne, J., 39
Deuel, R. K., 257
Deutsch, S. E., 188, 190
Dewey, D., 258, 260–262
Dhawan, V., 230
Dichiro, G., 57
Disimoni, F., 188
Divac, I., 208, 214, 221
Doar, B. P., 257
Domich, L., 219
Doody, R. S., 230
Drake, C., 254–256
Dreyer, D. R., 178
Dronkers, N. F., 186–187
Duffy, J. R., 30, 78, 113–114, 178, 184, 191
Duffy, R. J., 30, 78, 113–114
Dugas, C., 115–116, 120, 128–129
Duhamel, J., 270–271
Dum, R. P., 210
Duncan, G. W., 185–186
Dunlop, J. M., 177
Durand, M., 188–190

Eidelberg, D., 230
Elliott, D., 127
Ellis, A. W., 149, 151, 153–156, 165–166, 208–209
Ellis, E., 246
Eslinger, P. J., 231, 233, 237
Essens, P. J., 130
Eustache, F., 156–157
Evarts, E. V., 184, 191
Exner, S., 159

Faglioni, P., 14, 32, 41, 75, 131, 192–193, 225, 227, 232, 276
Fahn, S., 230
Falanga, A., 53
Farah, M. J., 53–54
Farmer, A., 177
Farringdon, J., 283
Fedio, P., 57
Feeser, H. R., 219
Feldman, G., 94
Feldman, R. G., 227
Ferro, J. M., 3, 75–76, 78, 224–225, 227, 233–234
Feyereisen, P., 78
Figel, T., 94
Finkelnburg, F., 8, 30
Finkelstein, S., 185–185
Finset, A., 76
Fisk, J. D., 122, 124
Fischer, R. S., 157–158
Fitts, P. M., 127
Fitzpatrick, P. M., 78
Fitzpatrick-DeSalme, E. J., 278–282, 285
Flaherty, D., 112–114, 133, 142–143, 194
Fleet, W. S., 13, 32, 35–36, 160, 193
Fletcher, J. M., 247
Flude, B. M., 153–154, 165–166

Foix, C., 188
Folger, W. N., 184, 191
Folstein, M. F., 227
Fookson, O., 94
Ford, F. R., 246
Foster, N. L., 57
Foundas, A. L. , 2–3, 23–24, 58, 63, 66, 70–71, 76–77, 81, 277–278, 285
Fragassi, N. A., 53
Franks, I. M., 117
Freeman, F., 178–179, 194
Freidenberg, D. L., 227–231, 233
Freund, H. J., 184, 188
Friedman, R. B., 156–158
Friedman, R. J., 78
Friesen, H., 113, 262
Freund, C. S., 53
Fromm, D., 194
Fry, D. B., 153, 177
Fujimura, O., 177
Fukusako, Y., 194
Funkenstein, H., 185–186
Funnell, E., 43–44, 51, 53
Fust, R., 78–79

Galaburda, A. M., 178, 185
Gates, J., 161
Gaunt, C., 78, 86
Gazzaniga, M. S., 11–12, 14, 31–32, 34, 161
Gelb, I. J., 149
Gentil, M., 185
Georges, J. B., 79, 88
Georgopoulos, A. P., 118
Gerfen, C. R., 215
Gersh, F., 161
Gerstman, L. J., 78
Geschwind, N., 5, 9–14, 16, 24–26, 75, 93, 107, 112, 158, 160–161, 187, 192–193, 249
Gibson, J. J., 270
Gilman, S., 163–164
Ginos, J. Z., 230
Glosser, G., 78
Godersky, J., 12, 24, 160
Godschalk, M., 184
Goldberg, G., 164, 184, 186–187
Goldberg, M. E., 21–22
Goldenberg, G., 160, 228–229, 231, 233
Goldman-Rakic, P. S., 218
Goldstein, K., 159, 192
Goldstein, L. H., 113–114
Gollomp, S., 229, 231, 233
Gonyea, E. F., 9, 25–26, 112, 158
Gonzalez, F., 218
Goodale, M. A., 116, 118, 120, 122, 124, 142
Goodglass, H., 8, 25, 76, 78, 86, 112, 114, 178, 185, 189, 250
Goodman, R. A., 156–157
Goossens, M., 78
Gordinier, H. C., 158–159
Gordon, A. M., 130
Gordon, N. S., 249
Gordon, P. C., 117
Gracco, V. L., 181
Graff-Radford, N. R., 12, 24, 160, 231, 233, 237
Grafman, J., 277
Gray, F., 161
Greene, P., 230
Greenwald, M. L., 41, 42–43, 54, 57, 62–70, 81–82, 86
Greitz, T., 140
Grice, J. W., 117, 120, 124
Grillner, S., 175
Grossi, D., 53
Grossman, M., 229, 231, 233
Growdon, J. H., 228
Gubbay, S. S., 235–246, 248–249, 259

Haaland, K. Y., 112–114, 117–118, 120, 122–124, 128–129, 131–140, 142–143, 194
Haaxman, R., 192
Hacker, P., 149, 156–157
Hajgaard, K., 223
Hall, C., 175
Hall, D., 248
Hallet, M., 23
Halligan, P., 281
Hammond, G., 137
Hanlon, R. E., 78
Hanson, W. R., 181, 186, 189
Hardcastle, W. J., 179, 194
Hardie, M., 195
Hardison, D., 177
Harlock, W., 186
Harrington, D. L., 117–118, 120, 122–124, 128–129, 131–140
Harris, E., 195
Harris, K., 179
Harris, K. S., 177–178
Hart, J., 54
Hartman, D. E., 184, 191–192
Hasher, L., 283
Hawkins, S. R., 130
Hayden, D., 196, 258
Hayden (Chumpelik), D., 195–196
Head, H., 29
Hécaen, H., 8, 165, 177, 274
Hedley-Whyte, T., 228
Heilbronner, K., 157
Heilman, K. M., 2–3, 5, 9, 11–16, 21–26,

31–33, 35–41, 42–44, 46, 51–52, 54–58, 62–71, 75–79, 81–86, 93–106, 108, 112–115, 138, 154–158, 160–161, 173, 184, 192-192, 194–195, 207–209, 217, 219, 223, 225–227, 230, 232–238, 256, 262, 264–265, 271–272, 276–278, 281–283, 285
Heiss, W. -D., 223
Helm, N. A., 196
Helm-Estabrooks, N., 61, 78, 189, 196
Henderson, S. E., 248–249
Henriksen, L., 164
Henschen, S. E., 159
Hermann, M., 78
Hillis, A. E., 51, 53–54
Hirayama, K., 161
Hirose, H., 178–179, 194
Hodges, J. R., 157–159
Hogg, S., 113–114, 143
Hogg, S. C., 30, 174, 178, 195
Holmes, G., 107, 164
Holtzman, J. D., 161
Holzner, F., 160
Hong, K., 23–24
Hooper, P., 78
Horenstein, S., 217
Hornykiewicz, O., 230
Horowitz, S. J., 258
Huber, S. J., 217–218, 228–231, 233
Hughlings Jackson, J., 2–3, 8
Hulme, C., 259
Humphreys, G. W., 32, 43–45, 51, 53–54, 86, 87
Hurtig, H. I., 229, 231, 233
Hyland, B., 116

Ikeda, H., 161
Imamura, T., 160
Inhoff, A. W., 117
Iorio, L., 53
Ireland, J. V., 177
Itoh, M., 178–179, 190, 194
Ivry, R. I., 116, 118, 129, 138, 140, 262–263
Iwasa, H., 31, 158

Jackson, A., 215, 234
Jackson, C. A., 181, 186, 189
Jackson, J. P., 252
Jaffee, D. W., 178
Jankovic, J., 230
Janowsky, J. S., 131
Jason, G. W., 129, 131
Jeannerod, M., 115–116, 118, 120, 128–129, 208, 235
Jennings, P. J., 94
Jesudowich, B., 177
Johannsen-Horbach, 78
Johns, D. L., 176–178, 194
Johnstone, S., 213
Jonas, S., 187
Jones, E. G., 216, 219
Jurgens, U., 187

Kalmus, H., 153
Kanazawa, Y., 161
Kaplan, B. J., 260–261
Kaplan, E., 8, 11–12, 14, 25, 31, 76, 78, 84, 112, 114, 142, 160, 178, 189, 223, 250–253, 255, 264
Kaplan, R. F., 40
Kapur, N., 156–157
Kartounis, L. D., 155–156
Kasai, Y., 179
Kawamura, M., 161
Kayne, H. L., 229–230
Kazui, S., 161
Kearns, K. P., 78, 81, 83
Keele, S. W., 116–118, 120, 129, 138, 262
Kegl, J., 194–195
Keith, R., 196
Kelso, J. A. S., 120, 140
Kempler, D., 30, 181, 186, 189
Kent, R. D., 175–176, 179–180, 190, 194
Kepes, J. J., 218, 236
Kerr, B., 117
Kerschensteiner, M., 194
Kertesz, A., 3, 24, 75–76, 78, 186, 193, 224–225, 227, 233–234, 258
Kim, A. K., 218, 236
Kimura, D., 9, 113–114, 129, 131, 136, 142, 174, 178, 190, 192–194, 247
Kinsbourne, M., 155–157
Kirshner, H. S., 78–79
Kleist, K., 7, 141, 188, 192
Klich, R. J., 177
Klima, E., 63
Knight, R. T., 186–187
Knoll, R. L., 117, 129
Knopman, D. S., 227, 230–231, 233–234, 236
Knuckey, N. W., 248
Kolb, B., 115, 131
Kooistra, C. A., 225–227, 234–236
Kools, J. A., 252–253
Koomar, J., 256
Kornblum, S., 116–117
Kornhuber, H. H., 215
Kothari, A., 41, 93–101, 271–272
Kuhl, D. E., 189
Kurata, K., 186
Kuypers, H. G., 184, 192

Laiacona, M., 76, 226–227, 232

Lambert, J., 156–157
Lang, A. E., 230
LaPointe, L. L., 176, 195–197
Larsen, B., 35, 140, 186
Lashley, K. S., 116, 129
Lassen, N. A., 35, 140, 164, 186, 223
Laurent, B., 157, 160
Lauritzen, M., 164
Lawton, N. F., 156–157
Leavitt, J. L., 117
Lebrun, Y., 153, 165
Lechtenberg, R., 163–164
Lecours, A. R., 185
Lehmkuhl, G., 112–114, 142–143, 274, 281–282
Leiguarda, R., 160
LeMay, A., 78
Lemme, M. L., 195
Lemon, R. N., 184
Lennox, L., 256
Lenti, C., 230–231, 233
Leonard, C. M., 23–24
Lesny, I. A., 249
Lesser, R., 149
Levine, D. N., 154–155, 157, 185–186
Levine, H. L., 189
Levita, E., 217
Levy, J., 162
Lewis, A., 230
Lezine, I., 251, 255
Lhermitte, F., 54, 185, 271
Li, C. L., 234
Lichtheim, L., 188
Liepmann, H., 1, 5–7, 9–13, 15–16, 19, 24–26, 29, 31–33, 36, 38, 41–42, 52, 61–62, 80, 112, 138–141, 142–143, 160–161, 184, 191, 271
Liles, B. Z., 30
Liss, J., 180
Lodesani, M., 75, 131
Lombardino, L., 264
Lopez, O. L., 276
Louarn, F., 161
LoVerme, S. R., 224, 231, 233, 237
Lozano, R. A., 178
Lucchelli, F., 15, 32, 52, 112, 142–143, 275–276, 282
Lucius-Hoene, G., 78
Lund, G., 161
Lundy-Ekman, L., 262–263
Luria, A. R., 139, 188, 190, 192, 255, 277
Luzzatti, C., 193, 223–224, 227, 233
Lyard, G., 185

Maas, O., 9–11, 16, 24–25, 31, 160
Macauley, B., 2–3, 41, 58, 66, 70–71, 77, 93, 95–101, 271–272, 277–278, 285
Macdonnell, R. A. L., 23
Mack, L., 14–15, 37–41, 57–58, 63–65, 76, 79, 81–83, 93, 95–96, 101–103, 113–114, 194–195, 225–227, 234–236, 264–265, 271–272
MacKay, D. G., 117, 130, 138
MacKenzie, C. L., 115–117, 120, 128–129, 194
MacNeilage, P. F., 175
Macko, K. A., 142
Maher, L. M., 2–3, 42, 62–71, 76–77, 81–86, 232–233, 235–238, 277–278, 285
Mahowald, M. W., 160
Mandell, A. M., 227, 230
Mani, R. B., 154–155, 157
Mansi, L., 57
Marci, P., 165
Margolin, D. I., 149, 152–158, 163–165
Marie, P., 177, 188–189
Marin, O. S. M., 42, 55
Marit, J., 113–114
Marquardt, T. P., 177, 179
Marsden, C. D., 163–164, 213–215, 222, 227, 230–231, 235, 271
Marteniuk, R. G., 115–117, 120, 128–129, 194
Martin, A. D., 177
Martin, J. G., 130, 175
Martin, R. E., 178, 181–184, 186, 192, 195
Martone, M., 221
Mastri, A., 160
Mateer, C., 193–194
Maxwell, R., 161
Mayer, E., 162
Mayer, N. H., 77, 277, 282
Mazzocchi, F., 186
Mazziota, J., 181
Mazziotta, J. C., 164
McCarthy, R., 39, 54–55
McClelland, J. L., 237
McCormick, D. A., 219
McCrary, J. W., 153
McDonald, S., 34, 76
McFarling, D., 217
McHugh, P. R., 227
McKinlay, I., 259
McLennan, J. E., 163–164
McNeil, M. R., 179–180, 192, 194
McReynolds, L. V., 81, 83
McShane, L. M., 23
Mehler, M. F., 36, 39
Meininger, V., 54
Menzer, L., 229–230
Merians, A. S., 41, 93–101, 271–272
Merton, P. A., 152

Metter, E. J., 181, 186, 189, 192
Meyer, D. E., 116–117, 130
Miceli, G., 149
Michel, D., 157, 160
Michel, F., 118
Miller, E., 210, 212–213
Miller, N., 248
Milner, A. D., 118, 142
Milner, B., 26, 116, 131
Mink, J. W., 212, 214
Mintz, T., 193
Mishkin, M., 142
Mitani, Y., 160
Mitchell, I. J., 215, 234
Mizon, J. P., 158
Mlcoch, A. G., 176, 189–191
Moberg, P. J., 221, 237
Moeller, J. R., 230
Mohr, J. P., 185–186
Monsell, S., 117, 129
Montagna, C. G., 78, 86
Montgomery, K., 229–230
Montgomery, M., 77
Montgomery, M. W., 277, 282
Moran, G., 259
Morasso, P., 99–100
Moreau, O., 185
Morgan, R. A., 179, 194
Morin, P., 156–157
Morlaas, J., 275
Morningstar, D., 195–196
Morris, R. G., 276
Morrison, D., 258
Morton, J., 154
Motomura, N., 42, 142
Motti, F., 84, 114, 225–226
Moyes, F. A., 249
Mozaz, M. J., 113–114
Mulligan, M., 180
Murakami, S., 194
Mycielska, K., 284

Nadeau, S. E., 210, 212–213, 219–221, 231–234, 237–238
Naeser, M. A., 142, 178, 185, 189, 196, 210, 212, 234
Nagafuchi, M., 187
Nagashima, T., 161
Nakano, K., 163–164
Nass, R., 253–254
Nebes, R. D., 162, 180
Newcombe, F., 54
Newell, K. M., 123
Nichelli, P., 84, 114, 225–226
Nielsen, J. M., 159–160, 175
Niizuma, H., 187
Nimmo-Smith, I., 149
Noll, D., 176
Norman, D. A., 130, 283–284

Oakson, G., 219
Obayashi, T., 31, 158
Oberg, G. E., 208, 214, 221, 223
Ochipa, C., 5, 14–16, 26, 36–37, 42–44, 46, 51–52, 55–58, 62–70, 75, 77, 82–86, 93–95, 138, 160, 173, 184, 193, 209, 271–273, 276, 281–283
O'Connell, P., 177
Odell, K. H., 192
O'Donnell, B., 229–230
Ogle, J. W., 157
Ohigashi, Y., 186
Ojemann, G. A., 214, 234
Okita, N., 187
Ombredane, A., 188–190
Orlando, G., 224, 226, 231–232
Orton, S. T., 246, 249
Osumi, Y., 160–161
Overton, W. F., 252

Paillard, J., 32, 207–208, 215–216, 235, 271, 274
Paivio, A., 53–54, 77
Palmer, C., 277
Palumbo, C., 142, 185, 210, 212, 234
Pandya, D. N., 161
Papagno, C., 157–158, 165, 210, 215, 218, 226–227, 232
Parent, A., 210, 216
Parker, J. C., 218, 236
Patronas, N. J., 57
Patterson, K., 155–157
Payne, M., 221
PDP Research Group, 237
Pearson, K., 30, 78
Pelisson, D., 120
Pellat, J., 185
Pena, J., 113–114
Peña-Casanova, J., 40–41
Penfield, W., 13, 34, 186, 217
Penney, J. B. Jr., 212, 215
Perez, R., 218
Pessin, M. S., 185–186
Peterson, H. A., 177–178
Petrides, M., 131
Petrovici, J. N., 187, 193
Phelps, M., 181, 189
Phelps, M. E., 164
Picard, N., 210
Pick, A., 14–15
Pieczuro, A., 223–224, 252–253, 257, 272–273
Pilgrim, E., 32, 43–45, 52, 86–87
Pitres, A., 159–160

Poeck, K., 15, 52, 76, 112–115, 186, 194, 274–275, 281–282
Poirier, J., 161
Poizner, H., 14, 32, 41–42, 76, 93–106, 194–195, 265, 271–272
Poncet, M., 270–271
Porac, C., 20–21, 23–24
Porrino, L. J., 218
Porter, R., 13, 34
Povel, D. J., 129–130
Powers, J. M., 230
Prablanc, D., 118, 120
Price, C. J., 43, 45, 87
Putnam, J. J., 149
Qualizza, L., 179
Quintana, L. A., 81

Raade, A. S., 193, 262
Rapcsak, S. Z., 26, 42, 57, 158, 161, 271, 276
Rapp, B. C., 51, 53–54
Ratcliff, G., 54
Raymer, A. M., 2–3, 41, 43–44, 54–55, 58, 62–63, 66–71, 76–77, 81–82, 86, 93–95, 221, 237, 277–278, 285
Reason, J. T., 283–284
Redfern, B. B., 186–187
Reed, E., 77
Reed, E. S., 277
Reichle, T., 78
Reinhart, J. B., 178
Reinvang, I., 76–77
Resch, L., 230
Restle, F., 129–130
Reuben, R. N., 248
Ribera, V. A., 217
Richardson, M. E., 41, 43, 54, 70
Riddoch, M. J., 43–45, 51, 53–54, 87
Riege, W., 181, 189
Rigby, J., 34, 76
Rigrodsky, S., 177
Riklan, M., 217
Riley, D. E., 230
Risse, G. L., 161
Roberts, L., 217
Robinson, D. L., 21–22
Rochon, E. A., 175
Roeltgen, D. P., 39, 157–158, 220–221, 232–233, 237–238, 246, 248–249, 262
Roig-Rovira, T., 41
Roland, E., 140
Roland, P. E., 35, 186
Rolls, E. T., 213
Romani, C., 51, 53–54, 149
Ronday, H. K., 184
Rondot, P., 8
Rosa, A., 158
Rosenbaum, D. A., 117, 130
Rosenbek, J. C., 176, 180, 190, 192, 194–197
Rosenfield, D. B., 155–157
Rosenkilde, C. E., 208, 214, 221
Ross, B., 103
Rothi, L. J. G., 2–3, 5, 12–16, 26, 31–33, 35–41, 42–44, 46, 51–52, 62–71, 54–58, 75–86, 93–106, 108, 112, 113–115, 138, 173, 184, 193–195, 256, 262, 264–265, 271–273, 276–278, 281–283, 285
Roy, E. A., 15, 30, 41, 52, 56, 75, 113–114, 127, 131, 138, 142–143, 152, 156–158, 160, 207–209, 217, 221, 223, 225–227, 230, 232, 234–237, 258, 261–262, 270, 272, 282
Rubens, A. B., 57, 160–162, 173–176, 178–180, 193–195, 271, 276
Rubow, R. T., 196
Rudel, R. G., 264
Rumelhart, D. E., 130, 237

Sabouraud, O., 187
Saffran, E. M., 40, 42, 54–55, 282
Saillant, B., 43, 53–54
Sakai, T., 142
Sala, S. D., 76
Salot, D., 94
Sambrook, M. A., 215, 234
Samra, K., 217
Sands, E., 178–179
Sanguineti, I., 226
Sargent, G. I., 130
Sasanuma, S., 177–179, 190, 194
Sassoon, R., 149
Sawada, T., 161
Sax, D. S., 221, 229–230
Scarpa, M., 32, 225, 227, 232
Schaltenbrand, G., 214, 234
Schank, R. C., 283
Schell, G. R., 184
Schiff, H. B., 178, 185
Schinsky, L., 78–79
Schnaberth, G., 228–229, 231, 233
Schneider, W., 283
Schon, F., 271, 284
Schwab, R. S., 163–164
Schwartz, M. F., 42, 55, 77, 271, 278–283, 285
Seemuller, E., 215
Seitz, R. J., 140
Selzer, B., 161
Semenza, C., 54
Seo, T., 142
Sevush, S., 220–221, 232–233, 237–238
Seymour, P. H. K., 53

Shallice, T., 43, 51, 53–54, 271, 277, 283–284
Shankwieiler, D., 177–178
Shapiro, J. K., 186–187
Shelton, P. A., 227, 229–231, 233–234, 236
Shewan, C. M., 3, 76, 78
Shiffrin, R. M., 283
Shimamura, A. P., 131
Shriver, E., 223
Shuren, J. E., 232–233, 235–238
Shuttleworth, E. C., 227–231, 233
Sidtis, J. J., 161, 230
Siguru, A., 270–271
Silvers, G., 153
Simmons, N. N., 78, 196
Sisterhen, C., 78
Sittig, O., 141
Skelly, M., 3, 78–79
Skinhoj, E., 35, 140, 186
Skinner, J. E., 219
Skyhoj-Olsen, T., 223
Smetanin, B., 94
Smith, J. E. K., 116–117
Smith, K. U., 153
Smith, W. M., 153
Smith, R., 78–79
Smyrnis, N., 118
Smyth, M. M., 153
Soechting, J. F., 32, 76, 93–95, 99, 103
Sokoloff, L., 234
Sommers, R. K., 178, 180, 188–191, 193
Sorgato, P., 14, 32, 41
Southwood, H., 196
Sparks, R. W., 196
Sperry, R. W., 11–12, 14, 162
Spinnler, H., 76, 193, 223–224, 227, 233
Square, P. A., 15, 41, 52, 56, 75, 113–114, 138, 173–175, 180–184, 186, 188–194, 196, 262, 272, 282
Square-Storer, P. A., 30, 113–114, 142–143, 174, 176, 178–180, 188–191, 193–195, 258
Squire, L. R., 131, 220–221
Starkstein, S., 160
Steinthal, P., 7–8, 29
Stellar, S., 217
Stelmach, G. E., 262
Steriade, M., 219
Stern, M. B., 229, 231, 233
Sternberg, S., 117, 129
Stichfield, S., 195
Stokoe, W. C., 63
Stone-Elander, S., 140
Stott, D. H., 249
Strick, P. L., 163–164, 184, 186, 210, 230
Strother, S. C., 230
Stuss, D. T., 185, 187
Suger, G., 215
Sugishita, M., 171
Summers, J. J., 130
Sundet, K., 76–77
Suzuki, T., 194
Sweet, W. H., 160

Tagliabue, M., 226
Taira, M., 118
Takeda, K., 179
Talland, G., 221
Tamaki, N., 161
Tanabe, M., 186
Tanaka, Y., 31, 161
Tarel, V., 185
Tate, R. C., 76
Tate, R. L., 34
Tatsumi, I. F., 194
Taylor, G., 25
Taylor, H. G., 247
Taylor, M. L., 178
Teetson, M., 192
Terzuolo, C., 152
Terzuolo, C. A., 99–100, 103
Thach, W. T., 212, 214
Thal, D., 258
Thelan, E., 175
Thomas, A. P., 78
Thompson, C. K., 79, 88
Tissot, R., 1
Tobias, S., 258
Tognolo, G., 193
Tolosa–Sarro, E., 41
Tonkonogy, J., 185
Touwen, B. C. L., 262
Toyokura, Y., 161
Tranel, D., 2
Tredici, G., 224, 226, 231–232
Trentini, P., 78, 86
Triggs, W. J., 23
Trillet, M., 157, 160
Trost, J. E., 177, 194
Tseng, C. H., 180, 194
Tuch, B. E., 175
Tuller, B., 140
Tulley, R. C., 218, 236
Tweedie, D., 252–253
Tyler, H. R., 163–164
Tzeng, O. J. L., 137

Uchida, Y., 179
Ungerleider, L. G., 142
Ushijima, T., 178–179, 190, 194

Valenstein, E., 12, 21–22, 33, 75, 115, 157–158, 219, 230, 232, 271
Van Allen, M. W., 114
Van Bergeijk, W. A., 153
Van Buren, J. M., 214, 234
Van Den Abell, T., 22
Van Donkelaar, P., 117
Van Galen, G. P., 149, 152
Van Lancker, D., 186, 189
Van Riper, C., 195
Vecchi, A., 75, 131
Verfaellie, M., 14–15, 21–22, 40, 63–65, 81–83, 93, 95–96, 101–103, 113–114, 194–195, 225–226, 265, 271–272
Vernon, G., 229, 231
Viader, F., 156–157
Vignolo, L. A., 52, 186, 193, 223–224, 231, 233, 252–253, 257, 272–273
Vila, A, 185
Villa, G., 149
Viviani, P., 99–100, 152
Voeller, K. K. S., 264
Volpe, B. T., 161
Volterra, V., 264
von Keyerlingk, D. G., 186

Wallace, M. A., 54
Wallace, S. A., 123
Wallesch, C. -W., 78, 187, 210, 215, 218
Walton, J. N., 246
Waltz, J. M., 217
Wang, W. S. Y., 137
Wapenski, J. A., 164
Ward, E. A., 258
Ward, L. M., 258
Warren, R. L., 218, 236
Warrington, E. K., 39, 54–55, 157–158, 282
Washino, K., 179, 194
Wassermann, E. M., 23
Waters, G., 175
Watson, J. B., 20
Watson, R. T., 11–13, 21–22, 24, 31–33, 35, 160–161, 193, 219–221, 232–233, 237–238
Watt, F., 30
Webb, W. G., 78–79
Weidner, W. E., 177
Weiner, M., 78
Weismer, G., 180, 194
Welch, K., 13, 34, 160
Wells, S. E., 228
Welsh, K., 12, 24, 186
Welt, C., 184–185
Wernicke, K., 8, 16, 24, 29, 31, 175, 177, 188
Wertz, R., 176, 195–197
Wessely, P., 160
Whishaw, I. Q., 115
Wickens, J., 116
Wilcox, M. J., 86
Willis, A. L., 227
Willmes, K., 112–114
Wilson, B., 281
Wilson, D. H., 161
Wimmer, A., 160, 228–229, 231, 233
Wing, A. M., 149, 155–158, 163–164, 210, 212–213, 262
Witelson, S. F., 258
Woltring, H. J., 94
Woollacott, M., 262–263
Wooten, E., 94
Worringham, C. J., 262
Wright, C. E., 116–117, 129, 152

Yamada, R., 161
Yamada, T., 231, 233, 237
Yamadori, A., 160–161
Yamamoto, H., 161
Yanagisawa, N., 215, 235
Yingling, C. D., 219
Young, A. W., 153–154, 165–166
Yoshioka, M., 161
Yosioka H., 178–179, 194
Younge, E. H., 195
Young, A. B., 212, 215
Young, A. W., 208–209
Yule, G., 180
Yushigima, T., 194

Zacks, R. T., 283
Zanella, R., 31, 161
Zangwill, O. L., 5, 15, 154, 157–158
Zesiger, P., 162
Zimmerman, J., 217
Zola-Morgan, S., 220

Subject Index

Action lexicon,
  definition, 36
  input, 36–38
  output, 36–38
Action semantics, 40
  category-specific aphasia, 53–55
  modality-specific aphasia, 53, 54
  optic aphasia, 53, 54
Activities of daily living, see Apraxia, ADLs
Afferent agraphia, 165–166
  sensory feedback, 165–166
  parietal lobe, 165–166
Agraphia, peripheral, 154–168
  afferent, see Afferent agraphia
  allographic, see Allographic agraphia
  apraxic, see Apraxia agraphia
  nonapraxic, see Nonapraxic agraphia
Aiming, 118–121
Allographic agraphia, 154–157
  anagram spelling, 155
  letter imagery, 155
  oral spelling, 155
  typing, 155
Amerind, 3, 78
Apraxia
  ADLs, 77–79
  agnosia, 8, 30
  Alzheimer's disease, 228–229, 276
  anosoagnosia, see Apraxia, error awareness
  aphasia, 3, 9, 26, 30, 78, 112
  asymbolia, 8, 30
  basal ganglia, 140, 207–238
  buccofacial, 8
  callosal, 9, 10, 12, 24, 31, 77, 80
  cerebellum, 140
  conceptual, 15, 16, 42–44, 52, 56–58, 77, 112, 139–140
  conduction, 70
  consequences, 76–77
  cortical–basal ganglionic degeneration, 230–231
  crossed, 9, 26
  definition, 1, 2, 3, 75, 112
  developmental, see Developmental dyspraxia
  disconnection theory, 31–32
  elemental motor disorders, 8
  error awareness, 79
  executive control, 77
  frontal apraxia, 77
  frontal lobes, 9, 13
  gesture, 78
  gesture learning, 78–79
  hand preference, 24–26
  history of, 7–16
  Huntington's disease, 229–231, 233–234
  ideational, 7, 15, 42–44, 52, 112, 139–140, 272–276
  ideomotor, 7, 139–140

incidence, 62, 75–76
limb kinetic, 7, 8
language, 12
left hemisphere, 30, 62, 66,
natural context, 3, 15, 58, 269–286
optic, 70
parietal lobe, 12, 13, 33–34
Parkinson's disease, 228–229, 233–234
prefrontal cortex, 140
right hemisphere, 62
sequencing, 129–137, 140
significance, 75–77
speech, see speech apraxia
subcortex, 223–238
subcortical dementia, 227
subcortical white matter, 223–227
supplementary motor area, 13, 140
supramarginal gyrus, 11, 12, 33–34
sympathetic, 11, 13
syndrome evolution, 76, 81
relevance,3, 4, 76–77
testing, 61, 71–72, 112, 235
thalamus, 207–238
tool selection, 79
tool use, 79, 268–286
treatment, 75–88
unilateral, 8, 80
Apraxic agraphia 157–162
anagram spelling, 157
case, 158
frontal lobe, 160
letter copying, 158
letter imagery, 157
limb apraxia, 157
oral spelling, 157
parietal lobe, 160
supplementary motor area, 160
typing, 157
unilateral, 160–162
writing style, 158
Automatic movement, 2, 3
Basal ganglia, 210–216, 222, 233–236
alternate selections, 215–216, 235–236
language, 210–213
movement combination, 214–235
praxis, 210–213, 233–236
release of action segments, 234–235
Caregiver counseling, 58
Corpus callosum, 9, 13
Developmental dyspraxia, 245–265
definition, 246–248
developmental learning disabilities, 254–258
etiology, 248–249
incidence, 248
language disorders, 258–259
motor deficits, 259–263
prevalence, 248
speech disorders, 258–259
testing, 249–250
treatment, 251
unilateral lesions, 253–254
Error types, 63–65, 81, 113–114
content, 56
FAB, see Florida Apraxia Battery
FAST, see Florida Apraxia Screening Test
FAST-R, see Florida Apraxia Screening Test-Revised
Florida Apraxia Battery, 62, 66
Florida Apraxia Screening Test, 62
Florida Apraxia Screening Test–Revised, 62–63, 66, 70
Gesture
auditory command, 40, 62–63, 76
imitation, 41, see also Gesture imitation testing
intransitive, 62, 81, 113–114
production, see Gesture production testing
reception, see Gesture reception testing
semantics, 40, 51–58
tool selection testing, 69, 77
transitive, 62, 81, 113–114
visual command, 40
Gesture imitation testing
real gesture imitation testing, 68
nonsense praxis imitation testing, 68, 113–114
Gesture production testing
gesture to verbal command, 68
gesture to visual tool, 68
gesture to tactile tool, 40, 68
Gesture reception testing
gesture naming, 67
gesture decision, 67
gesture recognition, 67
Gesture semantics testing, 69
tool-object association, 56–57
Hand preference, 19–27
alien hand syndrome, 21
aphasia, 26
apraxia, 26
attention, 22, 23
callosal disconnection, 21
crossed aphasia, 21
cultural influences, 20
definition, 19
elemental motor asymmetries, 23–24
environmental influences, 20
hand precision, 23
hand skill, 23
hand strength, 23
hemispace, 22–23
intention to act, 21, 22, 23

and language, 20, 26,
motor learning, 25, 26
movement formula, 25
and writing, 20
Innervatory patterns, 13, 139–140
definition, 31
premotor cortex, 34–36
supplementary motor area, 34–36
Kinesthetic engrams, 139
Kinetic memories, 139–140
definition, 31
Left hemisphere damage
aiming, 122–129
closed-loop processing, 122–123
open-loop processing, 122–123
Management, 79
Meaningless movement production, 39
Motor control, 111–143
Motor program, 116–121
Motor programming, 111–143
Movement formulae, 25, 33–34, 138–143
definition, 31
Movement sequencing, 9, 15, 42, 52, 55–56, 77, 116–117, 129–138
also see naturalistic action
Naturalistic action, 269–286
activities of daily living, 285–286
Alzheimer's disease, 276, 286
closed head injury, 278–282, 285–286
error type, 272, 273, 279
executive function, 271, 283, 284
frontal apraxia, 277, 285
frontal lobes, 277, 284
ideational apraxia, 272–276, 284–285
ideomotor apraxia, 271–272, 277–281, 285
kinematic analyses, 271
parietal lobe, 284
schema, 277, 283–286
sequencing, 274
supervisory attention, 284–286
testing, 70–71, 272–274
utilization behavior, 271
working memory, 284–286
Nonapraxic agraphia, 162–163
cerebellar dysfunction, 163–165
micrographia, 163–165
Parkinson's disease, 163–165
Nonfamiliar gesture production, 37–39
Nonsense gesture production, 37–39
Optic aphasia, 43
Parallel distributed processing, 237
Peripheral writing process, 150
afferent control systems, 153
allographic conversion, 150–152
graphic innervatory patterns, 152–153
graphic motor programming, 152
Praxis
acquisition of gesture in children, 251–253
attention, 270
auditory command, 40, 96
basal ganglia, see Basal ganglia
gesture reception, 36–38, 39, 40
imitation, 11, 12, 32, 37–38, 40, 41
language, 11, 32, 55–56, 208–238
learning, 4, 33
left hemisphere, 9
nonsense gesture, 37–39
production, 37, 52,
subcortex, 207–222, 233–238
testing, 61–72
to tactile command, 40, 96
to visual command, 40–41, 96
Quantitative movement analyses, 93–108
computer graphic analyses, 94
joint coordination, 95, 101–107
space-time relations, 95, 99–101
spatial orientation, 95–99
three dimensional data analyses, 94
Reflexes, 2,3
Speech apraxia, 173–197
anatomic substrates, 190–192
aphasia, 174, 177–180
Broca's area, 174
definition, 173–174
developmental, 258–259
dysarthria, 174–175
error type, 176
frontal lobe, 174
motor programming, 177–178
parietal lobe, 174
phonemic paraphasia, 179–180
pure, 180
treatment, 195–197
Speech motor control
frontal lobe, 181–183
lenticular zone, 188–189
parietal lobe, 181, 187–188
pars opercularis, 185–186
posture production, 175
premotor cortex, 181, 184–185
primary motor area, 181–184
seriation, 175
supplementary motor area, 181, 186–187
Thalamus, 216–221, 222, 236–238
declarative memory, 220–221, 237–238
pre-execution feedback, 218–219, 236
selective engagement, 219–220, 237
Tool affordances, 32, 270
Tool manufacture, 4, 57
Tool use, 2, 12, 15, 32, 52, 55, 58, 76–77
Treatment of apraxia, 79

aphasia, 86
error type, 82–86
restitution, 80–87
substitution-vicariative, 80, 87
substitution-compensatory, 80, 87
Verbal apraxia, see speech apraxia
Volitional movement, 1, 2, 3
Writing, 149–168
anagram spelling, 150
central processes, 150
copying, 153–154
peripheral writing processes, see Peripheral writing processes
physical letter codes, 150
sensory feedback, 153
typing, 150

DEPAUL UNIVERSITY LIBRARY
3 0511 00632 8371